PROCESS PIPING:

THE COMPLETE GUIDE TO ASME B31.3

Fourth Edition

by

Charles Becht IV

ASME PRESS

Library of Congress Cataloging-in-Publication Data

Names: Becht, Charles, IV, author.

Title: Process piping : the complete guide to ASME B31.3 / by Charles Becht, VI.

Description: Fourth edition. | New York : ASME Press, [2021] | Includes bibliographical references and index. | Summary: "This book is based on the 2020 Edition of ASME B31.3, Process Piping [Code]. Because changes, some very significant, are made to the Code every edition, the reader should refer to the Code for any specific requirements. This book should be considered as providing background information and not specific current Code rules. The equations in this book are numbered sequentially in each chapter. When equations from ASME B31.3 are reproduced herein the latter equation numbers are given as well"-- Provided by publisher.

Identifiers: LCCN 2020046152 | ISBN 9780791883792 (hardcover) | ISBN 9780791883808 (epub)

Subjects: LCSH: Piping--Standards.
Classification: LCC TJ930 .B347 2021 | DDC 660/.283--dc23
LC record available at https://lccn.loc.gov/2020046152

To My Wife

Mary

And Children

George, Charles, Derek, John, Kristen

And Grandchildren

Nathanial, Kokopelli, Charles, Audrey, Kate, Amelia, Ethan, Preston

About the Author

Dr. Becht is a recognized authority in pressure vessels, piping, expansion joints and elevated temperature design. He is Chief Executive Office of Becht, a consulting engineering company that provides both process and equipment engineering services as well as project, turnaround, training and other services to the energy sector, including refining, petrochemical, chemical, power and renewables; CEO of Helidex, LLC; and Director of Sonomatic Ltd., a company that provides advanced NDE services. He has performed numerous expert troubleshooting and failure investigation, design reviews and construction inspections for capital projects into the billion dollar range, and consulting to manufacturers on design, development and code compliance for new and existing equipment. He was previously with Energy Systems Group, Rockwell International and Exxon Research and Engineering where he was a pressure equipment specialist.

Dr. Becht is a member and former Chair of the ASME B31.3, Process Piping Committee and a member of ASTM F-17 Plastic Piping Systems Main Committee. Other committees he was previously on include: ASME Board on Pressure Technology Codes and Standards, ASME B31 Code for Pressure Piping Standards Committee, ASME B31 Executive Committee, ASME Post Construction Committee (former Chair), ASME Post Construction Subcommittee on Repair and Testing (founding Chair), ASME Post Construction Executive Committee (former Chair), ASME B&PV Code Subgroup on Elevated Temperature Design (former Chair), ASME B&PV Code Subcommittee on Transport Tanks, ASME B31 Mechanical Design Committee, ASME B&PV Code Subcommittee on Design, ASME B&PV Code TG on Class 1 Expansion Joints for Liquid Metal Service, and the ASME PVP Division Design and Analysis Committee.

Dr. Becht has developed and introduced many new concepts into ASME B31.3 including weld joint strength reduction factors, new stress limits for occasional loads at elevated temperatures, extension of *f* factors to unlimited numbers of cycles, expansion joint rules in Appendix X (now largely revised with the changes in the tenth Edition of EJMA), alternative rules for flexibility analysis in Appendix P (now largely incorporated into the base Code), and impact test exemption rules.

He received a PhD from Memorial University in Mechanical Engineering (dissertation: Behavior of Bellows), a MS from Stanford University in Structural Engineering, and BSCE from Union College, New York (Outstanding Civil Engineering Alumni Gold Award). He is a licensed engineer in 26 states and provinces, an ASME Fellow since 1996, recipient of the ASME Dedicated Service Award in 2001, recipient of the ASME Pressure Vessel and Piping Medal in 2009, recipient of the ASME J. Hall Taylor Medal in 2014, and has more than 70 publications and seven patents.

ACKNOWLEDGEMENTS

Thank you to Don Frikken who extensively updated the Third Edition to create this Fourth Edition, updating for all the changes from the 2008 edition of ASME B31.3 to the 2020 edition of ASME B31.3.

Mr. Frikken is an internationally recognized authority in piping systems, valves, and the development of standards and specifications. His work with Becht includes designing new piping systems, analyzing existing piping systems, assisting manufacturers with respect to code compliance issues, and providing instruction on process piping and piping flexibility analysis courses. Don has received several awards including the ASME B31 Forever Medal for Excellence in Piping, and the ASME Melvin R. Green Codes and Standards Medal in 2002, which recognizes outstanding contributions to the development of documents used in ASME programs of technical codification, standardization and certification.

He received the ASME Fellow Award in 1994 and the ASME Dedicated Service Award in 1995. Prior to joining Becht Engineering, Don spent 34 years with Solutia, Inc. and Monsanto Company working on a wide range of activities including piping and mechanical design, project engineering, and engineering standards. Mr. Frikken is a registered PE in Missouri.

ASME Standards and Certification activities include several positions in ASME B16 and B31 committees, and past membership is several supervisory boards. Activities in other organizations include ISO Technical Advisory Committees, API Refinery Department and AIChE.

Hundredth Meeting, Williamsburg, Virginia, October 1994

Front Row (Left to Right): Ron Haupt, Robert Meyerand, Dick Hinckley, Dave Moody, Dick Hudson, Pete Gardner, John Wier, Dick Straiton, Helmut Thielsch, Tom Estilow, Don Frikken, Warren Thomas, Ken Kluge, Bill McLean, Bob McKee, Ed Bane, Merven Pfeiffer

Center Row (Left to Right): Bruce Bassett, Al Simmons, Roy Grichuk, Phil Ellenberger, Art Post, Glynn Woods, Jon Labrador, Roger Bradshaw, Jerry Byers, Bob Nichols, Don Glover, Chris Ziu, Rick Fairlamb, Bill Koves, Jack Gott, Zoltan Romoda, Lalgudy Balasundaram, Quy Truong, Gary Nariani, Chuck Becht, Bijan Khamanian, Larry Smith, Jim Lang, Dave Fetzner, Walt Canham

Back Row (Left to Right): Bill Short, Jim Meyer, Bryan Harris, Steve Costa, Will Parrish, Barry Agee, Ken Oswald, Chandan Nath, Fred Tatar, Bob Hookway, Rex Engle, Carlos Davila, David Coym, Dan Christian, Ralph Rapp, Jerry D'Avanzo, Mike Jeglic, Joel Andreani, Don Edwards, Robert Silvia

American Society of Mechanical Engineers, B31.3 Chemical Plant and Petroleum Refinery Piping Section Committee

Still members in 2020:

Ron Haupt	Glynn Woods	Dave Fetzner	Carlos Davila
Don Frikken	Don Glover	Jim Meyer	Dan Christian
Roy Grichuk	Chris Ziu	Barry Agee	Don Edwards
Phil Ellenberger	Chuck Becht	Fred Tatar	Bob Silvia

Committee members and visitors at the 125th meeting of the ASME B31.3 Section Committee

Row 1 (Left to Right): Melo, Cynthia; Kadakia, Dixit; Engle, Rex Ch IRG and SG C; Meyer, *Jimmy Vice Chair;* Becht, Charles Chair; Xiaoli (Shelly) Tang; Truong, Quy; Medvick. Richard; Nalbandian, Ara; Shyu, John

Row 2 (Left to Right): D'Avanzo, Jerry Ch SG A; Rangus, Anthony; Hodgins, Jim; Tatar, Fred Ch SG G; Silvia, Robert Ch SG E; Babuder, Jerry; Koves, William Past B31.3 Ch; McLean, William; Woods, Glynn; Molina, III, Vicencio Ch SG H; Nam, Chandan

Row 3 (Left to Right): McLeod, Robert; Wendler, Joseph *Secretary*; Fraser, Douglas; Diehl, David Ch SG B; Smith, J. Larry; Widder, Chris; Ellenberger, James; Reinhardt, Gregory; Luf, John

Row 4 (Left to Right): Haupt, Ronald; Campbell, Rich; Shipley, Kraig; Grichuk, Roy Ch SG D; Verone. Lou; May, Maurice; Glover, Donald Ch SG O; Edwards, Don; (Unidentified)

Row 5 (Left to Right): Moore, Craig; Frikken, Donald Past B31.3 Ch; Burnett, Gordon; Fetzner, David; Clark, Mark; Katcher, Mike; Prawdzik, Joe; Bojarczuck, Ray

Committee members at the 150th meeting of the ASME B31.3 Section Committee

Row 1 (Left to Right): Shyu, Chaur Yue (John); Diehl, David (Chair SC); Nalbandian, Ara

Row 2 (Left to Right): Melo, Cynthia (Chair SG-A); Davila, Carlos (Vice Chair SG-A); D'Avanzo, Jerry; Pham, Kha; Mohamed, Riad (Secretary); Eskridge, Chip (Vice Chair SC); Larsen, Chris (Vice Chair SG-E)

Row 3 (Left to Right): Wills, Terry; Trinker, Gary*; Rahoi, Dennis; Scrivner, Trace (Vice Chair SG-B); Christian, Daniel; Greulich, Owen; Jaouhari, Matt

Row 4 (Left to Right): Swartz, Ben*; Agee, Barry (Chair SG-D); Arnett, David; Grichuk, Roy; Kalnins, John; McDaniel, Michael (Chair SG-F); Bounds, Brandon (Chair SG-G); Fraser, Douglas; Shipley, Kraig (Chair SG-B); Stewart, Matthew*

Row 5 (Left to Right): Huitt, William (Chair SG-H); Campbell, Richard (Chair SG-E); Swezy, John; Haupt, Ronald; Tatar, Fred; Walker, Brad; Gordon, Billee Shane; Ianiro, Dino; Tonkins, Scott (Vice Chair SG-D)

Row 6 (Left to Right): Davio, Jason (Vice Chair SG-E); Flenner, Philip; Meyer, Jimmy; Moore, Craig (Vice Chair SG-F);

Row 7 (Left to Right): Frikken, Donald; Reamey, Raymond; Bojarczuk, Raphael; Leishear, Robert

* These three individuals were not Committee members at the time of the photo.

Contents

List of Figures

List of Tables

Table Number

CHAPTER

1

Background and General Information

This book is based on the 2020 Edition of ASME B31.3, *Process Piping [Code].* Because changes, some very significant, are made to the Code every edition, the reader should refer to the Code for any specific requirements. This book should be considered as providing background information and not specific current Code rules.

The equations in this book are numbered sequentially in each chapter. When equations from ASME B31.3 are reproduced herein the latter equation numbers are given as well.

1.1 HISTORY OF ASME B31.3

In 1926 the American Standards Association initiated Project B31 to develop a piping code. ASME was the sole administrative sponsor. The first publication of this document, *American Tentative Standard Code for Pressure Piping,* was in 1935. From 1942 through 1955 the Code was published as the *American Standard Code for Pressure Piping,* ASA B31.1. It was composed of separate Sections for different industries.

These Sections were split off, starting in 1955 with *Gas Transmission and Distribution Piping Systems,* ASA B31.8. ASA B31.3, *Petroleum Refinery Piping Code,* was first published in 1959. A number of separate Sections have been prepared, most of which have been published. The various Section designations are as follows:

- B31.1 *Power Piping*
- B31.2 *Fuel Gas Piping* [withdrawn in 1988]
- B31.3 *Process Piping*
- B31.4 *Liquid Transportation Systems for Hydrocarbons, Liquid Petroleum Gas, Anhydrous Ammonia, and Alcohols*
- B31.5 *Refrigeration Piping*
- B31.6 *Chemical Plant Piping* [never published]
- B31.7 *Nuclear Piping* [moved to *ASME Boiler and Pressure Vessel Code,* Section III]
- B31.8 *Gas Transmission and Distribution Piping Systems*
- B31.9 *Building Services Piping*
- B31.10 *Cryogenic Piping* [never published]
- B31.11 *Slurry Piping* [withdrawn and incorporated into B31.4 in 2012]
- B31.12 *Hydrogen Piping and Pipelines*

A draft of the Section for Chemical Plant Piping, B31.6, was completed in 1974. However, it was decided to merge this Section into ASME B31.3, because the two Code Sections were closely related. A joint Code

Section, Chemical Plant and Petroleum Refinery Piping, was published in 1976. At this time items such as fluid service categories (e.g., Category M), Nonmetallic Piping, and Safeguarding were introduced into ASME B31.3. In 1980 the Nonmetal portions of the B31.3 Code were gathered and combined into one Chapter, Chapter VIII.

A draft Code for Cryogenic Piping was prepared by Section Committee B31.10 and was ready for approval in 1981. Again, since the coverage overlapped with ASME B31.3, it was decided to merge the Section Committees and develop a single, inclusive Code. This Code was issued in 1984. In the same year another potentially separate Code was added as a new chapter to ASME B31.3, High-Pressure Piping, Chapter IX.

The resulting document is a Code that is very broad in scope. It covers fluids as benign as water and as hazardous as mustard gas. It covers temperatures from cryogenic conditions to 815°C (1500°F) and beyond, and pressures from vacuum and atmospheric to 340,000 kPa (50,000 psi) and higher. Part of the philosophy of the Code stems from this broad coverage. There is a great deal of responsibility placed with the owner and latitude to use good engineering.

The acronym that appears in front of B31.3 in the title of the Code has changed from ASA to ANSI to ASME. The initial designation, ASA, referred to the American Standards Association. Between 1967 and 1969 ASA became first the United States of America Standards Institute and then the American National Standards Institute (ANSI). In 1978 the Standards Committee was reorganized as a committee operating under the American Society of Mechanical Engineers (ASME) with ANSI approval. Thus, it is currently correct to refer to the Code as ASME B31.3. These changes have not changed the Committee structure or the Code.

1.2 SCOPE OF ASME B31.3

ASME B31.3 is the generally accepted standard for process piping, worldwide. It is the de facto standard for piping for the oil, petro-chemical and chemical industries, worldwide. To facilitate its international use, ISO 15649, **Petroleum and natural gas industries—piping,** was approved. This ISO standard makes a normative reference to ASME B31.3. It is an evergreen reference, as it is not tied to a specific edition. The ISO Standard also provides additional technical requirements, such as provisions for buried piping.

To facilitate use of ASME B31.3, as referenced by ISO 15649, in Europe, a guide was prepared by EEMUA. This guide, entitled "Guide to the Use of ISO 15649 and ANSI/ASME B31.3 for Piping in Europe in Compliance With the Pressure Equipment Directive," provides specific guidance on how to comply with the European Pressure Equipment Directive (PED) while using ISO 15649. Following this guide does not provide a presumption of conformity with the PED, therefore, the conformity assessment requirements of the PED must be followed in full.

Within the United States, ASME B31.3 is the recognized and generally accepted good engineering practice for process piping. It is specifically referenced by legislation in the following states (Alaska, California, Maryland and Oklahoma) and provinces of Canada (Alberta, British Columbia, Northwest Territories, Ontario, Prince Edward Island, Quebec and Saskatchewan). However, it is also referenced by other Codes, Standards and regulations, such as NFPA and CFR. As such, it is often required by law, although not directly specified in state regulations.

The ASME B31.3 Code was written with process piping in mind. It is very broad in scope, reflecting the scope of process piping services. Although ASME B31 allows owners to select the piping code most appropriate to their piping installation, the following are typical examples of the types of facilities that ASME B31.3 is intended to cover:

- Chemical plants
- Oil refineries
- Loading terminals
- Bulk processing plants
- Cryogenic piping plants

- Petroleum and natural gas production facilities
- Pharmaceutical plants
- Textile plants
- Paper plants
- Ore processing plants
- Semiconductor plants
- Food and beverage processing facilities

The Code was previously intended to cover all piping within the property limits of a process plant, with certain exclusions. The ASME B31 Codes now state that it is the owners' responsibility to decide which Code is most applicable to their piping system. However, these previously listed exclusions show the intent of the writers of the Code. The following items were excluded from coverage by earlier editions of ASME B31.3:

- Fire protection systems
- Sanitary and storm water systems
- Plumbing
- Piping in property set aside for transportation piping (ASME B31.4, and B31.8.)
- Piping within the jurisdiction of Section I of the *ASME Boiler and Pressure Vessel Code* (BPV) (i.e., boiler external piping, which is require to conform to B31.1)

Certain piping was considered optional:

- Steam piping from a boiler that is outside the jurisdiction of ASME BPV, Section I, which could be either ASME B31.1 or ASME B31.3
- Refrigeration piping that is part of a packaged system (factory-assembled), which could be either ASME B31.5 or ASME B31.3
- Hydrogen piping that could be either ASME B31.12 or ASME B31.3

Boiler external piping is the piping between a boiler casing and generally the first block valve outside of the boiler. This is within the jurisdiction of Section I of ASME BPVC, although Section I refers to the rules of ASME B31.1 for this piping.

Although the Code included building systems piping (e.g., heating systems) that are on the process plant property, other than items specifically excluded (e.g., sanitary), these are better covered by ASME B31.9. In practice, it is doubtful that many people have used ASME B31.3 for their building systems piping in any case.

When deciding which piping codes to use for a given facility, the owner should consider the ability of the facility engineering staff to properly apply the requirements of multiple codes. In general, the fewer number of codes that govern the piping at a facility the better.

The Code excludes low pressure piping systems from its scope, if they meet all of the following conditions:

(1) the pressure is less than 105 kPa (15 psi);
(2) the pressure is not less than zero (i.e., no vacuum condition);
(3) the fluid is nonflammable, nontoxic and not damaging to human tissue;
(4) the temperature is not less than -29°C (-20° F); and
(5) the temperature is not greater than 186°C (366°F).

1.3 WHAT IS PIPING?

ASME B31.3 covers process piping, but what is within the scope of piping? This is defined in the Code in the definitions part of Chapter I.

A piping system is defined as follows.

Interconnected piping subject to the same set or sets of design conditions.

Piping is defined as follows.

Assemblies of piping components used to convey, distribute, mix, separate, discharge, meter, control, or snub fluid flows. Piping also includes pipe-supporting elements, but does not include support structures, such as building frames, bents, foundations, or any equipment excluded from this Code (see para. 300.1.3).

Piping includes piping components, which are defined as follows.

Mechanical elements suitable for joining or assembly into pressure-tight fluid-containing piping systems. Components include pipe, tubing, fittings, flanges, gaskets, bolting, valves, and devices such as expansion joints, flexible joints, pressure hoses, traps, strainers, in-line portions of instruments, and separators.

Piping also contains pipe, which is defined as follows.

A pressure-tight cylinder used to convey a fluid or to transmit a fluid pressure, ordinarily designated pipe in applicable material specifications. Materials designated tube or tubing in the specifications are treated as pipe when intended for pressure service.

These definitions should be reviewed when determining whether something is within the scope of the Code. Some examples follow.

There is often some confusion as to whether instrumentation is covered by the Code. However, the definition of piping components makes it clear that in-line portions of instruments are included in the scope and must comply with ASME B31.3 rules. Often, they are qualified as unlisted components (see Section 4.1 and 4.15).

With respect to piping supports, the supporting elements such as shoes, spring hangers, hanger rods, sway braces etc. and their attachments to the structure (the definition of pipe-supporting element in the Code includes the structural attachment) are included in the scope of ASME B31.3. However, the structures to which they are attached are not.

1.4 INTENT

The Code specifies engineering requirements deemed necessary for safe design and construction of pressure piping. While safety is the primary consideration, this factor alone will not necessarily govern the final specifications for any piping installation. The Code is not a design handbook. The Code relies on the users of the Code to make the additional decisions that must be made to produce a sound piping installation. Furthermore, it is for new piping. Although the intent of ASME B31.3 may be considered in evaluating existing piping, piping that has been in service is not within the scope of the Code. Other standards, such as API 570, *Piping Inspection Code,* should be considered as the governing code for piping after it has been placed in service. The ASME B31.3 Code does not address operation or maintenance of piping systems.

The scope of the Code is new piping; it does not include repair. However, the issue of replacement is less clear. While an earlier interpretation indicated that replacement was covered by ASME B31.3, a recent interpretation stated that the subject of replacement was not addressed. The wording of the Code scope had been changed between those two interpretations, and the committee is perhaps more rigorous now in issuing interpretations that are clearly supported by specific requirements in the Code.

From a practical standpoint, if an entire piping system is to be replaced, it should be constructed to the current Code and if a small portion of a piping system is to be replaced, it should be replaced in kind (as a repair). Where to draw the line between these two extremes is a matter of judgment. However, in the opinion of the author, it is prudent in any case to assess the nature of the Code changes that would impact the design of a portion of a piping system being replaced as a repair.

When a new piping system is being attached to an existing piping system, the demarcation is at the connection to the existing system. The new piping, exclusive of the attachment to the existing system, is governed by ASME B31.3. The connection to the existing system is not considered new construction, but is rather subject to the requirements of a post construction code, such as API 570. It is for this reason that inquiries with respect to hot taps have been have received responses that ASME B31.3 does not apply (see, for example, Interpretation 13-04). With respect to leak testing, the new piping is required to be pressure tested; it can be leak tested prior to tying it in to the existing system. For the connection to the existing system, alternatives to leak testing are provided in API 570.

The following paragraphs, quoted from para. 300(c) at the beginning of the Code, provide important insights into its basic intent:

(3) The Code generally specifies a simplified approach for many of its requirements. A designer may choose to use a more rigorous analysis to develop design and construction requirements. When the designer decides to take this approach, the designer shall provide to the owner details and calculations demonstrating that design, construction, examination, and testing are consistent with the design criteria of this Code. These details shall be adequate for the owner to verify the validity and shall be approved by the owner. The details shall be documented in the engineering design.

(4) Piping elements shall conform to the specifications and standards listed in this Code or, if not prohibited by this Code, shall be qualified for use as set forth in applicable Chapters of this Code.

(5) The engineering design shall specify any unusual requirements for a particular service. Where service requirements necessitate measures beyond those required by this Code, such measures shall be specified by the engineering design. Where so specified, the Code requires that they be accomplished.

These statements help in understanding the philosophy of the ASME B31.3 Code. The Code is not intended to rigorously set forth every procedure, approve every component, and approve every material. Instead, procedures are set forth for evaluating the use of unlisted components and unlisted materials. This differs, for example, from the *ASME Boiler and Pressure Vessel Code* requirement that Code Cases be prepared to approve any material for use that is not listed by the Code.

Paragraph 300(c)(3) states that more rigorous analysis methods may be used. Finite-element analysis is an example of one such method. However, the designer must be able to demonstrate (to the owner) the validity of the more rigorous analysis method. The paragraph was revised in the 2000 Addendum of the Code. It was clarified to specifically state that the validity must be accepted by the owner and the approach be documented in the engineering design. This change was made to address concerns that 300(c)(3) could be interpreted too liberally, and to more specifically state the intent.

The same approach can be found throughout the Code. For example, heat treatments other than those specified may be used, components in listed standards may be rerated, and the temperatures for which allowable stresses are provided may be exceeded. There is a great deal of freedom for good engineering practice and much responsibility for owners.

1.5 RESPONSIBILITIES

1.5.1 Owner

The owner's first responsibility is to determine which Code Section should be used. The owner has overall responsibility within the ASME B31.3 Code for compliance with the Code and for establishing the requirements for design, fabrication, examination, inspection, and testing. Owners are also responsible for designating Category D, Category M, High Pressure and High Purity fluid services when desired.

The owner is the individual or organization that will own and operate the facility after it is constructed. For example, in turnkey construction, the owner is not the contractor, but the organization to which the facility will be turned over. The Code gives the owner permission to designate a representative to carry out selected responsibilities, but the owner retains ultimate responsibility for the actions of the representative.

1.5.2 Designer

The Designer is responsible to the owner for assurance that the engineering design of piping complies with the requirements of the Code and with any additional requirements established by the owner. Qualifications for the designer were added in the 2000 Addendum, as para. 301.1. Note that the Designer is the person in charge of the engineering design of the piping system. This is not the job classification typically called "piping designer" nor the design firm.

The qualifications are stated as minimum experience, including some combination of education or professional registration, and experience in the design of related pressure piping. Experience that includes design calculations for pressure, sustained and occasional loads, and piping flexibility is considered to satisfy this experience requirement.

Different systems may require more or less experience, depending on their complexity and criticality. A Designer with less experience may be acceptable in some circumstances; this is permissible with the approval of the owner.

The number of years of experience is not specified for engineers with Professional Engineering registration. Reliance is instead placed on jurisdictional laws that prohibit Professional Engineers from taking responsible charge of work for which they do not have the necessary competence.

1.5.3 Manufacturer, Fabricator, and Erector

The manufacturer, fabricator, and erector of piping are responsible for providing materials, components, and workmanship in compliance with the requirements of the Code and of the engineering design. Remember that additional requirements specified in the engineering design become requirements of the Code for that piping installation.

1.5.4 Owner's Inspector

The owner's Inspector is responsible to the owner for ensuring that the requirements of the Code for inspection, examination, and testing are met. The Inspector is the owner's representative and oversees the examination and testing work performed by the contractor. The Inspector, therefore, cannot be an employee of the contractor, but could be a third party retained by the owner. The Inspector is a representative of the owner, not an independent third party, such as an authorized inspector. Required qualifications of the Inspector are provided in Chapter V of the Code and are discussed below in Chapter 13. Note that the Inspector is not required to perform examinations or check design.

1.6 HOW IS ASME B31.3 DEVELOPED AND MAINTAINED?

ASME B31.3 is a consensus document. It is written by a Committee intended to be made up of a balanced representation from a variety of interests. These include members with the following perspectives:

- Manufacturer (AK)
- User (AW)
- Designer (AB)
- Constructor (AA)
- Regulatory (AT)
- Insurance Inspection (AH)
- General Interest (AF)

The members of the Committee are not intended to be representatives of specific organizations; their membership is considered based on the qualifications of the individuals and the desire for a balanced representation of various interest groups.

ASME B31.3 is written as a consensus code and is intended to reflect industry practice. This differs from a regulatory approach, in which rules may be written by a government body.

Changes to the Code are prepared by the B31.3 Section Committee. Within the Section Committee, responsibility for specific portions of the Code are split among the following Subgroups:

• SG-A, General Requirements:	Chapters I, II (Parts 3 and 4), IV, and VIII Appendices G, M, and Q Coordination of Appendices E and F Technical Oversight of Code
• SG-B, Design:	Chapter II (Parts 1,2, 5, and 6) Appendices H, S, V, W and X
• SG-C, Edit:	
• SG-D, Materials:	Chapters II (302.3.1 through 302.3.4 only) and III Appendices A and C (metals only)
• SG-E, Fabrication, Examination and Testing	Chapters V and Appendix R
• SG-F, Non-Metallic Piping:	Chapter VII Appendices B and C (nonmetals only)
• SG-G, High Pressure Piping:	Chapter IX Appendix K
• SG-H, Ultra-High Purity Systems:	Chapter X

An International Review Group (IRG) was formed in 2007 to facilitate participation of interested parties outside of the USA. The purpose of the group is to provide a direct means of technical input into the development of ASME B31.3 rules *so* that the Code reflects an ever more global perspective.

To make a change to the Code, the responsible Subgroup prepares the proposed change and sends out a letter ballot for the Section Committee as a whole to vote on the change. Anyone who votes against the change, i.e., votes negatively, must state their reason for doing so, which is sent to the entire Section Committee. The responsible Subgroup usually makes an effort to incorporate comments and resolve any negatives. In fact, it is rare for any change to go forward with more than a couple of negative votes out of a Section Committee that consists of 50 or more individuals. Only two-thirds approval is required for the change to move forward to the next level of approval, the B31 Standards Committee.

Proposed changes to the Code are forwarded to the B31 Standards Committee along with the written reasons for any negative votes and the Section Committee responses to those negatives. In this fashion, the Standards Committee is given the opportunity to see any opposing viewpoints. If anyone on the B31 Standards Committee votes negatively on a change on first consideration, the item is returned to the Section Committee with written reasons for the negative vote. The Section Committee must consider and respond to any negatives, either by withdrawing or modifying the proposed change or by providing explanations that respond to the negative. If the item is returned to the Standards Committee for second consideration, it requires a two-thirds approval to pass.

Once an item is passed by the Standards Committee, it is forwarded to the Board on Pressure Technology Codes and Standards. While an item is being considered by the B31 Standards Committee, the Board members have the opportunity to provide technical comments. Once the item passes the B31 Standards Committee, the Board considers whether or not the procedures were followed correctly. The Board is the final level at which the item is voted on within ASME. Again, any negative vote at this level returns the item to the Section Committee and a second consideration requires two-thirds approval to pass.

Although the Board on Pressure Technology Codes and Standards reports to the Council on Standards and Certification the Council does not vote on changes to the Code.

The final step is a public review process. The availability of drafts is announced on the ASME and ANSI websites. Any comments from the public are considered by the Section Committee. These procedures provide for careful consideration and public review of any proposed change to the Code.

1.7 CODE EDITIONS, ERRATA AND CODE CASES

Starting with the 2002 Edition, the long standing practice of issuing addenda was eliminated. Further, the editions were changed from a three year cycle to a two year cycle. This book is based on the 2020 Edition. The primary reason for the change was that the addenda service was incompatible with electronic publishing for ASME.

Significant changes can occur in each Edition. An engineer whose practice includes process piping should keep a current copy of the Code. New Editions may be obtained from ASME.

The elimination of addenda service can add an additional year between the passing of a Code change and its publication. To address this, if a Code change is considered to be sufficiently urgent, it may be issued as a Code Case. Code Cases may be used as soon as they are approved; there is no need to wait for publication. They will be posted on the ASME B31.3 Section Committee website Instructions for accessing this website can be found in ASME B31.3 on the page following the title page. Note that Code Cases are essentially optional rules.

Errata are corrections to typographical errors and the like. Errata are also posted on the ASME B31.3 Section Committee website.

1.8 HOW DO I GET ANSWERS TO QUESTIONS ABOUT THE CODE?

The B31.3 Section Committee responds to all questions on the Code via an inquiry process. Instructions for writing a request for an interpretation are provided in ASME B31.3, Appendix Z. The Committee will provide a strict interpretation of the existing rules. However, as a matter of policy, the Committee will not approve, certify, rate, or endorse anything, nor will it act as a consultant on specific engineering problems or the application of Code rules. Further, it will not provide explanations as to background or reasons for Code rules. If you need one of these to be done, you should take a course, read this book, or hire a consultant, as appropriate.

The Section Committee will answer any request for interpretation with a literal interpretation of the Code. It will not create rules that do not exist in the Code, and will state that the Code does not address an item if it is not specifically covered by rules written into the Code. Even if the Section Committee disagrees with how the Code is written, it will answer in accordance with how it is written, and then possibly consider changes in the Code.

This strict procedure is considered to be necessary, because interpretations are often asked due to disagreements that may be subject to contractual terms or litigation. If the Code is the binding document, it must be taken as it was written at the time, not how one may prefer it to have been written.

Answering requests for interpretations is one of the highest priorities in the Section Committee meetings. The response is almost always prepared at the first Section Committee meeting following receipt of the request, and is then forwarded to the inquirer by ASME staff. Interpretations are posted on the website for the benefit of all Code users.

Instructions for accessing published interpretations on the ASME website can be found in ASME B31.3 Appendix Z.

1.9 HOW CAN I CHANGE THE CODE?

The simplest means of trying to change the Code is to propose a change to the Section Committee. Any requests for revision to the Code are considered by the Subgroup responsible for the respective Section of the Code. Note that the response of the Subgroup must be approved by the full Section Committee. To increase the likelihood of a suggested change being adopted, the individual should come to the meeting at which the item will be discussed. ASME B31.3 Section Committee meetings are open to the public, and participation of interested parties is generally welcomed. Having a person explain the proposed change and the need for it is even more effective than sending a letter. If you become an active participant and have appropriate professional and technical qualifications, you could be invited to become a B31.3 Subgroup or Section Committee member.

Your request for a Code change may be passed to one of three Technical Committees under ASME B31. These are the Fabrication and Examination Technical Committee, the Materials Technical Committee and the Mechanical Design Technical Committee. These Technical Committees exist to provide technical advice and ensure consistency among the various Code Sections.

CHAPTER

2

GENERAL CONSIDERATIONS

2.1 FLUID SERVICES

The Code includes six categories of fluid service, which provides a means of discriminating among possible degrees of hazard. Less stringent design, examination, and testing are permitted for fluid service of lower hazard (Category D) and more stringent requirements are applied for more hazardous fluid service (Normal, Category M, Elevated Temperature and High Pressure). High Purity fluid service provides requirements needed for piping systems requiring a certain level of cleanness.

Except for Elevated Temperature Fluid Service, it is the owner's responsibility to select the fluid service category. The fluid service is assumed to be Normal Fluid Service unless a different selection is made by the owner. Because some owners may not be familiar with this responsibility, the piping designer should inform the owner of the responsibility and may offer recommendations regarding categories for various fluid services.

Category D Fluid Service is for less hazardous service. It includes fluids that are nontoxic, nonflammable, and not dangerous to human tissue and are at a pressure less than 1035 kPa (150 psi) and temperature colder than 186°C (366°F). If the minimum temperature is caused by anything other than ambient conditions, the temperature must be warmer than -29°C (-20°F). These criteria can be found in the definitions in Chapter I of the Code, under "fluid service." Cooling water, plant air, inert gas and low-pressure steam piping systems are frequently designated Category D Fluid Service. An important additional distinction is that it is the condition of the fluid on leakage, not in the pipe, that must be considered. For example, a steam system with a relief pressure less 1035 kPa (150 psi) can be classified as Category D Fluid Service even though an individual in a container filled with such steam would obviously suffer tissue damage. Rules are provided in the Code that result in lower installed costs for these less hazardous services.

Normal Fluid Service is used for most process piping systems, some of which are considered highly hazardous. Before the 1976 edition of B31.3, this was the only choice. Except for Elevated Temperature Fluid Service, it is the default designation for all piping systems unless the owner makes another choice.

Category M Fluid Service is used for extremely hazardous fluid services. Examples of fluid services that would usually be designated as Category M include systems containing methyl isocyanate, phosgene, and nerve gas. Systems containing fluids such as chlorine, H2S and hydrogen cyanide are sometimes designated as being in Category M Fluid Service. It is not possible to create a list of Category M fluids, because the conditions of the installation must be considered in making the classification. For a fluid service to be Category M, the potential for personnel exposure must be judged to be significant. If a piping is double contained, for example, it could be judged that even highly toxic fluids such as phosgene do not make the system Category

M because the potential for personnel exposure is not significant. For Category M fluid service, the rules for Normal Fluid Service are not applicable. Instead, provisions designed to enhance piping system tightness result in higher installed costs are provided in Chapter VIII.

Elevated Temperature Fluid Service is a fluid service in which the piping metal temperature is in the creep range for long periods of time. The Code provides additional design, fabrications and examination requirements for such services.

High Pressure Fluid Service is a fluid service for which the owner specifies the use of Chapter IX for piping design and construction. The Code does not provide specific pressure for designation as High Pressure Fluid Service. In practice, the Chapter IX rules make sense for small bore piping with pressures in excess of 170 MPa (25,000 psi). Larger bore piping systems with lower pressures may prove to result in a lower cost installations.

High Purity Fluid Service is a fluid service which requires a controlled level of cleanness. The term thus applies to piping systems defined for other purposes as high purity, ultra-high purity, hygienic, or aseptic. Industries include food and beverage, pharmaceutical, and silicon wafer manufacturing. The Code provides alternative methods of fabrication, inspection, examination, and testing for such systems in Chapter X.

The Code definitions for these fluid services are subjective. Because of that and because piping systems are installed in different environments and operated by companies with different philosophies, owners designate different fluid service categories for similar fluid services. For example, if higher integrity piping is desired by the owner, even though the fluid service may not meet the definition for Category M Fluid Service, the owner may designate the piping as Category M so that the additional design, construction, examination, and testing requirements that are provided in Chapter VIII are used for the installation. Alternately, the owner may choose to designate the piping as being in Normal Fluid Service and specify specific additional requirements to make the installation safer.

A flowchart intended to assist fluid service selection is provided in ASME B31.3, Appendix M, Guide to Classifying Fluid Services. This Appendix is considered by the Code to provide guidance, not Code requirements. The actual Code requirements for fluid service classification are the definitions in Chapter I.

2.2 SEVERE CYCLIC CONDITIONS

The Code defines severe cyclic conditions as conditions applying to specific piping components or joints for which the owner or the designer determines that construction to better resist fatigue loading is warranted. This designation invokes a set of requirements in the Code that will result in a more robust piping system against fatigue. Examples of piping where such a designation might be warranted include piping expected to be subjected to vibration like piping near a reciprocating compressor.

Some Code requirements invoked when piping is designated for severe cyclic conditions:

- Only fittings that are forged, wrought (seamless or 100% RT/UT), or cast with factor $E_c \geq 0.90$ may be used
- Fabricated branch connections must be the "stub-in" type
- Welding neck flanges
- High strength bolting
- Threaded joints prohibited except for non-moment bearing connections
- Socket welding prohibited above NPS 2
- Tubing joints may be used only if safeguarded
- Gray and malleable iron are prohibited
- Welding procedure shall provide a smooth, regular, fully penetrated inner surface
- 100% radiographic examination with more stringent weld acceptance criteria
- Backing rings should be removed

2.3 CODE ORGANIZATION

The ASME B31.3 Code has certain organizational features that, when understood, make it easier to follow. The Code consists of ten chapters. The first six chapters are what the committee calls the base Code. It provides the basic piping Code requirements for Normal, Elevated Temperature and Category D fluid service metallic piping. Chapter VII contains rules for nonmetallic piping, including requirements for metallic piping lined with nonmetals. Chapter VIII contains rules for piping in Category M Fluid Service.

Chapter IX contains rules for high-pressure piping and Chapter X provides rules for high-purity piping.

The paragraphs in the Code follow a specific numbering scheme: Paragraphs are designated in the 300 range. The 300 series paragraphs are exclusive to the ASME B31.3 Code Section of the ASME B31 *Code for Pressure Piping.* The paragraph numbers in the base Code are repeated in Chapters VII through IX. Those paragraphs are differentiated by a letter placed in front of the paragraph numbers in each case. Chapter VII uses the letter A, Chapter VIII uses the letter M for metallic piping and MA for nonmetallic piping, Chapter IX uses the letter K and Chapter X uses the letter U.

Using the same paragraph numbers permits easy cross-referencing between each of the last four chapters and the base Code requirements. The last four chapters refer back to base Code requirements whenever they are applicable. In this arrangement, the same requirements need not be repeated multiple times. Also, these four chapters tell users whenever they need to refer back to base Code requirements.

2.4 LISTED COMPONENTS

Chapter IV of the Code, Standards for Piping Components, tabulates listed standards. Listed piping components are those that are in accordance with the standards listed in Table 326.1. Metallic pipe and tube listed in Appendix A are also considered to be listed components. Components that do not comply with those standards are unlisted components. Note that listed components for nonmetallic and high-pressure piping are different as described below.

Fluid Service	Chapters	Listed Components in
Normal, Category D, Elevated Temperature	I-VI	Table 326.1 and Appendix A
Nonmetallic	VII	Table A326.1
Category M	VIII	Table 326.1 and Appendix A
High Pressure	IX	Table K326.1 and Appendix K
High Purity	X	Table 326.1 and Appendix A

Listed components are manufactured in accordance with standards that have been reviewed and accepted by the code committee for use in ASME B31.3 piping systems. The requirements for piping systems in ASME B31.3 do not apply to listed components, unless specifically stated in the Code. As stated in paragraph 326.3, "The design, materials, fabrication, assembly, examination, inspection, and testing requirements of this Code are not applicable to components manufactured in accordance with the documents listed in Table 326.1, unless specifically stated in this Code, or the listed document."

2.5 LISTED MATERIALS

Listed materials are those that are listed in the allowable stress tables of the Code and those permitted in component standards listed in the Code, as in Table 326.1, Table A326.1, or Table K326.1, as appropriate.

2.6 SAFEGUARDING

ASME B31.3 incorporates the concept of safeguarding. Safeguarding involves consideration of factors beyond the simple design of the pipe in the overall safety of the piping installation. It brings in the concepts of consequences of failure and probable sources of damage, which essentially considers risk.

Safeguarding is a concept that works well in the context of ASME B31.3, because the owner has overall responsibility for all aspects of the piping system. This differs from the much more limited scope of responsibilities in ASME B31.1 and ASME BPVC, Section VIII, Division 1. Because the owner has complete responsibility for both design and operation, the owner also has the ability to effectively specify and implement safeguarding provisions.

ASME B31.3 permits the use of certain components, joining methods, and other procedures when appropriate safeguards are provided. For example, brazed joints are prohibited from use in piping systems containing flammable or toxic fluids, unless safeguarded. When the concern with brazed joints is failure of the joint on fire exposure due to melting of the brazing material, appropriate safeguards could involve protecting the joints from fire exposure.

A partial list of conditions where safeguarding is required includes the following:

- Paragraph 305.2.2, use of ASTM A 134 pipe made from ASTM A 285 plate and A 139 pipe for other than Category D Fluid Service
- Paragraph 308.2.4, use of flanges other than weld neck flanges meeting certain criteria for severe cyclic conditions
- Table 314.2.1, use of threaded joints in sizes larger than DN 50 (NPS 2) when the fluid is flammable, toxic, or damaging to human tissue
- Paragraph 314.2.2, use of straight-threaded joints under severe cyclic conditions when the joint is subject to external moment loadings
- Paragraph 315.2, use of flared, flareless, or compression-type tubing fittings covered by listed standards, in normal fluid service, subject to severe cyclic conditions
- Paragraph 317.2, use of brazed and braze-welded joints in fluid services that are flammable, toxic, or damaging to human tissue
- Paragraph 323.2.2(c), exemption of safeguarded piping from impact testing based on the stress state from external loads
- Paragraph 323.4.2, use of gray iron other than ductile iron permitted for specified conditions only when safeguarded against excessive heat and thermal shock and mechanical shock and abuse
- Paragraph A323.4.1, safeguarding against excessive temperature, shock, vibration, pulsation, and mechanical abuse when nonmetallic materials are used in any fluid service
- Paragraph A323.4.2(a), use of thermoplastics in other than Category D fluid service
- Paragraph A323.4.2(b), use of reinforced plastic mortar (RPM) in other than Category D Fluid Service
- Paragraph A323.4.2(c), use of reinforced thermosetting resin (RTR) piping in toxic or flammable fluid service
- Paragraph A323.4.2(d), use of safeguarding against large, rapid temperature changes when borosilicate glass and porcelain are used in any fluid service; additionally, use of general safeguarding when they are used in toxic or flammable fluid services
- Paragraph U315.3 for face seal or hygienic clamp-type joints used under severe cyclic conditions

The following safeguards are also mentioned in the Code:

- Paragraph M300(d) requires consideration be given to additional engineered safeguards for Category M fluid service.
- Paragraph FA323.4(b) (Appendix F) recommends safeguards for thermoplastic piping in aboveground compressed gas (including air) service.

Appendix G of ASME B31.3 discusses the concept of safeguarding and provides examples of safeguarding by plant layout and operation, and engineered safeguards.

Safeguarding requires addressing the potential consequences of failure. Therefore, the hazardous properties of the fluid, the quantity of the fluid that could be released by a piping failure, the consequences of such a release with respect to personnel exposure and equipment damage (with potential additional consequences), and the conditions of the environment and their effect on the hazards caused by a possible piping failure are all considerations of safeguarding. These address the consequence aspect of risk. The safety inherent in the piping by virtue of materials of construction, methods of joining, and history of service reliability, also mentioned, addresses the probability-of-failure aspect of risk.

Safeguarding by plant layout and operation includes the following examples from Appendix G of ASME B31.3:

"(a) Plant layout features, such as open-air process equipment structures; spacing and isolation of hazardous areas; slope and drainage; buffer areas between plant operations and populated communities; and control over plant access.

(b) Protective installations, such as fire protection systems; barricades or shields; ventilation to remove corrosive or flammable vapors; instruments for remote monitoring and control; and containment and/or recovery facilities or facilities (e.g., incinerators) for emergency disposal of hazardous materials.

(c) Operating practices, such as restricted access to processing areas; work permit system for hazardous work; and special training for operating, maintenance, and emergency crews.

(d) Means for safe discharge of fluids released during pressure relief device operation, blowdown, cleanout, etc.

(e) Procedures for startup, shutdown, and management of operating conditions, such as gradual pressurization or depressurization and gradual warmup or cooldown, to minimize the possibility of piping failure, e.g., brittle fracture."

Examples of engineered safeguards include the following:

"(a) Means to protect piping against possible failures, such as (1) thermal insulation, shields, or process controls to protect from excessively high or low temperature and thermal shock; (2) armor, guards, barricades, or other protection from mechanical abuse; (3) damping or stabilization of process or fluid flow dynamics to eliminate or to minimize or protect against destructive loads (e.g., severe vibration pulsations, cyclic operating conditions).

(b) Means to protect people and property against harmful consequences of possible piping failure, such as confining and safely disposing of escaped fluid by shields for flanged joints, valve bonnets, gages, or sight glasses; or for the entire piping system if of frangible material; limiting the quantity or rate of fluid escaping by automatic shut-off or excess flow valves, additional block valves, flow-limiting orifices, or automatic shutdown of pressure source; limiting the quantity of fluid in process at any time, where feasible."

The above serve as examples of what is meant by safeguarding. They do not, however, represent all the possibilities of how piping systems can be safeguarded.

CHAPTER

3

Design Conditions and Criteria

3.1 DESIGN CONDITIONS

Design conditions in ASME B31.3 are specifically intended for pressure design. The design pressure and temperature are the most severe *coincident* conditions, defined as the conditions that result in the greatest pipe wall thickness or highest required pressure class or other component rating. Design conditions are not intended to be a combination of the highest potential pressure and the highest potential temperature, unless such conditions occur at the same time.

The code defines a piping system as "interconnected piping subject to the same set or sets of design conditions." Although it is possible for one operating condition to govern the design of one part of interconnected piping and another condition to govern the design of another part, this is relatively rare. If this situation is encountered, the two different parts of interconnected piping system should be designated as separate piping systems.

3.1.1 Design Pressure

In determining the design pressure, all expected conditions of internal pressure must be considered. These include thermal expansion of trapped fluids, surge, and failure of control devices. The determination of design pressure can be significantly affected by the means used to protect the pipe from overpressure. Process piping systems are permitted to be used without protection of safety relief valves. However, in the event that none are provided on the pipe (or on attached equipment that would also protect the pipe), the piping system must be designed to safely contain the maximum pressure that can reasonably be expected to occur in the piping system, including consideration of failure of any and all control devices. Such a pressure could be, for example, the dead-head pressure of a pump. Unlike some industry standards, there is no consideration given to whether the source of failure depends on single- versus double-contingency events.

Although all these pressures must be considered, they do not necessarily become the design pressure. The Code permits pressure and temperature variations per para. 302.2.4. If the event being considered complies with the Code requirements of 302.2.4, the allowable stress and/or component pressure rating may be exceeded for a short time, as discussed in Section 3.5. Although this is often considered to be an allowable variation above the design condition, the variation limitations are related to the maximum allowable working pressure of the piping, not the design conditions, which could be lower than the maximum allowable pressure at temperature.

3.1.2 Design Temperature

It is the metal temperature that is of interest in establishing the design temperature. Thus, the design temperature does not necessarily coincide with the temperature of the process fluid. In addition to fluid temperature, other considerations include ambient cooling, ambient heating, solar radiation, and maximum heat tracing temperature.

For an insulated pipe, the metal temperature is generally considered to be the fluid temperature, unless some other means of heating is a consideration, such as the presence of steam tracing, which could potentially heat the pipe to temperatures much warmer than normal during a no flow condition.

For uninsulated pipe, the metal temperature is taken as the fluid temperature if the temperature is below 65°C (150°F), unless some other effect, such as solar heating of the pipe, could result in a greater temperature. A commonly assumed potential temperature due to solar heating for piping exposed to the sun is 50°C (120°F); local conditions may dictate a lower or higher temperature.

For an uninsulated pipe, the Code permits consideration of ambient cooling effects if the fluid temperature is 65°C (150°F) or greater. Of course, if solar radiation can bring the metal temperature to higher than this temperature, as in Saudi Arabia, ambient cooling effects could only be considered at some higher temperature. The metal temperature of the pipe can be determined via heat transfer calculations or testing, such as measurement of the temperature on a similar system, or based on presumptive values given in the following table, which provides conservatively high estimates of wall temperature:

Assumed Metal Temperature	Component
95% of fluid temperature	Components having a wall thickness comparable to pipe, such as valves, pipe, lapped ends, and welding fittings
90% of fluid temperature	Flanges (except lap joint), including those that are on fittings and valves
85% of fluid temperature	Lap joint flanges
80% of fluid temperature	Bolting

For an internally insulated pipe, such as a refractory-lined pipe, the metal temperature must be determined by heat transfer calculations or test. Note that the heat transfer calculations must consider the appropriate range of conditions of wind and ambient temperatures to which the pipe will be subjected.

Insert 3.1 Calculation of Wall Temperature for Refractory-Lined Pipe

In lines for high-temperature fluid solids (e.g., in fluid catalytic cracking units), it is sometimes necessary to calculate the metal temperature with an internal, insulating refractory lining. In this case, the metal temperature may be significantly less than the fluid temperature. The refractory lining resists erosion while at the same time permitting use of carbon steel with a fluid that would otherwise be prohibitively hot for its use. Note that the metal temperature is often deliberately kept warm, often above 150°C (300°F), to prevent condensation against the inside of the steel pipe (of gasses passing through refractory pores and cracks) and resultant corrosion (e.g., due to sulfuric acid and sulfurous acid in flue gas) of the metal under the refractory. The following equations and a chart of film coefficient are provided as means for calculating the metal temperature. Note that, for design temperature, conditions that maximize the metal temperature are assumed (zero wind and hottest ambient temperature).

The heat flow through the pipe, per square foot per hour, is given by

$$Q = \frac{2\pi\Delta T}{\dfrac{1}{r_1 h_i} + \dfrac{\ln(r_2/r_1)}{K_r} + \dfrac{\ln(r_3/r_2)}{K_s} + \dfrac{1}{r_3 h_o}} \tag{3.1}$$

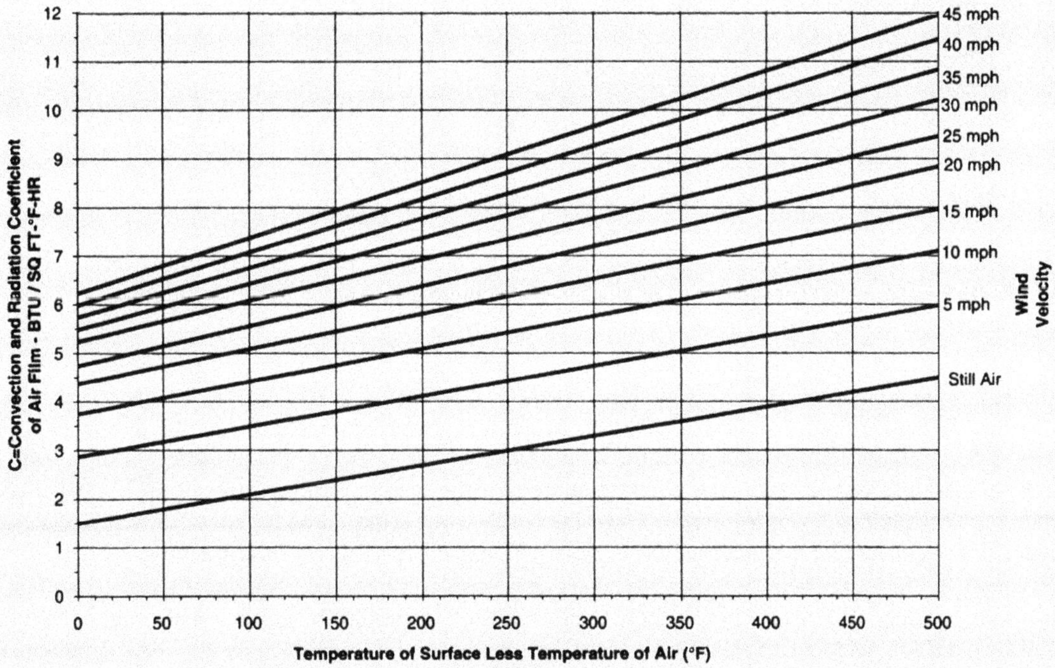

FIG. 3.1
EFFECTIVE OUTSIDE FILM COEFFICIENT

where

Q = heat flow, Btu/ft-h
ΔT = temperature difference between fluid inside the pipe and ambient temperature, °F
h_i = film coefficient on the inside of the refractory (often assumed to be very high, so the term including this in the denominator may be neglected), Btu/ft2-°F-h
h_o = film coefficient on the outside of the pipe (see Fig. 3.1), Btu/ft2-°F-h
r_1 = radius inside the refractory, in.
r_2 = radius outside of the refractory, in.
r_3 = outside radius of the pipe, in.
K_r = conductivity of the refractory material, Btu-in./ft2-°F
K_s = conductivity of the pipe material, Btu-in./ft--°F

The change in temperature through any layer is equal to the heat flow times the resistance. Thus, the temperature on the outside surface of the pipe is given by the equation

$$T_o = Q\frac{1}{2\pi r_3 h_o} + T_a \qquad (3.2)$$

where
T_o = temperature on the outside surface of the pipe, °F
T_a = ambient temperature, °F

The temperature on the inside surface of the pipe is given by the equation

$$T_i = Q\frac{\ln\left(r_3 / r_2\right)}{2\pi K_s} + T_o \tag{3.3}$$

where

T_i = temperature on the inside surface of the metal pipe, °F

This is normally essentially the same as the temperature on the outside surface, because the conductivity of the pipe material is generally high relative to the refractory and outside film coefficient.

3.1.3 Design Minimum Temperature

The design minimum temperature is required to determine the impact test requirements of ASME B31.3. As the temperature is decreased, the notch toughness of the material drops. This effect is particularly pronounced for carbon and low alloy steels. Thus, impact tests are required under conditions of sufficiently low temperature to ensure that the material has adequate toughness to avoid a brittle fracture.

In establishing the minimum temperature, one must consider the lowest component temperature that can be expected in service. This can include effects such as autorefrigeration (the cooling that occurs due to depressurization) and low ambient temperatures. Note that some piping is subjected to very low temperatures, but not while in service. For example, a water line cannot be in service at temperatures below 0°C (32°F). So the design minimum temperature for the water line would be 0°C (32°F) even though it might be subjected to -40°C (-40°F) during a winter shutdown.

With the inclusion of stress as a consideration in the need to perform impact testing, it is possible that more than one minimum temperature, each with different stress conditions, will require consideration. This is because the loads may be different at different low-temperature conditions, and lower temperatures without impact testing may be permitted at lower stress conditions.

3.2 ALLOWABLE STRESS

3.2.1 General

The Code provides allowable stresses for metallic piping based on criteria listed in para. 302.3. These criteria are not applicable to nonmetallic piping (Chapter VII) and high pressure piping (Chapter X). These are, with certain exceptions, the lowest of the following:

- One-third of the specified minimum tensile strength (which is at room temperature)
- One-third of the tensile strength at temperature (times 1.1)
- Two-thirds of the specified minimum yield strength (which is at room temperature)
- Two-thirds of the "minimum" yield strength at temperature
- Average stress for a minimum creep rate of 0.01%/1000 hours
- Two-thirds of the average stress for creep rupture in 100,000 hours
- 80% of the minimum stress for a creep rupture in 100,000 hours

Specified minimum values are those described in the material specifications. For unlisted material, the "minimum" at temperature is required in para. 302.3.2 to be determined by multiplying the specified (room temperature) values by the ratio of the average strength at temperature to that at room temperature. The allowable stresses listed in the Code are determined by the ASME Boiler and Pressure Vessel Code Committee on Materials (BPV II), and are based on trend curves that show the effect of temperature on yield and tensile strengths.

(The trend curve provides the aforementioned ratio.) An additional factor of 1.1 is used with the tensile strength at temperature.

An exception to the above criteria is made for austenitic stainless steel and nickel alloys with similar stress-strain behavior, for which one can use as high as 90% of the yield strength at temperature. This is due not to a desire to be less conservative, but to a recognition of the differences in behavior of these alloys. The quoted yield strength is determined by drawing a line parallel to the elastic loading curve on a stress-strain diagram with a 0.2% offset in strain. The yield strength is the intercept of this line with the stress-strain curve. Such an evaluation provides a good yield strength value of carbon steel and alloys with similar behavior, but it does not represent the strength of austenitic stainless steel, which has considerable hardening and additional strength beyond this strain value. However, the additional strength is achieved at the cost of additional deformation. Thus, the higher allowable stresses relative to yield are only applicable to components that are not deformation-sensitive. Thus, although one might use the higher allowable stress for pipe, it should not be used for flange design. The US Customary (USC) allowable stress tables use a style that lets the Code user know when the allowable stress exceeds two-thirds of the yield strength at temperature, as will be described in Section 11.8.

Use of the higher allowable stress can create some complications, because the Code has limitations in the form of stress in the piping relative to yield strength. For example, variations above what would otherwise be considered to be the maximum allowable working pressure of the piping are permitted, to as high as 33% above the allowable stress. However, the Code requires that the stress not exceed yield stress. Thus, if the pipe was designed using 90% of yield stress, the allowable variation would be much less than 33%. Similar circumstances occur for occasional loads (see para. 302.3.6; stress is limited to yield).

The allowable stress for bolting is similar, except that it is based on one-fourth of the tensile strength rather than one-third. Special consideration is given to bolting materials for which the strength has been enhanced by heat treatment or strain hardening. They are limited to one-fifth of the specified minimum tensile strength and one-fourth of the specified minimum yield strength, unless these values are lower than the values of allowable stress for the annealed material, in which case the annealed material values of allowable stress are used.

For gray and malleable cast iron materials, the behavior is brittle and the allowable stress differs accordingly. For gray iron, the basic allowable stress is the lower of one-tenth of the specified minimum tensile strength (at room temperature) and one-tenth of the "minimum" strength at temperature, also based on the trend of average material strength with temperature. For malleable iron, a factor of one-fifth is used rather than a factor of one-tenth.

3.2.2 Occasional Loads

The Code limits stresses due to sustained plus occasional loads to approximately 90% of the yield strength. When materials with an allowable stress at temperature based on 2/3 of yield strength is involved, the Code directs the user to set the limit at 1.33 times the allowable stress. [1.33 times 2/3 Sy = 0.89 Sy.] For allowable stresses that are based on 90% of yield and those based on creep properties, determining the allowable stress for sustained plus occasional loads is more complicated.

In the 2004 edition, alternative rules were added for occasional loads for piping at elevated temperatures, in the creep regime. In Elevated Temperature Fluid Service, the allowable stress is governed by long term creep rupture/rate properties, which are not particularly relevant to whether the piping can withstand short term loading. In the alternative rules for Elevated Temperature Fluid Service, the basic allowable is 90% of yield strength, reduced by a strength reduction factor which takes into account that some metals have reductions in yield strength over time when in elevated temperature service. This factor is set at 0.8 (which is a lower bound for relevant materials) or 1.0 for austenitic stainless steel which does not lose strength with time at elevated temperature. In a study documented in an ASME Research Report STP-PT-024, it was recommended that an additional limit of four times the basic allowable stress be used to avoid more than 20% creep damage if the duration of the occasional loads is ten hours or less, which is also in the rules. If the duration is greater than ten hours, the designer is required to calculate the limit using Appendix V.

In the drafting of the rules for later editions including 2018, a couple of errors were introduced which may be corrected in a future edition of ASME B31.3. First, the word "weld" was added in (a) (2) (-a) in front of "strength reduction factor" which can lead the Code user to misinterpret this as "weld joint strength reduction factor" which is something entirely different. The "strength reduction factor" referred to in (-a) is defined further down in the text as either 1.0 of 0.8. The second error is that paragraph that requires casting quality factors be included and to also limit the allowable stress to yield strength at temperature was inadvertently incorporated into the alternative rules for elevated temperature in 302.3.6 (a) (2). These rules preceded the incorporation of the alternative rules for elevated temperatures and apply to all evaluations of occasional loads. It should be a standalone paragraph (e.g. (b) apart from the elevated temperature service rules.

3.3 QUALITY FACTORS

Quality factors for welded joints are used in pressure design and are applied at pipe longitudinal and spiral weld joints. Quality factors for castings are applied for pressure design, design for occasional loads, and for evaluating displacement stress range.

The quality factors for weld joints are provided in Table A-1B of the Code, and vary from 0.6 to 1.0. The basis for these factors is summarized in Figure 3.2. The quality factor for furnace butt weld pipe is the lowest, at 0.60. Note that this quality factor does not generally effect the thickness of furnace butt weld pipe, since it is only permitted for Category D Fluid Service, and furnace butt welded pipe is typically supplied in standard wall, DN 100 (NPS 4) and smaller.

Electric resistance welded pipe has a quality factor of 0.85. This cannot be improved with additional examination.

The quality factor for electric fusion welded pipe varies from 0.80 to 1.0, depending upon whether it is a single or double sided weld, and the degree of radiographic examination (none, spot, or 100%).

The quality factors for castings are described in para. 302.2.3 of the Code. Due to the conservative stress basis, the casting quality factor for gray and malleable iron is 1.0. The casting quality factor for other castings is 0.8, unless supplemental requirements are met. These supplemental requirements involve additional examination and/or machining of surfaces to enhance examination. Compliance with these supplemental requirements can increase the quality factor to from 0.85 to 1.0, depending upon which supplemental examinations are performed.

3.4 WELD JOINT STRENGTH REDUCTION FACTORS

Weld joint strength reduction factors were added to ASME B31.3 in the 2004 edition. These apply at temperatures above 427°C (800°F). and are based on consideration of the effects of creep. They apply to longitudinal and spiral weld joints in pressure design, and to circumferential weld joints in evaluation of stresses due to sustained loads, S_L. They were added because weldment creep rupture strength has been determined to be lower than base metal creep rupture strength in some circumstances. The designer may determine the weld joint strength reduction factor for the specified weldments based on creep rupture test data. This is encouraged to develop factors specific to the base material/weld material combinations used in the design. However, a simplified factor was provided for use by the designer, in the absence of more applicable data. Because it was impractical at the time to establish factors for specific materials, a general factor was used. The values were fine tuned for various material groups and multiple factors were added in the 2008 ediition.

FIG. 3.2
LONGITUDINAL WELD JOINT QUALITY FACTOR, Ej
(ASME B31.3-2020, TABLE 302.3.4)

No.	Type of Joint		Type of Seam	Examination	Factor, E_j
1	Furnace butt weld, continuous weld	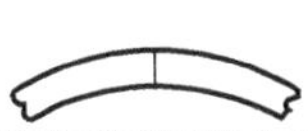	Straight	As required by listed specification	0.60 [Note (1)]
2	Electric resistance weld	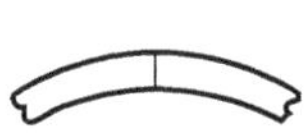	Straight or spiral (helical seam)	As required by listed specification	0.85 [Note (1)]
3	Electric fusion weld				
	(a) Single butt weld (with or without filler metal)		Straight or spiral (helical seam)	As required by listed specification or this Code	0.80
				Additionally spot radiographed in accordance with para. 341.5.1	0.90
				Additionally 100% radiographed in accordance with para. 344.5.1 and Table 341.3.2	1.00
	(b) Double butt weld (with or without filler metal)	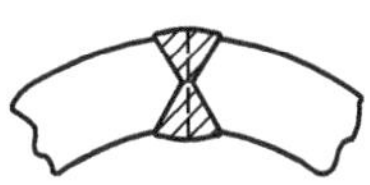	Straight or spiral (helical seam) (except as provided in 4 below)	As required by listed specification or this Code	0.85
				Additionally spot radiographed in accordance with para. 341.5.1	0.90
				Additionally 100% radiographed in accordance with para. 344.5.1 and Table 341.3.2	1.00
4	Specific specification				
	API 5L, electric fusion weld, double butt seam	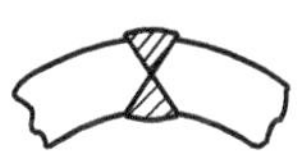	Straight (with one or two seams) or spiral (helical seam)	As required by specification	0.95
				Additionally 100% radiographed in accordance with para. 344.5.1 and Table 341.3.2	1.00

NOTE: (1) It is not permitted to increase the joint quality factor by additional examination for joint 1 or 2.

The factor is applied to the allowable stress used when calculating the required thickness for internal pressure and may be used when evaluating stresses due to sustained loads.

The weld joint strength reduction factor is not applied to the allowable stress range for displacement stresses, S_A, because these stresses are not sustained. The displacement stresses relax over time. The allowable stress criteria for displacement stress range is designed so that the piping system will self-spring so that the highest level of displacement stresses only occurs at the hot condition once over the lifetime of the piping system.

Background on how those factors were developed is provided in Insert 3.2. A general factor was used for all materials initially because, as its developer, this author, did not have sufficient data to propose material specific factors covering the wide range of materials permitted in elevated temperature service in ASME B31.3. However, the factor was introduced to address a known issue and safety concern, with the expectation that further work would lead to improvements.

Following ASME B31.3, the Power Piping Section Committee (B31.1) and Subcommittee on Power Boilers (Section I) began work to include weld joint strength reduction factors into their respective codes. In order to develop consistent requirements in the ASME Codes, including ASME B31.3, a project team was formed to develop consistent rules for all three codes. The starting point was the rules in ASME B31.3; they were then significantly enhanced to provide requirements that were more material specific.

The factor is used when calculating the required thickness of longitudinal and spiral welded pipe and fittings in pressure design. In the 2004 and 2006 edition, it was also used in evaluating stress due to sustained loads at girth weld locations. In the 2008 edition, this was changed from a requirement, to stating that application of weld joint strength reduction factors to girth welds is the responsibility of the designer. This change was made as a compromise to avoid inconsistency between ASME B31.3 and ASME B31.1; there was not general agreement that it should be required for girth welds.

The factor does not apply to the following conditions.

1. It is not used to reduce the allowable displacement stress range, S_A, because these stresses are not sustained. The displacement stresses relax over time.
2. It is not used for evaluating stresses due to occasional loads, as such loads have short durations.
3. It is not used when considering the allowable stress for permissible variations, as provided in para. 302.2.4, as such loads have short durations.

As mentioned above, the Code does not specify whether to use the factor or not when evaluating stress due to sustained loads at girth welds.

In the 2008 edition, for most materials, the same slope was used for weld joint strength reduction factor versus temperature, but the starting temperature was changed to T_{cr} which is material specific. T_{cr} is 25°C (50°F) below the temperature at which point the allowable stress is governed by creep properties in the allowable stress tables. The stress level at which creep governs the allow stress is indicated in the metric stress tables and in Section VIII, Div 1 allowable stress table. The material specific factors are provided in Table 302.3.5 (Table 3.1).

A few highlights of the table follow:

1. Factors are provided for CrMo (through 9Cr-lMo), CSEF, and austenitic stainless grade 3xx, and N088xx nickel alloy materials. The prior ASME B31.3 factors are used for other materials, when required.
2. CSEF material weldment performance is highly dependent upon the heat treatment; if a subcritical PWHT is performed, the factor is dropped to 0.5 throughout the creep range.
3. Autogenous (i.e., no filler metal welds) austenitic stainless grade 3xx and N088xx and N066xx nickel alloys are assigned a *W* factor of 1.0 if they are solution annealed after welding.
4. Reference is made to permit use the Section III, Subsection NH, Tables 1-14.10 A-xx, B-xx and C-xx factors specified for 100,000 hour duration. Observing Figure 3.3, and in particular the 304B line (304 welded with 316 weld metal), this can give higher factors for some combinations of base metal and weld metal.

TABLE 3.1
WELD JOINT STRENGTH REDUCTION FACTOR, W (ASME B31.3-2020, Table 302.3.5)

Steel Group	Component Temperature, T_i, °C (°F) ≤427 (≤800)	454 (850)	482 (900)	510 (950)	538 (1,000)	566 (1,050)	593 (1,100)	621 (1,150)	649 (1,200)	677 (1,250)	704 (1,300)	732 (1,350)	760 (1,400)	788 (1,450)	816 (1,500)
Carbon Steel	1	1	1	1	1	1	1	...	...	...	...	...	...	...	...
CrMo [Notes (1)–(3)]	1	0.95	0.91	0.86	0.82	0.77	0.73	0.68	0.64	...	...	...	...	...	...
CSEF (N + T) [Notes (3)–(5)]	...	...	...	1	0.95	0.91	0.86	0.82	0.77	...	...	...	...	...	...
CSEF [Notes (3) and (4)] (Subcritical PWHT)	...	...	1	0.5	0.5	0.5	0.5	0.5	0.5	...	...	...	...	...	...
Autogenous welds in austenitic stainless grade 3xx, and N088xx and N066xx nickel alloys [Note (6)]	...	...	...	1	1	1	1	1	1	1	1	1	1	1	1
Austenitic stainless grade 3xx and N088xx nickel alloys [Notes (7) and (8)]	...	...	...	1	0.95	0.91	0.86	0.82	0.77	0.73	0.68	0.64	0.59	0.55	0.5
Other materials [Note (9)]	...	...	...	...	...	...	...	...	...	...	...	...	...	...	...

GENERAL NOTES:

(a) Weld joint strength reduction factors at temperatures above the upper temperature limit listed in Appendix A for the base metal or outside of the applicable range in Table 302.3.5 are the responsibility of the designer. At temperatures below those where weld joint strength reduction factors are tabulated, a value of 1.0 shall be used for the factor W where required; however, the additional rules of this Table and Notes do not apply.

(b) T_{cr} = temperature 25°C (50°F) below the temperature identifying the start of time-dependent properties listed under "NOTES - TIME-DEPENDENT PROPERTIES" (Txx) in the Notes to ASME BPVC, Section II, Part D, Tables 1A and 1B for the base metals joined by welding. For materials not listed in Section II, Part D, T_{cr} shall be the temperature where the creep rate or stress rupture criteria in paras. 302.3.2(d)(4), (5), (6), and (7) governs the basic allowable stress value of the metals joined by welding. When the base metals differ, the lower value of T_{cr} shall be used for the weld joint.

(c) T_i = temperature, °C (°F), of the component for the coincident operating pressure–temperature condition, i, under consideration.

(d) CAUTIONARY NOTE: There are many factors that may affect the life of a welded joint at elevated temperature and all of those factors cannot be addressed in a table of weld strength reduction factors. For example, fabrication issues such as the deviation from a true circular form in pipe (e.g., "peaking" at longitudinal weld seams) or offset at the weld joint can cause an increase in stress that may result in reduced service life and control of these deviations is recommended.

(e) The weld joint strength reduction factor, W, may be determined using linear interpolation for intermediate temperature values.

NOTES:

(1) The Cr–Mo Steels include: $^1/_2$Cr–$^1/_2$Mo, 1Cr–$^1/_2$Mo, 1$^1/_4$Cr–$^1/_2$Mo–Si, 2$^1/_4$Cr–1Mo, 3Cr–1Mo, 5Cr–$^1/_2$Mo, 9Cr–1Mo. Longitudinal and spiral (helical seam) welds shall be normalized, normalized and tempered, or subjected to proper subcritical postweld heat treatment (PWHT) for the alloy. Required examination is in accordance with para. 341.4.4 or 305.2.4.

(2) Longitudinal and spiral (helical seam) seam fusion welded construction is not permitted for C–$^1/_2$Mo steel above 454°C (850°F).

(3) The required carbon content of the weld filler metal shall be ≥0.05 C wt. %. See para. 341.4.4(b) for examination requirements. The basicity index of SAW flux shall be ≥1.0.

(4) The CSEF (Creep Strength Enhanced Ferritic) steels include grades 91, 92, 911, 122, and 23.

(5) N + T = Normalizing + Tempering PWHT.

(6) Autogenous welds without filler metal in austenitic stainless steel (grade 3xx) and austenitic nickel alloys UNS Nos. N066xx and N088xx. A solution anneal after welding is required for use of the factors in the Table. See para. 341.4.3(b) for examination requirements.

(7) Alternatively, the 100,000 hr Stress Rupture Factors listed in ASME BPVC, Section III, Division 1, Subsection NH, Tables I-14.10 A-xx, B-xx, and C-xx may be used as the weld joint strength reduction factor for the materials and welding consumables specified.

(8) Certain heats of the austenitic stainless steels, particularly for those grades whose creep strength is enhanced by the precipitation of temper-resistant carbides and carbonitrides, can suffer from an embrittlement condition in the weld heat-affected zone that can lead to premature failure of welded components operating at elevated temperatures. A solution annealing heat treatment of the weld area mitigates this susceptibility.

(9) For materials other than carbon steel, CrMo, CSEF, and the austenitic alloys listed in Table 302.3.5, W shall be as follows: For $T_i \le T_{cr}$, W = 1.0. For SI units, for $T_{cr} < T_i \le 816$°C, $W = 1 - 0.00164(T_i - T_{cr})$. For U.S. Customary units, for $T_{cr} < T_i \le 1{,}500$°F, $W = 1 - 0.000909(T_i - T_{cr})$. If T_i exceeds the upper temperature for which an allowable stress value is listed in Appendix A for the base metal, the value for W is the responsibility of the designer.

Use of factors other than those provided in Table 302.3.5 are permitted based on use of creep test data for weldments other than Creep Strength Enhanced Ferritic (CSEF) materials, which can have significant issues in the heat affected zone which require longer term testing to observe. Further, with the owner's approval, extensive successful service experience may be used to justify higher *W* factors. Succesful experience must include same or like material, weld metal composition, and welding process under equivalent, or more severe, sustained operating conditions.

In addition to changes to the weld joint factors, additional changes were made to the fabrication and examination rules for welds in Elevated Temperature Fluid Service.

1. Piping materials (this includes those in accordance with listed standards) with longitudinal or spiral welds in P-No. 4 or P-No, 5 materials are required to be examined by 100% radiography or 100% ultrasonic examination (para. 305.2.4).
2. Additional examination requirements are specified for elevated temperature piping in para. 341.4.4. These include;
   a. 100% visual examination rather lhan random visual examination
   b. Specific discussion of evaluating the installed system to ensure that movement of the piping under all conditions of startup, operation, and shutdown will be accommodated without undue binding or unanticipated constraint.
   c. Longitudinal welds for P-No. 4 and P-No. 5 material made as a part of fabrication (welds made in the manufacturing process were covered by I, above), also require 100% radiographic or ultrasonic examination.
   d. Socket welds and branch connection welds for P-No. 4 and P-No. 5 materials that are not radiographed or ultrasonically examined are required to be examined by magnetic particle or liquid penetrant methods.
3. Weld metal requirements for CSEF materials are provided in note 3 of Table 302.3,5.
4. Required heat treatment conditions for use of the factors in Table 302.3.5 are specified for some materials therein.

Insert 3.2 Background on Weld joint Strength Reduction Factors

The weld joint strength reduction factors added in the 2004 edition were based on those provided in ASME BPVC Section HI. Subsection NH. Weld metal, base metal and cross-weld creep rupture data had been used to develop weld joint strength reduction factors in Subsection NH for limited material combinations. The basic weld joint strength reduction factors provided in ASME B3I.3 were developed based on evaluating the factors provided in Subsection NH. Factors for 100,000 hour durations were used. These factors are plotted in the attached charts.

Figure 3.3 shows the data for 2-l/4Cr-lMo, 800H, 9Cr-lMo, and austenitic stainless steel with at least 0.04% minimum carbon content weld metal as well as the basic weld joint strength reduction factor, *W*, provided for use in the absence of more applicable data. Figure 3.4 shows data for type 304 and 316 base material, required by Subsection NH to have a minimum 0.04% carbon content. welded with lower creep strength low carbon (<0.04%) weld metal. Note that the sudden drop in the weld joint strength reduction factor for Type 316 stainless steel between 510°C (950°F) and538°C (1000°F) is due to a shift from tensile property to creep property control of the allowable stress. The weld joint strength reduction factor is 1.0 when tensile properties govern the allowable stress.

The data in Figure 3.4 were considered to represent a poor choice of weld material for high temperature service and were not included in the data used to develop the weld joint strength reduction factors. Selection of an appropriate weld material for elevated temperature service remains the responsibility of the Designer. The weld joint strength reduction factors will not protect against poor material selection, but are a step towards addressing creep effects in high temperature welds.

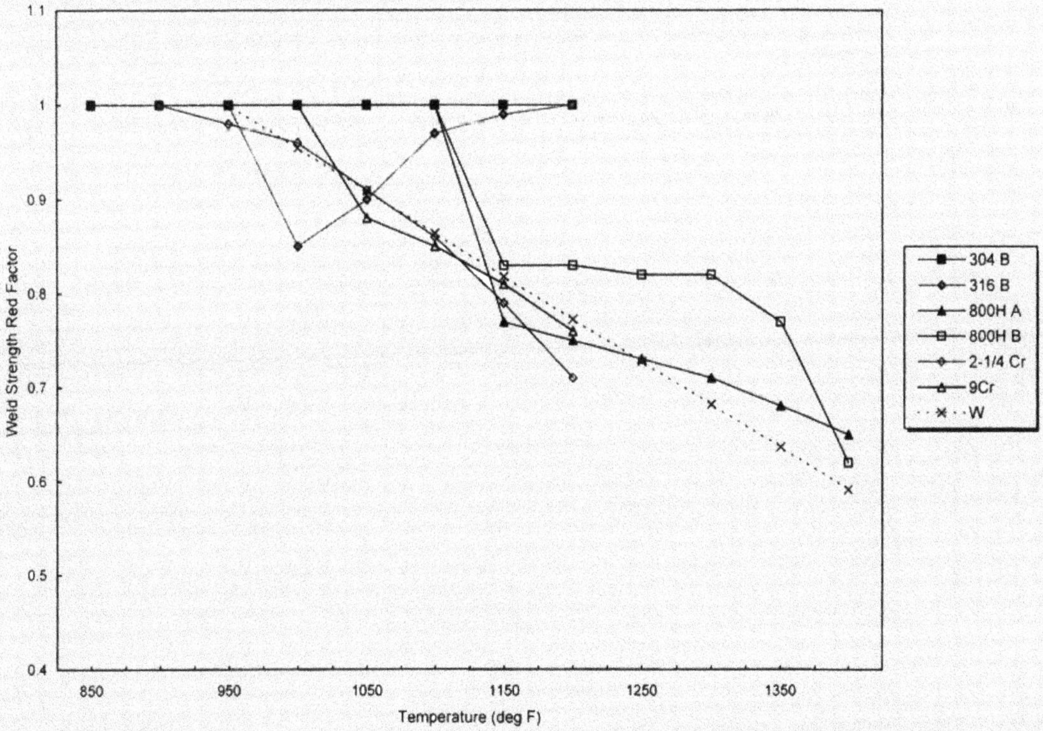

Key to Weldment Designation

304 B 304 stainless steel welded with SFA 5.22 EXXXT-G (16-8-2 chemistry); SFA 5.4 E 16-8-2; and SFA 5.9 ER 16-8-2

316 B 316 stainless steel welded with SFA 5.22 EXXXT-G [16-8-2 chemistry); SFA 5.4 e 16-8-2; and SFA 5.9 ER 16-8-2

800H A 800H welded with SFA 5.11 ENICrFe-2 (Inco A)

800H B 800H welded with SFA 5.14 ErNICr-3 (Inco 82)

2-1/Cr 2-1/4 CM Mo (60/30) welded with SFA 5.28 E 90C-B3; SFA 5.28 ER 90S-B3; SFA 5.5 E 90XX-B3 (>0.05C); SFA 5.23 EB 3; SFA 5.23 ECB 3 (>0.05C); SFA 5.29 E 90T1-B3 (>0.05C)

9Cr 9Cr-1Mo welded with standard 9Cr-1 Mo filler wire and modified 9Cr-1Mo filler wire GTA, SNA and TA welding processes

FIG. 3.3
BASIS FOR WELD JOINT STRENGTH REDUCTION FACTOR

Generally good experience with the strength of carbon steel weldments relative to carbon steel base material indicated that there was no need to use the strength reduction factor at temperatures below 510°C (950°F) for carbon steel. This is independent of other concerns, such as the potential for graphitization, that may limit use of carbon steel al elevated temperatures.

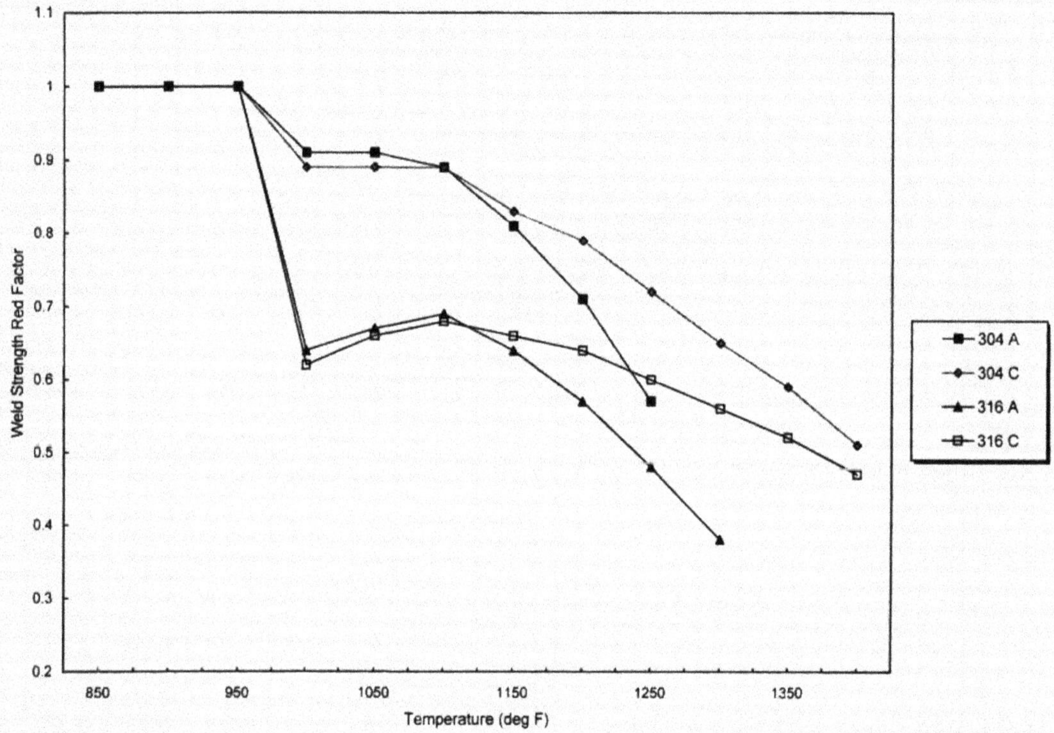

Key to Weldment Designation

304 A 304 stainless steel welded with SFA 5.22 E 308T and E 308LT; SFA 5.4 E 308 and E 308; and SFA 5.9 ER 308 and ER 308L

304 C 304 stainless steel welded with SFA 5.22 E 316T and E 316LT-1, -2, and -3; SFA 5.4 E 316 and E 316L; and SFA 5.9 ER 316 and ER 316L

316 A 316 stainless steel welded with SFA 5.22 E 308T and E 308LT; SFA 5.4 E 308 and E 308L; and SFA 5.9 ER 308 and ER 308L

316 C 316 stainless steel welded with SFA 5.22 E 316T and E 316LT-1, and -2; SFA 5.4 E 316 and E 316L; and SFA 5.9 ER 316 and ER 316L

FIG. 3.4
LOW CARBON WELDS WITH HIGHER CARBON BASE MATERIAL

3.5 ALLOWANCES FOR PRESSURE AND TEMPERATURE VARIATIONS

Although the Code does not use the term "maximum allowable working pressure (MAWP)," the concept is useful in a discussion of the allowances for variations. Pressure design of piping systems is based on the design conditions. Because piping systems are assemblies of standardized parts, there is quite often significant pressure capacity in the piping beyond that needed for operations. Let's call it the MAWP. The allowances for variations are relative to the MAWP. Because the variation is not permitted to exceed the test pressure and because the test pressure is a function of the design pressure, many designers and owners simplify project execution by setting the design conditions equal to the MAWP.

The allowances for variations are relative to the MAWP. The allowances for variations are not used in sustained, occasional, or displacement stress evaluations. They are only used in pressure design.

Increases in pressure and temperature above a certain percentage of the MAWP is permitted for short-term events when several conditions are satisfied.

- The piping system shall not have pressure-containing components of gray iron or other non-ductile metal.
- The increased pressure shall not exceed the test pressure.
- The nominal pressure stress (hoop stress for straight pipe or, for rated components, the pressure divided by the allowable pressure times two-thirds the yield strength) must be less than the yield strength of the material.
- The sustained stresses must be within the normally permitted limits.
- The total number of pressure-temperature variations above the design conditions must be less than 1000 over the life of the system (note that this is the number anticipated in the design of the system, not some count taken during operation of the system; the ASME B31.3 Code is for design of new piping systems).

If the above conditions are satisfied, and if the owner approves, the pressure rating or allowable stress (essentially the maximum allowable working pressure) may be exceeded by 33% for events that are not more than 10 hours at any one time nor more than 100 hours per year, and by 20% for events that are not more than 50 hours at any one time nor more than 500 hours per year. It is clear how a variation in pressure is handled. There is sometimes confusion relative to variations in temperature. The variation in temperature decreases the allowable stress or pressure rating. Thus, the stress or pressure may exceed the allowable value during a variation in temperature, without a change in pressure.

If the above variations are used, the designer must determine that the piping system, including the effects of the variations, is safe over the service life of the piping, using methods that are acceptable to the owner. Note that the pressure test provides such assurance for piping operating below the creep regime. For piping at elevated temperatures, within the creep regime for the material of construction, the pressure test does not ensure long-term pressure integrity. Therefore, Appendix V of ASME B31.3 was provided to evaluate the effect of short-term variations at elevated temperature on the life of the piping system.

As an alternative to the prior two paragraphs, if the variation is self-limiting, such as by accumulation during a pressure-relieving event, and lasts no more than 50 hours at any one time nor more than 500 hours per year, an allowable variation of 20% is permitted without the owner's approval and without requiring that the designer determine that the piping system is safe with the variations.

3.6 OVERPRESSURE PROTECTION

As discussed above in the section on design pressure, the piping system must either be designed to safely contain the maximum expected pressure considering such factors as failure of control devices and dynamic events such as surge, or be provided with overpressure protection, such as a safety relief valve. This basic requirement is provided in para. 301.2.2, whereas requirements for relief devices are provided in para. 322.6, Pressure Relieving Systems.

For example, consider a 4200 kPa (610 psi) system goes through a pressure-letdown valve to a 2100 kPa (305 psi) system and then through another letdown valve to a 1035 kPa (150 psi) system. The designer will have to consider possible failure of both letdown valves. One way to protect against failure of the low-pressure system is design it for at least 4200 kPa (610 psi). Another more common approach is to provide a relief device capable of handling the flow from the two failed letdown valves without exceeding the maximum pressure for the low-pressure system. Although not frequently used, the Code also allows protection of the piping using instrumentation that would, for example, close the letdown valves when the pressure was too high.

If a pressure-relieving device is used, ASME B31.3 refers to the requirements of ASME BPVC, Section VIII, Division 1. However, there are some exceptions. For example, design pressure is substituted for maximum allowable working pressure (MAWP), because the latter term is not used in the Piping Code.

More significantly, the set pressure requirements of ASME BPVC Section VIII, Division 1, are not followed. If they were, then the variations could never realistically be permitted, because any variation in pressure above the design pressure would result in opening of the pressure relief device. Rather than limit the set pressure to the design pressure, the Code allows the set pressure to be any value, as long as the maximum pressure during the relieving event, including consideration of potential accumulation (additional pressure buildup beyond the pressure at which the valve opens), does not exceed one of the following two alternatives a) the maximum relieving pressure permitted by BPVC Section VIII, Division 1, or b) the maximum pressure permitted in the allowances for variations provisions of ASME B31.3. However, the owner's approval is required for the set pressure to exceed the design pressure.

For liquid thermal expansion relief devices, the set pressure may simply be established at 120% of the design pressure, without any requirement for owner's approval nor any need to comply with the additional requirements necessary to use the allowances for variations. The set pressure can be set at higher than 120%; however, then the other requirements of justifications and approvals explained above must be satisfied.

Block valves are permitted in the inlet and discharge lines to the relief valve, with certain limitations. These requirements are outlined in para. 322.6.1.

CHAPTER

4

Pressure Design

4.1 METHODS FOR INTERNAL PRESSURE DESIGN

The ASME B31.3 Code provides four basic methods for the design of components for internal pressure, as described in para. 302.2:

(a) Components in accordance with listed standards for which pressure ratings are provided in the standard, such as ASME B16.5 for flanges, are considered suitable by ASME B31.3 for the pressure rating specified in the standard. Note that the other methods of pressure design provided in ASME B31.3 can be used to rerate such listed components and/or extend their temperature range.
(b) Some listed standards, such as ASME B16.9 for pipe fittings, state that the fitting has the same pressure rating as matching seamless pipe. ASME B31.3 modifies this slightly by stating that the fittings are accepted to have the same rating as the matching seamless pipe, considering the manufacturer's under tolerance for the fittings. This takes into consideration the typical mill tolerance for pipe. Note that design calculations are not usually performed for these components; design calculations are performed for the straight pipe, and matching fittings are simply selected.
(c) Design equations for some components, such as straight pipe and branch connections, are provided in para. 304 of ASME B31.3. These can be used to determine the required wall thickness with respect to internal pressure of components. Furthermore, some specific branch connection designs are assumed to be acceptable.
(d) Components that are not in accordance with a listed standard and for which design equations are not provided in the Code are treated in para. 304.7.2. This paragraph provides accepted methods, such as burst testing, for determining the pressure capacity of unlisted components.

The equations in the Code provide the minimum thickness required to limit the membrane, and in some cases bending stresses in the piping component, to the appropriate allowable stress. Mechanical and corrosion/ erosion allowances must be added to this thickness. Finally, the nominal thickness selected must be such that the minimum thickness that may be provided, per specifications considering mill tolerance, is at least equal to the required minimum thickness.

The pressure design rules in the Code are based on maximum normal stress, or maximum principal stress, versus maximum shear stress, or von Mises stress intensity. When the rules were developed in the 1940's, it was understood that stress intensity provided a better assessment of yielding, but it was felt that the maximum principal stress theory could generally provide a better measure of pressure capacity in situations where local yielding could simply lead to stress redistribution [Rossheim and Markl (1960)].

Mechanical allowances include physical reductions in wall thickness, such as due to threading and grooving the pipe. Corrosion and erosion allowances are based on the anticipated corrosion and/or erosion over the life of the pipe. This is based on estimates, experience, or literature, such as National Association of Corrosion

Engineers (NACE) publications. These allowances are added to the pressure design thickness to determine the minimum required thickness of the pipe or component when it is new.

For threaded components, the nominal thread depth (dimension *h* of ASME B 1.20.1; see Appendix I or equivalent) is used for the mechanical allowance. For machined surfaces or grooves where the tolerance is not specified, the Code requires that a tolerance of 0.5 mm (0.02 in.) on the depth of the cut be assumed.

Mill tolerances are provided in specifications. The most common tolerance on the wall thickness of straight pipe is 12.5%. This means that the wall thickness at any given location around the circumference of the pipe must not be less than 87.5% of the nominal wall thickness. Note that the tolerance on pipe weight is typically tighter, so that the volume of metal and its weight may be present although a thin region would control design for hoop stress due to internal pressure.

The appropriate specification for the pipe must be referred to in order to determine the specified mill tolerance. For example, plate typically has an undertolerance of 0.25 mm (0.01 in.). However, pipe formed from plate does not have this undertolerance; it can be much greater. The pipe specification, which can permit a greater undertolerance, governs for the pipe. The manufacturer of pipe can order plate that is thinner than the nominal wall thickness for manufacturing the pipe, as long as the pipe specification mill tolerances are satisfied. However, the weight tolerance could then govern. For example, the thickness tolerance for A53 pipe is 12.5%, but the weight tolerance is 10%. As a result, the minimum average thickness for A53 welded pipe made from plate material would be 10% under thickness because of the weight tolerance.

4.2 PRESSURE DESIGN OF STRAIGHT PIPE FOR INTERNAL PRESSURE

Equations for pressure design of straight pipe are provided in para. 304.1. The minimum thickness of the pipe selected, considering manufacturer's minus tolerance, must be at least equal to t_m, defined as[1]

$$t_m = t + c \tag{4.1}$$

where

c = sum of the mechanical allowances plus corrosion and erosion allowances
t = pressure design thickness
t_m = minimum required thickness including allowances

For pipe with $t<D/6$, the basic equation for determining pressure design thickness is provided in the Code,[2]

$$t = \frac{PD}{2(SEW + PY)} \tag{4.2}$$

where

D = pipe outside diameter (not nominal diameter)
E = quality factor
P = internal design gage pressure
S = allowable stress value
W = weld joint strength reduction factor per para. 302.3.5(e). (See Section 3.4)
Y = coefficient provided in Table 304.1.1 of the Code and Table 4.1 here

Note that Eq. (4.2) is based on the outside diameter, rather than the inside diameter, which is used in pressure vessel codes. This is for a very good reason: The outside diameter of pipe is independent of wall thickness. That is, an NPS 6 pipe will have an outside diameter of 168.3 mm (6.625 in.) regardless of the wall thickness. Therefore, the wall thickness can be directly calculated when the outside diameter is used in the equation.

[1]ASME B31.3, Eq. (2).
[2]ASME B31.3, Eq. (3a).

TABLE 4.1
VALUES OF COEFFICIENT *Y* FOR *t* < D/6 (ASME B31.3-2020, Table 304.1.1)

	Temperature, °C (°F)							
Material	482 (900) and Below	510 (950)	538 (1,000)	566 (1,050)	593 (1,100)	621 (1,150)	649 (1,200)	677 (1,250) and Above
Ferritic steels	0.4	0.5	0.7	0.7	0.7	0.7	0.7	0.7
Austenitic steels	0.4	0.4	0.4	0.4	0.5	0.7	0.7	0.7
Nickel alloys UNS Nos. N06617, N08800, N08810, and N08825	0.4	0.4	0.4	0.4	0.4	0.4	0.5	0.7
Gray iron	0.0	...	...	...	...	...	...	...
Other ductile metals	0.4	0.4	0.4	0.4	0.4	0.4	0.4	0.4

Equation (4.2) is an empirical approximation of the more accurate and complex Lamé equation (ca. 1833). The hoop or circumferential stress is higher toward the inside of the pipe than toward the outside. This stress distribution is illustrated in Fig. 4.1. The Lamé equation, provided below, can be used to calculate the stress as a function of location through the wall thickness. Equation (4.2) is the Boardman equation. Although it has no theoretical basis, it provides a good match to the more accurate and complex Lamé equation for a wide range of diameter-to-thickness ratios. It becomes increasingly conservative for lower D/t ratios (thicker pipe) if Y is held constant.

The Lamé equation for hoop stress on the inside surface of pipe follows. Note that for internal pressure, the stress is higher on the inside than the outside. This is because strain in the longitudinal direction of the pipe must be constant through the thickness, so that any longitudinal strain caused by the compressive radial stress (due to Poisson effects and considering that the radial stress on the inside surface is equal to the surface traction of internal pressure) must be offset by a corresponding increase in hoop tensile stress to cause an offsetting Poisson effect on longitudinal strain. The Lamé equation is

$$\sigma_h = P\frac{0.5(D/t)^2 - (D/t) + 1}{(D/t) - 1} \tag{4.3}$$

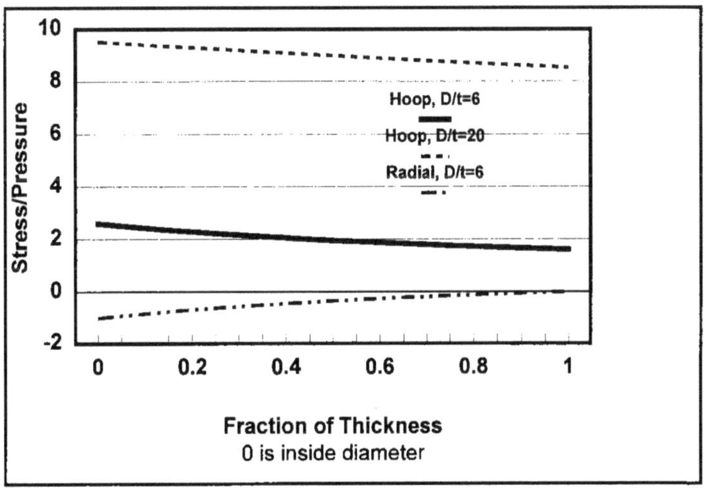

FIG. 4.1
STRESS DISTRIBUTION THROUGH PIPE WALL THICKNESS DUE TO INTERNAL PRESSURE

where

σ_h = hoop stress

The empirical Boardman representation of this simply bases the calculation of pressure stress on some intermediate diameter, between the inside and outside diameters of the pipe, as

$$\sigma_h = P\frac{D-2Yt}{2t} \tag{4.4}$$

where

$$Y = 0.4 \tag{4.5}$$

Simple rearrangement of the above equation and substitution of SEW for σ_h leads to Eq. (3a) of the code [Eq. (4.2) here]. Further, inside-diameter-based formulas add 0.6 times the thickness to the inside radius of the pipe rather than subtract 0.4 times the thickness from the outside radius. Thus, the inside-diameter-based formula in the pressure vessels codes and Eqs. (3a) and (3b) of ASME B31.3, the Piping Code, are consistent. The additional consideration in Eq. (3b) of ASME B31.3 is the addition of the allowances (internal corrosion increases the inside diameter in the corroded condition). With this additional consideration, Eq. (3b) of ASME B31.3 based on inside diameter provides the same required thickness as Eq. (3a) based on outside diameter.

A comparison of hoop stress calculated using the Lamé equation versus the Boardman equation (4.2) is provided in Fig. 4.2. Remarkably, the deviation of the Boardman equation from the Lamé equation is less than 1% for *D/t* ratios greater than 5.1. Thus, the Boardman equation can be directly substituted for the more complex Lamé equation.

For thicker wall pipe, ASME B31.3 provides the following equation for the calculation of the *Y* factor in the definition of *Y* in para. 304.1.1. Use of this equation to calculate *Y* results in Eq. (4.2), matching the Lamé equation for heavy wall pipe as well:

$$Y = \frac{d+2c}{D+d+2c} \tag{4.6}$$

The factor *Y* depends on temperature. At elevated temperatures, when creep effects become significant, creep leads to a more even distribution of stress across the pipe wall thickness. Thus, the factor *Y* increases, leading to a decrease in the calculated required wall thickness (for a constant allowable stress).

Three additional equations were formerly provided by the Code, but two were removed to be consistent with ASME B31.1 and simplify the Code. They may continue to be used, but SEW should be used in place of SE. The first of the removed equations is

$$t = \frac{PD}{2SE} \tag{4.7}$$

This equation is the simple Barlow equation, which is based on the outside diameter and is always conservative. It may be used, because it is always more conservative than the Boardman equation, which is based on a smaller diameter (except when $Y = 0$). The second removed equation is

$$t = \frac{D}{2}\left(1-\sqrt{\frac{SE-P}{SE+P}}\right) \tag{4.8}$$

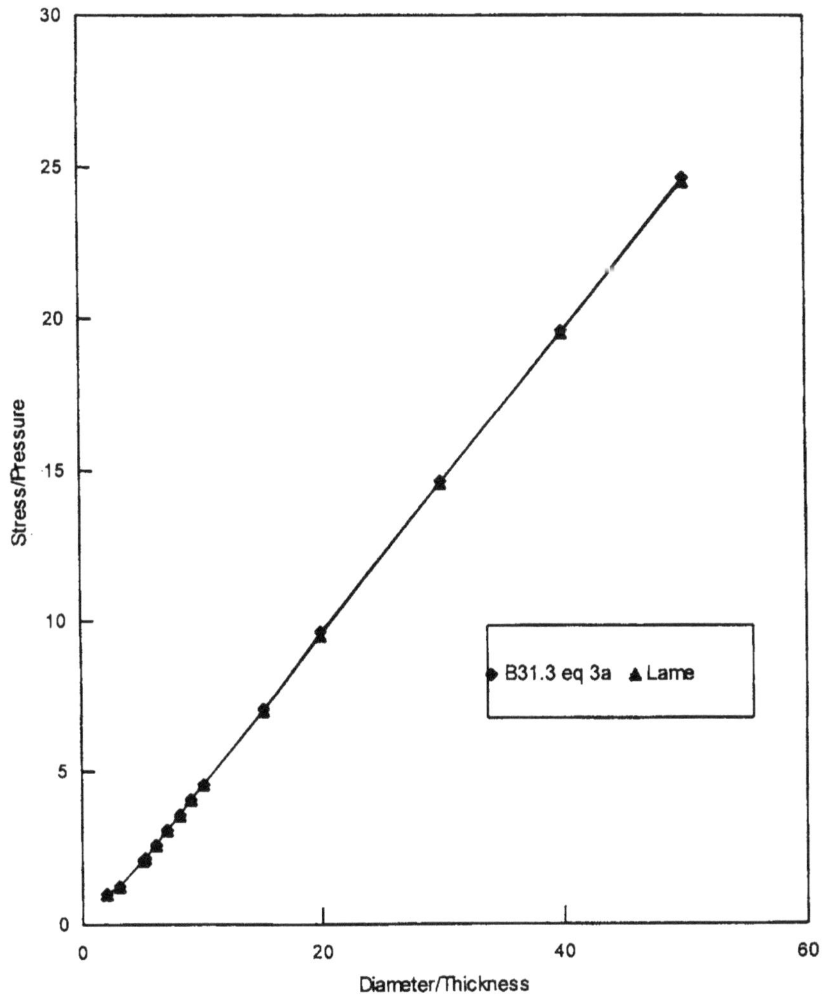

FIG. 4.2
COMPARISON OF LAMÉ AND BOARDMAN EQUATIONS

This equation is the Lamé equation rearranged to calculate thickness. Although it is not specifically included, it could be used, in accord with para. 300(c)3. However, it should not make a significant difference in the calculated wall thickness.

The following optional equation remains in ASME B31.33:

$$t = \frac{P(d+2c)}{2\left[SEW - P(1-Y)\right]} \tag{4.9}$$

where

d = inside diameter

[3]ASME B31.3, Eq. (3b).

Equation (4.9) is the same as (4.2), but with $(d+2c + 2t)$ substituted for D and rearranged to keep thickness on the left side.

Insert 4.1 Sample Wall Thickness Calculation

What is the required thickness of NPS 2 threaded A53 Grade B seamless pipe for the following conditions?:

- Design pressure = 150 psi
- Design temperature = 500°F
- Corrosion allowance (CA) = 1/16 in.
- SE = 18,900 psi
- W = 1.0
- D = 2.375 in.
- t = 150 (2.375)/[2(18,900 + 0.4 × 150)] = 0.0094 in.
- c = CA+thread depth = 0.0625 + 0.07 in.
- t_m = 0.0094 + 0.0625 + 0.07 = 0.14 in.

The minimum nominal pipe thickness, considering mill tolerance, is

$$\overline{T}_{\text{min.}} = \frac{t_m}{0.875} = 0.16\,\text{in.} \tag{4.10}$$

Schedule 80, XS pipe, with a nominal wall thickness of 0.218 in. is acceptable.

Insert 4.2 Basic Stress Calculations for Cylinders Under Pressure

The average (through-thickness) circumferential and longitudinal (axial) stresses in a cylinder due to internal pressure can be calculated from equilibrium considerations. The circumferential stresses can be calculated from a longitudinal section, as shown in Fig, 4,3. The forces acting on that section must equilibrate, or, per Newton's law, the parts on either side of the section will start accelerating away from each other.

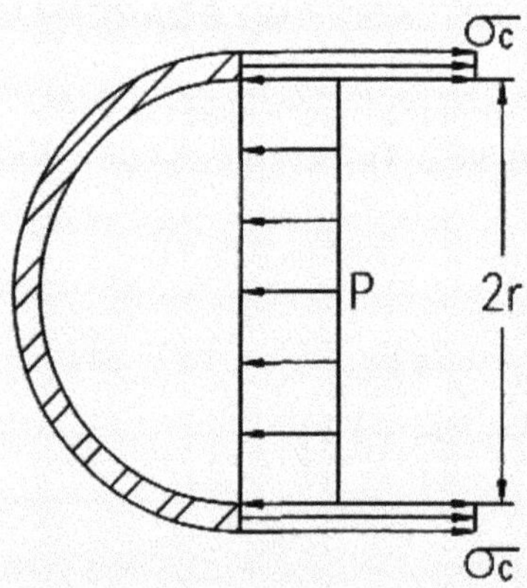

FIG. 4.3
EQUILIBRIUM AT A CIRCUMFERENTIAL CUT

The pressure force acting on the section is $2rP$, where r is the inside radius and P is the internal pressure. The circumferential force in the pipe wall resisting the pressure force is $2t\sigma_c$, where t is the thickness (times two, because there are two sides), and σc is the average circumferential stress. These must be equal, and solving for σc, one arrives at the following equation:

$$\sigma_c = \frac{Pr}{t} \tag{4.11}$$

The longitudinal stress can be determined by making a girth cut (as a guillotine cut) on the pipe, as shown in Fig. 4.4, The pressure force acting on the section is $\pi r^2 P$ and the longitudinal force in the pipe wall resisting this pressure force is $2\pi r t \sigma_l$. Note that, to be more precise, the mean radius of the pipe should be used to calculate the area of the pipe wall, but using the inside radius is generally close enough and conservative. The longitudinal stress can be calculated by equating these two forces and solving for σ_l:

$$\sigma_l = \frac{Pr}{2t} \tag{4.12}$$

Thus, the longitudinal stress in a cylinder due to internal pressure is about one-half of the circumferential stress. This is quite convenient in the design of piping, because the wall thickness is determined based on pressure design. This leaves at least one-half of the strength in the longitudinal direction available for supporting the pipe weight.

A common example of the fact that the stress in the circumferential direction is twice that in the longitudinal direction can be found when cooking a hot dog. A hot dog has a pressure-containing skin. When the internal temperature reaches the point where the fluids contained inside begin to vaporize, the hot dog skin is pressurized. When the skin is overpressurized and fails, the split is always longitudinal, transverse to the direction of highest stress, the circumferential direction. Hot dogs, at least in the experience of this author, never experience guillotine failures during cooking.

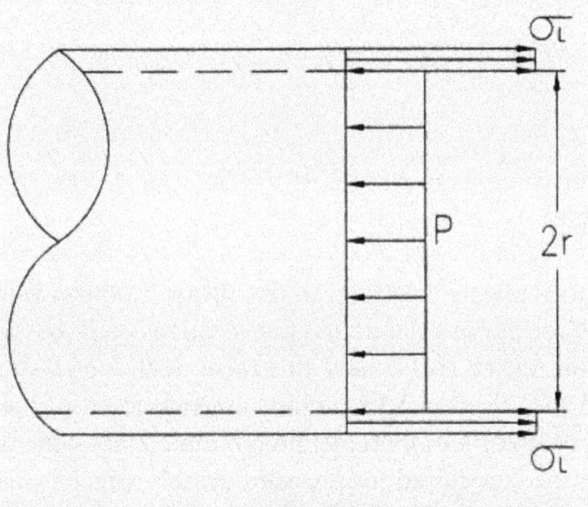

FIG. 4.4
EQUILIBRIUM AT A LONGITUDINAL CUT

4.3 PRESSURE DESIGN OF STRAIGHT PIPE UNDER EXTERNAL PRESSURE

For straight pipe under external pressure, there is a membrane stress check in accordance with Eq. (3a) [or (3b)] of ASME B31.3 [the equation for internal pressure; Eq. (4.2) or Eq. (4.9) here], as well as a buckling check in accordance with the external pressure design rules of ASME BPVC, Section VIII, Division 1.

Flanges, heads, and stiffeners that comply with ASME BPVC, Section VIII, Division 1, para. UG-29 are considered stiffeners. The length between stiffeners is the length between such components. The buckling pressure is a function of geometrical parameters and material properties.

Buckling pressure calculations in ASME BPVC, Section VIII, Division 1 require first the calculation of a parameter A, which is a function of geometry, and then a parameter B, which depends on A and a material property curve. The charts that provide the parameter B account for plasticity between the proportional limit of the stress-strain curve and the 0.2% offset yield stress. The chart for determination of A is provided in Fig. 4.5. A typical chart for B is provided in Fig. 4.6.

Two equations are provided for calculating the maximum permissible external pressure. The first uses the parameter B,

$$p = \frac{4B}{3\,D/t} \tag{4.13}$$

where

B = parameter from material curves in ASME BPVC, Section II, Part D, Subpart 3
D = Inside diameter (note that this differs from the ASME B31.3 definition of D, which is outside diameter; this is the Pressure Vessel Code definition of D)
p = allowable external pressure
t = pressure design thickness

The second equation is for elastic buckling, and must be used when the value of A falls to the left of the material property curves that provide B. This equation is

$$p = \frac{2AE}{3\,D/t} \tag{4.14}$$

where

A = parameter from geometry curves in ASME BPVC Section II, Part D, Subpart 3, Fig. G (see Fig. 4.5 here)

The second equation is based on elastic buckling, so the elastic modulus is used. A chart of B could be used, with the linear elastic portion of the curve extended to lower values of B, but this would unnecessarily enlarge the charts. The charts provided in ASME B31.5 have this form, with the elastic lines extended.

The procedures of ASME BPVC Section VIII include consideration of the allowable out-of-roundness in pressure vessels, and use a design margin of three. Although pipe is not generally required to comply with the same out-of-roundness tolerance as is required for pressure vessels, this has historically been ignored, and has not led to any apparent problems.

The basis for the approach of ASME BPVC Section VIII is provided in Bergman (1960), Holt (1960), Saunders and Windenburg (1960), Windenburg and Trilling (1960), and Windenburg (1960).

A new buckling evaluation procedure, provided in *ASME BPV Code Case 2286, Alternative Rules for Determining Allowable External Pressure and Compressive Stresses for Cylinders, Cones, Spheres, and Formed Heads Section VIII, Division 1,* is more relevant to piping, because it permits consideration of combined loads, including external pressure, axial load, and gross bending moment. Code Case 2286 also provides a more

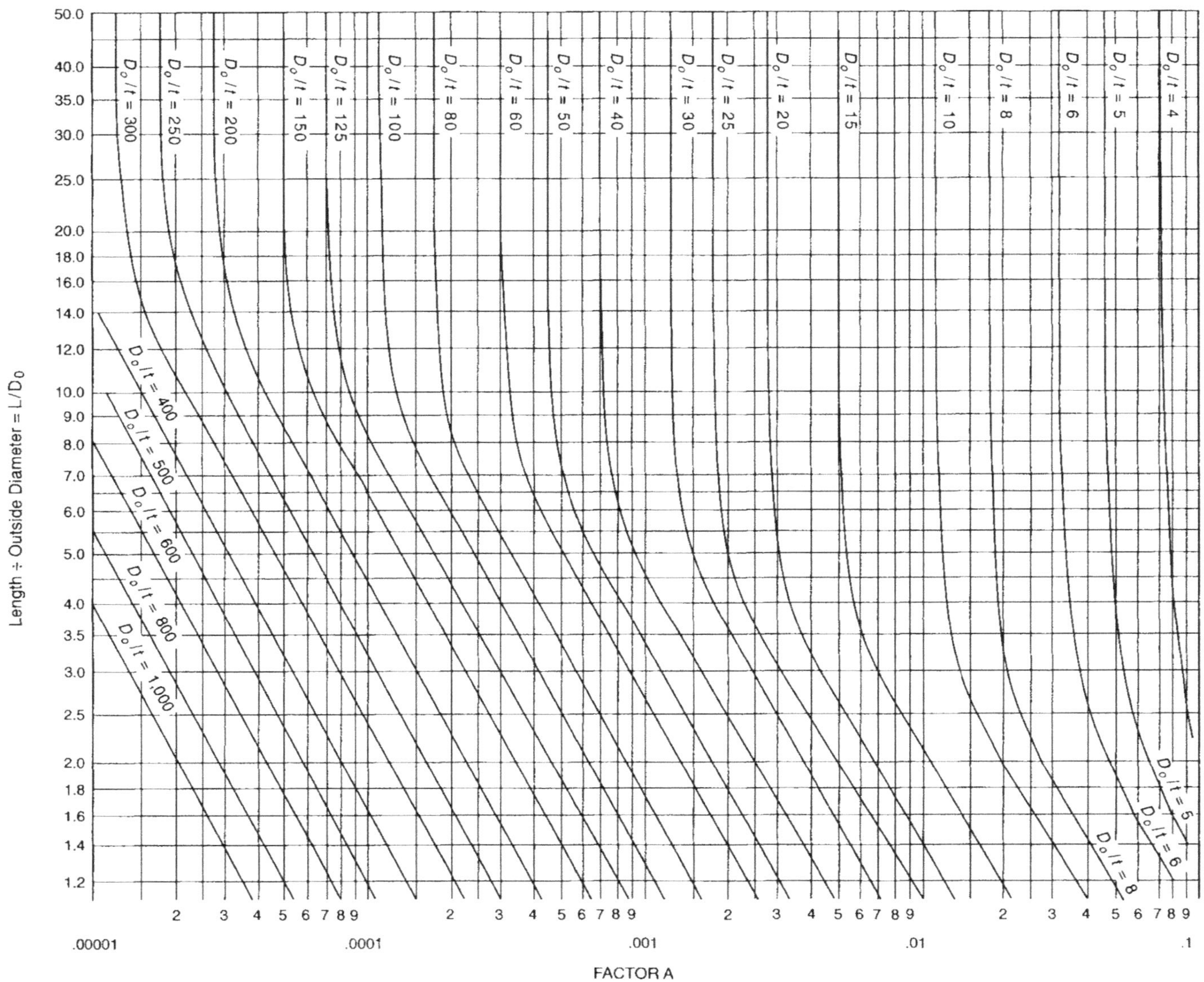

FIG. 4.5
CHART FOR DETERMINING A (ASME BPVC, SECTION II, PART D, SUBPART 3, FIG. G). TABLE G CITED IN THE FIGURE IS GIVEN IN ASME BPVC, SECTION II

The area A_1 of metal removed is defined as

$$A_1 = t_h d_1 \left(2 - \sin\beta\right) \tag{4.15}$$

where

d_1 = effective length removed from the run pipe at the branch connection
t_h = pressure design thickness of the header
β = the smaller angle between axes of branch and run

For a 90 deg. branch connection, d_1 is effectively, the largest possible inside diameter of the branch pipe; the inside diameter of the pipe if fully corroded and with the full mill tolerance removed from the inside of the pipe.

Figure 4.7 illustrates the nomenclature and process.

The angle β is used in the evaluation because a lateral connection, a branch connection with a β other than 90°, creates a larger hole in the run pipe. This larger hole must be considered in d_1. For a lateral, d_1 is the branch pipe inside diameter, considering mill tolerance and corrosion/erosion allowance, divided by sin β. The (2 – sin β) term is used to provide additional reinforcement that is considered to be appropriate because of the geometry of the branch connection.

The pressure design thickness is the pressure design thickness of the run pipe, with one exception. If the run pipe is welded and the branch does not intersect the weld, the weld quality factor E and weld joint strength reduction factor W should not be used in calculating the wall thickness. The weld quality factor and weld joint strength reduction factor only reduce the allowable stress at the location of the weld.

Only the pressure design thickness is used in calculating the required area since only the pressure design thickness is required to resist internal pressure. Corrosion allowance and mill tolerance at the hole are obviously of no consequence.

The area removed, A_1, must be replaced by available area around the opening. This area is available from excess wall thickness that may be available in the branch and run pipes as well as added reinforcement, and the welds that attach the added reinforcement. This metal must be relatively close to the opening of the run pipe to reinforce it. Thus, the metal must be within a certain limited area in order to be considered appropriate reinforcement of the opening.

The limit of reinforcement along the run pipe, taken as a dimension from the centerline of the branch pipe where it intersects the run pipe wall, is d_2, defined by

$$d_2 = \text{greater of } \left[d_1, (T_b - c) + (T_h - c) + d_1/2\right] \quad \text{but } d_2 \le D_h \tag{4.16}$$

where

D_h = outside diameter of header or run pipe
T_b = minimum thickness of branch pipe
T_h = minimum thickness of run pipe
c = allowance (mechanical, corrosion, erosion)

The limit of reinforcement along the branch pipe measured from the outside surface of the run pipe is L_4, defined as the lesser of 2.5 (T_h - c) and 2.5 (T_b - c) + T_r, where T_r is the minimum thickness of reinforcement.

The reinforcement within this zone is required to exceed A_1. This reinforcement consists of excess thickness available in the run pipe, A_2, excess thickness available in the branch pipe, A_3, and additional reinforcement, A_4. These can be calculated as follows:

$$A_2 = \left(2d_2 - d_1\right)\left(T_h - t_h - c\right) \tag{4.17}$$

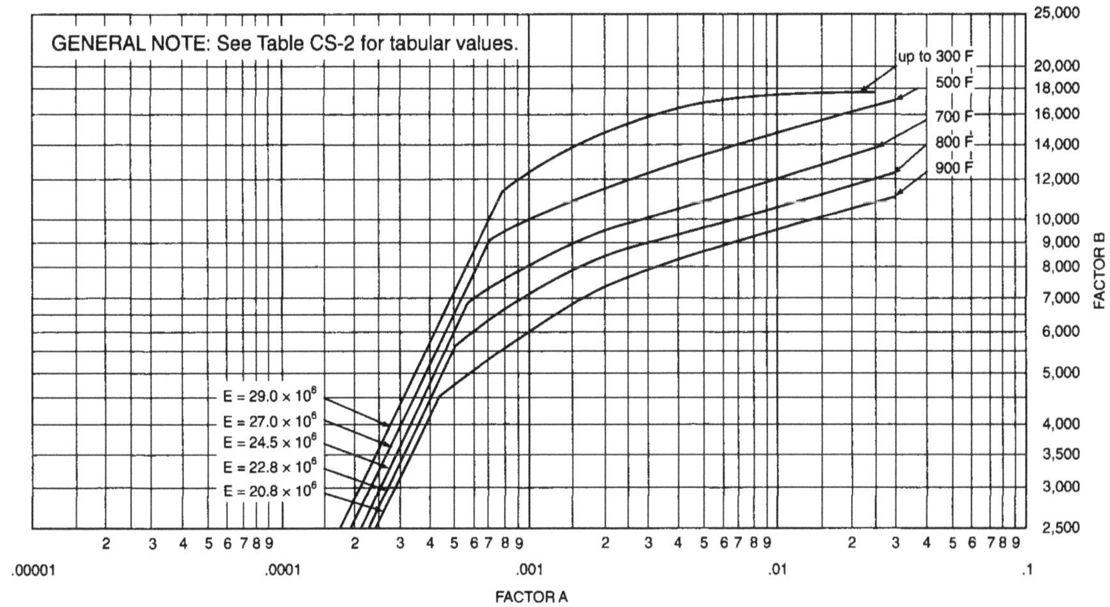

FIG. 4.6
TYPICAL CHART FOR DETERMINING B (ASME BPVC, SECTION II, PART D, SUBPART 3, FIG. CS-2). TABLE CS-2 CITED IN THE FIGURE IS GIVEN IN ASME BPVC, SECTION II

uniform margin of safety and a higher allowable external pressure. It is not at present explicitly recognized in ASME B31.3, but could be considered, given para. 300(c)3.

4.4 PRESSURE DESIGN OF WELDED BRANCH CONNECTIONS

The pressure design of branch connections is based on a rather simple approach, although the resulting design calculations are the most complex of the design-by-formula approaches for pressure design provided in the Code. A branch connection cuts a hole in the run pipe. The metal removed is no longer available to carry the forces due to internal pressure. An area replacement concept is used for those branch connections that do not either comply with listed standards or with certain designs (see Section 4.7). The area of metal removed by cutting the hole, to the extent that it was required for internal pressure, must be replaced by extra metal in a region around the branch connection. This region is within the limits of reinforcement, defined later.

The simplified design approach is limited with respect to the geometries to which it is considered applicable. These limitations are as follows:

- The run pipe diameter-to-thickness ratio is less than 100.
- The branch-to-run diameter ratio is not greater than one.
- The angle β (angle between branch and run pipe axes) is at least 45°.
- The axis of the branch intersects the axis of the run pipe.

Where the above limitations are not satisfied, the designer is referred to para. 304.7.2 (see Section 4.15). Alternatives in that paragraph include proof testing and finite-element analysis. Paragraph 304.3.5(e) suggests consideration of integral reinforcement or complete encirclement reinforcement for such branch connections.

The area A_1 of metal removed is defined as

$$A_1 = t_h d_1 (2 - \sin\beta) \tag{4.15}$$

where

d_1 = effective length removed from the run pipe at the branch connection
t_h = pressure design thickness of the header
β = the smaller angle between axes of branch and run

For a 90 deg. branch connection, d_1 is effectively, the largest possible inside diameter of the branch pipe; the inside diameter of the pipe if fully corroded and with the full mill tolerance removed from the inside of the pipe.

Figure 4.7 illustrates the nomenclature and process.

The angle β is used in the evaluation because a lateral connection, a branch connection with a β other than 90°, creates a larger hole in the run pipe. This larger hole must be considered in d_1. For a lateral, d_1 is the branch pipe inside diameter, considering mill tolerance and corrosion/erosion allowance, divided by sin β. The (2 – sin β) term is used to provide additional reinforcement that is considered to be appropriate because of the geometry of the branch connection.

The pressure design thickness is the pressure design thickness of the run pipe, with one exception. If the run pipe is welded and the branch does not intersect the weld, the weld quality factor E and weld joint strength reduction factor W should not be used in calculating the wall thickness. The weld quality factor and weld joint strength reduction factor only reduce the allowable stress at the location of the weld.

Only the pressure design thickness is used in calculating the required area since only the pressure design thickness is required to resist internal pressure. Corrosion allowance and mill tolerance at the hole are obviously of no consequence.

The area removed, A_1, must be replaced by available area around the opening. This area is available from excess wall thickness that may be available in the branch and run pipes as well as added reinforcement, and the welds that attach the added reinforcement. This metal must be relatively close to the opening of the run pipe to reinforce it. Thus, the metal must be within a certain limited area in order to be considered appropriate reinforcement of the opening.

The limit of reinforcement along the run pipe, taken as a dimension from the centerline of the branch pipe where it intersects the run pipe wall, is d_2, defined by

$$d_2 = \text{greater of } \left[d_1, (T_b - c) + (T_h - c) + d_1/2\right] \tag{4.16}$$
$$\text{but } d_2 \leq D_h$$

where

D_h = outside diameter of header or run pipe
T_b = minimum thickness of branch pipe
T_h = minimum thickness of run pipe
c = allowance (mechanical, corrosion, erosion)

The limit of reinforcement along the branch pipe measured from the outside surface of the run pipe is L_4, defined as the lesser of 2.5 (T_h - c) and 2.5 (T_b - c) + T_r, where T_r is the minimum thickness of reinforcement.

The reinforcement within this zone is required to exceed A_1. This reinforcement consists of excess thickness available in the run pipe, A_2, excess thickness available in the branch pipe, A_3, and additional reinforcement, A_4. These can be calculated as follows:

$$A_2 = (2d_2 - d_1)(T_h - t_h - c) \tag{4.17}$$

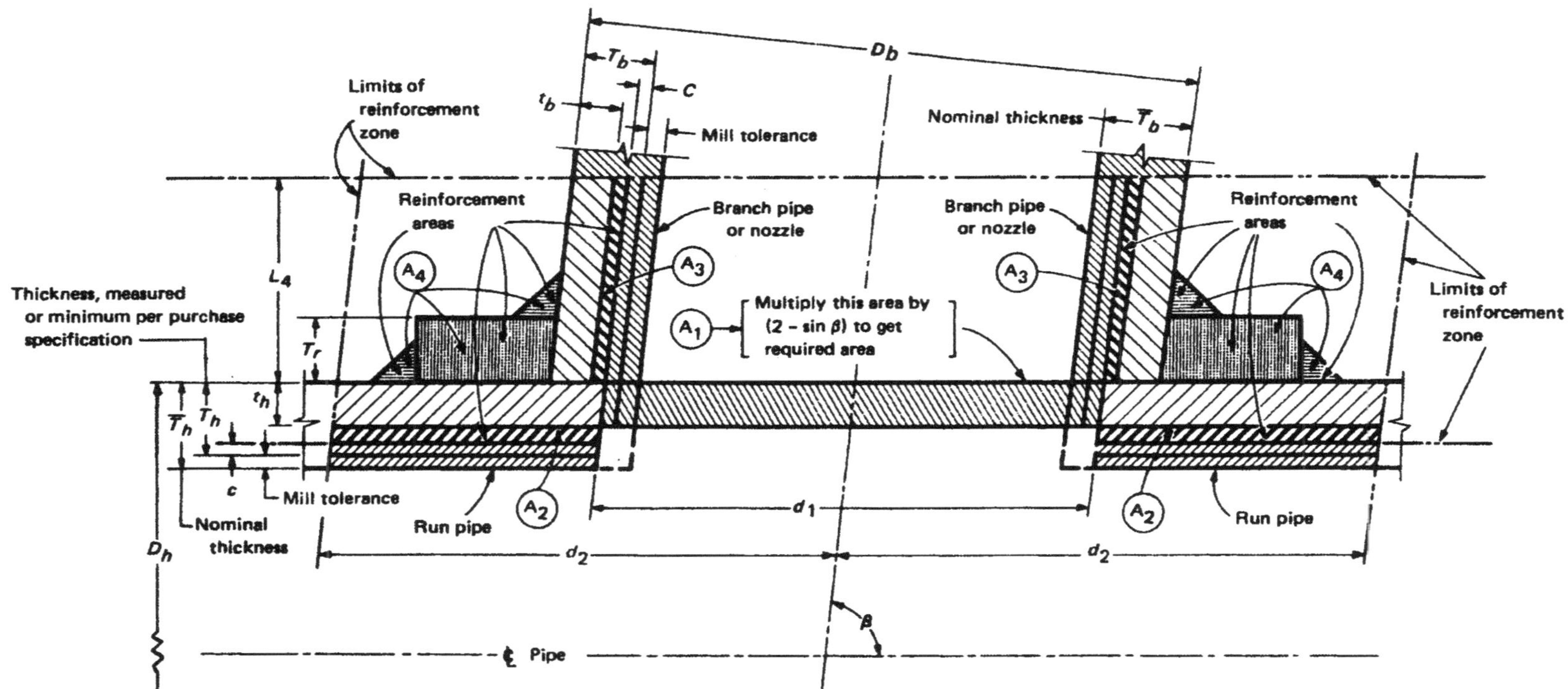

GENERAL NOTE: This Figure illustrates the nomenclature of para. 304.3.3. It does not indicate complete welding details or a preferred method of construction. For typical weld details, see Fig. 328.5.4D.

FIG. 4.7
ILLUSTRATION OF FABRICATED BRANCH CONNECTION SHOWING NOMENCLATURE (ASME B31.3, FIG. 304.3.3)

$$A_3 = 2L_4\left(T_b - t_b - c\right)/\sin\beta \tag{4.18}$$

The area $A4$ is the area of properly attached reinforcement and the welds that are within the limits of reinforcement. The Code specifies minimum weld sizes in para. 328.5.4. The designer is directed to assume that the minimum dimensions specified by the Code are provided, unless the welder is specifically directed to make larger welds. The ASME B31.3 Code does not require the designer to specify branch connection weld size, because generally acceptable minimum sizes are specified by the Code. Additionally, the ASME B31.3 Code differs from the Pressure Vessel Code in that strength calculations for load paths through the weld joints are not required.

In many cases, these calculations are avoided for fabricated branch connections by simply requiring the the reinforcing pad thickness to be the same as the run pipe and the width of the reinforcing pad to be half of the branch diameter. Such a reinforcing pad always provides adequate reinforcement. This approach allows the fabricator to use a scrap piece of run pipe for the reinforcing pad.

4.5 PRESSURE DESIGN OF EXTRUDED OUTLET HEADER

An extruded outlet header is a branch connection formed by extrusion, using a die or dies to control the radii of the extrusion. Paragraph 304.3.4 provides area replacement rules for such connections; they are applicable for 90-deg. branch connections where the branch pipe centerline intercepts the run pipe centerline, and where there is no additional reinforcement. Figure 4.8 (ASME B31.3, Fig. 304.3.4) shows the geometry of an extruded outlet header. Extruded outlet headers are subject to minimum external contour radius requirements, depending on the diameter of the branch connection.

A similar area replacement calculation is used as described in Section 4.4 for fabricated branch connections, except that the required replacement area is reduced for smaller branch-to-run pipe diameter ratios. The replacement area is from additional metal in the branch pipe, additional metal in the run pipe, and additional metal in the extruded outlet lip.

4.6 ADDITIONAL CONSIDERATIONS FOR BRANCH CONNECTIONS UNDER EXTERNAL PRESSURE

Branch connections under external pressure are covered in para. 304.3.6. The same rules described in Sections 4.4 and 4.5 are used. However, only one-half of the area described in Section 4.4, covering welded branch connections, requires replacement. In other words, only one-half of the area A_1 requires replacement. Also, the thicknesses used in the calculation are the required thicknesses for the external pressure condition.

4.7 BRANCH CONNECTIONS THAT ARE PRESUMED TO BE ACCEPTABLE

Some specific types of branch connections are presumed to be acceptable. This, of course, includes listed fittings (e.g., ASME B16.9 tees and MSS SP-97 branch connection fittings). It also includes the following:

- For branch connections DN 50 (NPS 2) or smaller that do not exceed one-fourth of the nominal size of the run pipe, threaded or socket welding couplings or half-couplings (Class 3000 or greater) are presumed to provide sufficient reinforcement as long as the minimum thickness of the coupling is at least as thick as the branch pipe.
- Branch connection fittings qualified per para. 304.7.2 are acceptable.

4.8 PRESSURE DESIGN OF BENDS AND ELBOWS

Bends were required to have, after bending, a wall thickness at least equal to the minimum required wall thickness for straight pipe in para. 304.2.1. However, this was changed in the 2000 Addendum. The Lorenz equation (ca. 1910) was included; it provides a means of calculating the required wall thickness. Note that the prior requirement that simply stated the thickness should be the same as required for straight pipe was deleted. The new requirement is more conservative for the intrados (inside curves) of bends and less conservative for the extrados (outside curve).

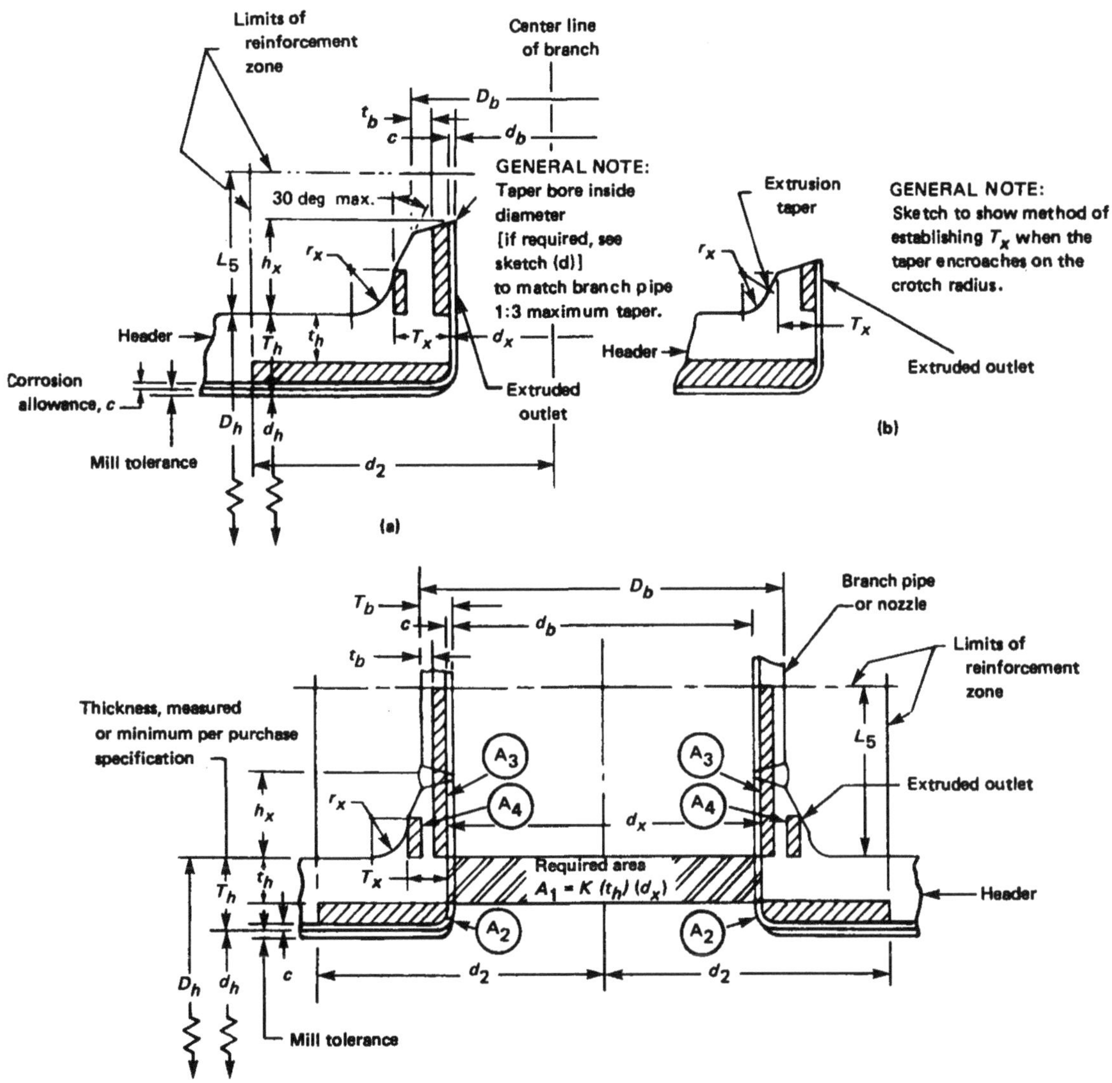

GENERAL NOTE: Sketch is drawn for condition where $K = 1.00$.

FIG. 4.8
ILLUSTRATION OF EXTRUDED OUTLET FITTING SHOWING NOMENCLATURE (ASME B31.3, FIG. 304.3.4). THIS FIGURE ILLUSTRATES THE NOMENCLATURE OF PARA. 304.3.4. IT DOES NOT INDICATE COMPLETE DETAILS OR A PREFERRED METHOD OF CONSTRUCTION

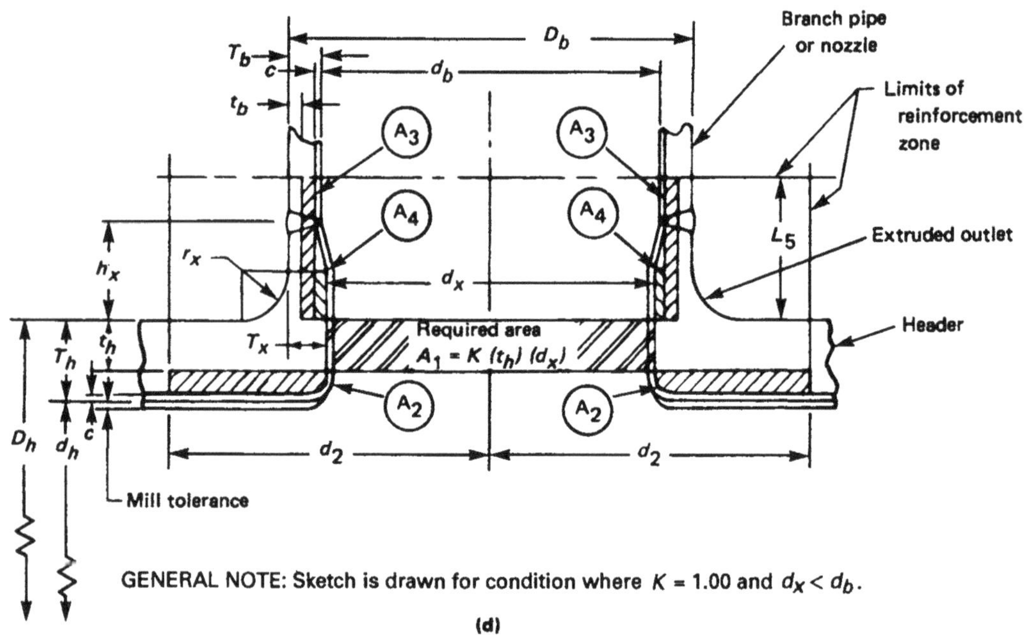

FIG. 4.8 CONTINUED

The Lorenz equation is basically the equation for a toroid. If the intrados and extrados had the same wall thickness, the inside would be subjected to higher hoop stress than straight pipe and the outside would be subjected to lower hoop stress than straight pipe. A simple way to envision this is that the inside has less metal over the curve and the outside has more metal over the curve. The Lorenz equation for an elbow or bend is given by[4]

$$t = \frac{PD}{2(SEW/I + PY)} \tag{4.19}$$

where the terms are as defined in Section 4.2 for Eq. (4.2), except for *I,* which is a stress index that accounts for the difference in hoop stress due to internal pressure in bends versus straight pipe.

On the inside curve of the bend, the intrados, we have[5]

$$I = \frac{4R_1/D - 1}{4R_1/D - 2} \tag{4.20}$$

Where

R_1 = radius of bend

[4]ASME B31.3, Eq. (3c).

[5]ASME B31.3, Eq. (3d).

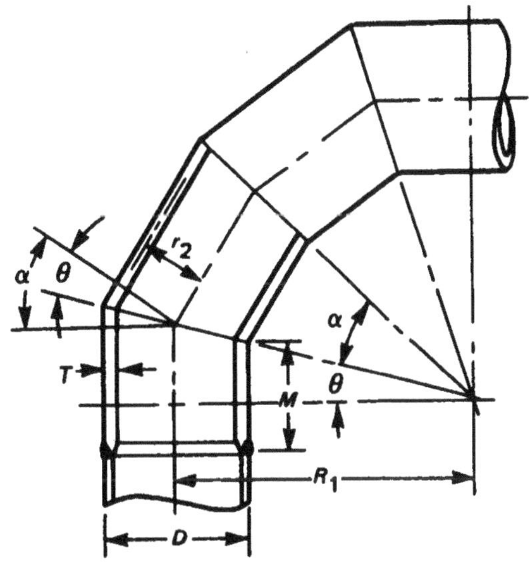

FIG. 4.9
ILLUSTRATION OF MITER BEND SHOWING NOMENCLATURE (ASME B31.3, FIG. 304.2.3)

On the outside of the bend, or the extrados, we have[6]

$$I = \frac{4R_1/D + 1}{4R_1/D + 2} \tag{4.21}$$

On the side of the elbow, or the crown, $I = 1.0$ (i.e., the hoop stress is the same as in straight pipe).

The thickness variation from the intrados to the extrados is required to be gradual, and the requirements are stated to apply at the midspan of the bend. The thickness at the ends is required to satisfy the required thickness for straight pipe per para. 304.1.

The normal process of making a bend by bending straight pipe produces this type of thickness variation. Part of the reason for providing these new rules is due to the practice of fabricating elbows by forming two "clamshells" out of plate and welding them together. This produces a bend of uniform thickness, and the thickness on the intrados would be too thin if it simply satisfied the required thickness for straight pipe.

Elbows in accordance with listed standards, or qualified by para. 304.7.2, are also permitted.

4.9 PRESSURE DESIGN OF MITER BENDS

Miter changes in direction with an angular offset of 3 deg. or less (angle α Fig. 4.9) do not require design consideration as a miter bend. The required wall thickness for this condition is the same as straight pipe. Design equations for multiple and single miter bends follow.

The design equation for determining the required wall thickness for internal pressure for multiple miter bends is the lesser of the following two equations, 4.22 and 4.23. Note that these equations are only valid when the angle θ does not exceed 22.5 deg. Equation 4c (4.24 herein) applies for angles θ greater than 22.5 deg.[7,8]

[6]ASME B31.3, Eq. (3e).
[7]ASME B31.3, Eq. (4a).
[8]ASME B31.3, Eq. (4b).

$$P_m = \frac{SEW(T-c)}{r_2}\left(\frac{T-c}{(T-c)+0.643\tan\theta\sqrt{r_2(T-c)}}\right) \tag{4.22}$$

$$P_m = \frac{SEW(T-c)}{r_2}\left(\frac{R_1-r_2}{R_1-0.5r_2}\right) \tag{4.23}$$

Single miters with an angle θ not greater than 22.5 deg. are calculated per equation 4.22. Otherwise, they are calculated per equation 4.24.[9]

$$P_m = \frac{SEW(T-c)}{r_2}\left(\frac{T-c}{(T-c)+1.25\tan\theta\sqrt{r_2(T-c)}}\right) \tag{4.24}$$

The miter wall thickness, T, used in Eqs. (4.22)–(4.24) is required to extend a distance at least M from the inside crotch of the miter end welds.

In the above equations, the following definitions apply:

E = quality factor
M = greater of $2.5(r_2T)^{0.5}$ and $(\tan\theta)(R_1-r_2)$
P_m = maximum allowable internal pressure for miter bends
R_1 = effective radius of miter bend, defined as the shortest distance from the pipe centerline to the intersection of the planes of adjacent miter joints
S = allowable stress from Table A-l in Appendix A of ASME B31.3
T = miter pipe wall thickness (measured or minimum per purchase specification)
W =weld joint strength reduction factor per para. 302.3.5(e)
c = sum of mechanical allowances plus corrosion and erosion allowances
r_2 = mean radius of pipe using nominal wall
θ = angle of miter cut
α = angle of change in direction at miter joint
$=2\theta$

4.10 PRESSURE DESIGN OF CLOSURES

Closures are covered in para. 304.4.1. Listed components, such as ASME B16.9 pipe caps, can be used for closures. The other two options provided in ASME B31.3 are to design the closure in accordance with ASME BPVC Section VIII, Division 1, or to qualify it as an unlisted component in accordance with para. 304.7.2 (see Section 4.15). Specific references to ASME BPVC, Section VIII, Division 1 paragraphs are provided for ellipsoidal, torispherical, hemispherical, conical, toriconical, and flat heads.

Openings in closures are covered in para. 304.4.2. These requirements are summarized below:

- If the opening is greater than one-half of the inside diameter of the closure (as defined in ASME BPVC Section VIII, Division 1, para. UG-36), it should be designed as a reducer per para. 304.6 if the closure is dished and as a flange per para. 304.5 if the closure is flat.

[9]ASME B31.3, Eq. (4c).

- Small openings and connections using branch connection fittings that comply with para. 304.3.2(b) or 304.3.2(c) are considered to be inherently adequately reinforced.
- The required area of reinforcement is determined per the relevant ASME BPVC Section VIII, Division 1 requirements [UG-37(b), UG-38, or UG-39], which depend on the type of closure. For example, only one-half of the area requires replacement for a flat head.
- The available area of reinforcement is calculated per the rules in ASME B31.3, specifically para. 304.3.3 or para. 304.3.4. Note that ASME B31.3 requires that boundaries for a curved closure follow the contour of the closure (versus a chord dimension).
- Rules for multiple openings follow para. 304.3.3 and para. 304.3.4 rules for multiple openings.

4.11 PRESSURE DESIGN OF FLANGES

Most flanges are in accordance with listed standards, such as ASME B16.5, and for larger flanges, ASME B 16.47. Appendix L is provided in ASME B31.3 to cover aluminum alloy flanges because they are not included in ASME B16.5. When a custom flange is required, design by analysis is permitted by para. 304.5.1. ASME B31.3 refers to the rules for flange design contained in ASME BPVC Section VIII, Division 1, Appendix 2 or ASME BPVC, Section VIII, Division 2, para. 4.16, but using the allowable stresses and temperature limits of ASME B31.3. For flanges with metal-to-metal contact outside of the bolt circle, the rules of ASME BPVC Section VIII, Division 1, Appendix Y are referenced.

4.12 PRESSURE DESIGN OF BLIND FLANGES

Most blind flanges are in accordance with listed standards, such as ASME B16.5. When designing a blind flange, the rules of ASME BPVC Section VIII, Division 1, para. UG-34 apply. The procedure in ASME BPVC Section VIII includes consideration of the moment due to the flange bolts, so the required bolt load for boltup and operation must be determined per the flange design rules of Appendix 2 of ASME BPVC Section VIII, Division 1.

4.13 PRESSURE DESIGN OF BLANKS

Blanks are flat plates that get sandwiched between flanges to block flow. A design equation for permanent blanks is provided in para. 304.5.3, as[10]

$$t_m = d_g \sqrt{\frac{3P}{16SEW}} + c \tag{4.25}$$

where

d_g = inside diameter of gasket for raised or flat face flanges, or the gasket pitch diameter for ring joint and fully retained gasketed flanges

Other terms are as defined in Section 4.2.

ASME B 16.48, Steel Line Blanks, was added as a listed standard in the 2004 edition. Blanks used for test purposes are not subject to these design rules and are often designed to higher allowable stress levels (e.g., 90% of the specified minimum yield strength).

[10]ASME B31.3, Eq. (15).

4.14 PRESSURE DESIGN OF REDUCERS

Most reducers in piping systems are in accordance with listed standards. However, pressure design per para. 304.7.2 and detailed design using rules for conical or toriconical closures are permitted. The rules for closures reference ASME BPVC Section VIII, Division 1, as described above in Section 4.10.

4.15 PRESSURE DESIGN OF UNLISTED COMPONENTS

If a component is not in accordance with a listed standard and the other design rules provided in para. 304 are not applicable, para. 304.7.2 is applicable. This paragraph requires that the design be based on the pressure design criteria of the Code and be substantiated by one of several methods. The meat of this paragraph is considered to be the substantiation; The methods to substantiate the calculations, and thereby the design, include the following:

- Extensive, successful service experience under comparable conditions with similarly proportioned components of the same or like material.
- Experimental stress analysis, such as described in ASME BPVC Section VIII, Division 2, Annex 5.F.
- Proof test in accordance with either ASME B16.9, MSS SP-97, or ASME BPVC Section VIII, Division 1, para. UG-101. Note that of these standards, those of B16.9 and MSS SP-97 are more applicable to piping components and have a margin of safety consistent with other components in the Piping Code (factor of three on burst rather than the factor of four in the Pressure Vessel Code).
- Detailed stress analysis (e.g., finite-element method) with results evaluated as described in ASME BPVC Section VIII, Division 2, Part 5. These are the design-by-analysis rules in the Pressure Vessel Code. Note that the allowable stress from ASME B31.3 is used in the assessment.

Of the above, the methods normally used to qualify new unlisted components are proof testing and detailed stress analysis.

It should be noted that the Code permits interpolation between sizes, wall thicknesses, and pressure classes, and also permits analogies among related materials. Extrapolation is not permitted.

The issue of how to determine that the above has been done in a satisfactory manner has been discussed in detail in B31.3 Section Committee meetings. Earlier editions of the Code required only that proof testing be approved by the Inspector. However, this created concerns that it may be interpreted that the Inspector must witness the proof test, which is not practical when the manufacturer performs proof tests to qualify a line of piping components. Obviously, all the potential future owner's Inspectors could not be gathered for this event. Furthermore, the other methods are of at least equal concern, and their review may be more appropriately done by an engineer rather than an Inspector. As a result of these concerns, the requirement was added that documentation showing compliance with the above must be available for the owner's approval. The review would be by an Inspector or some other qualified individual for the owner.

Although MSS SP-97 and ASME B16.9 provide a clear approach for determining that the rating of a component is equivalent to or better than matching straight pipe, they do not provide clear procedures for determining a rating for a component that may have a unique rating which may differ from matching straight pipe. The procedure generally used here is to establish a pressure-temperature rating by multiplying the proof pressure by the ratio of the allowable stress for the test specimen to the actual tensile strength of the test specimen. An example of this approach is provided by Biersteker et al (1991).

CHAPTER

5

Limitations on Components and Joints

5.1 OVERVIEW

ASME B31.3 includes limitations on components and joints in the design chapter, Chapter II. These are contained in Part 3, Fluid Service Requirements for Piping Components, and Part 4, Fluid Service Requirements for Piping Joints. Many of these are restrictions for severe cyclic conditions. The present chapter combines the limitations with pressure design and other considerations on a component-by-component basis.

The Code has limitations on the use of components in piping systems where the fluid is flammable, toxic or damaging to human tissue. The concern being addressed with these limitations is the consequence when the fluid escapes the piping pressure containment. The Code provides definitions for "flammable" and "damaging to human tissue".

flammable: for the purposes of this Code, describes a fluid that under ambient or expected operating conditions is a vapor or produces vapors that can be ignited and continue to burn in air. The term thus may apply, depending on service conditions, to fluids defined for other purposesas flammable or combustible.

damaging to human tissues: for the purposes of this Code, this phrase describes a fluid service in which exposure to the fluid, caused by leakage under expected operating conditions, can harm skin, eyes, or exposed mucous membranes so that irreversible damage may result unless prompt restorative measures are taken. (Restorative measures may include flushing with water, administration of antidotes, or medication.)

5.2 VALVES

Most valves in ASME B31.3 piping systems are in accordance with the following listed standards:

- ASME B16.10, *Face-to-Face and End-to-End Dimensions of Valves*
- ASME B 16.34, *Valves—Flanged, Threaded, and Welding End*
- API 6D, *Pipeline Valves*
- API 526, *Flanged Steel Pressure Relief Valves*
- API 594, *Check Valves: Flanged, Lug, Wafer and Butt-welding*
- API 599, *Metal Plug Valves—Flanged, Threaded and Welding Ends*
- API 600, *Bolted Bonnet Steel Gate Valves for Petroleum and Natural Gas Industry*

- API 602, *Gate, Globe, and Check Valves for Sizes DN 100 and Smaller for the Petroleum and Natural Gas Industries*
- API 603, *Corrosion-Resistant, Bolted Bonnet Gate Valves — Flanged and Butt-Welding Ends*
- API 608, *Metal Ball Valves—Flanged, Threaded, and Welding Ends*
- API 609, *Butterfly Valves: Double Flanged, Lug- and Wafer-Type*
- AWWA C500, *Metal-Seated Gate Valves for Water Supply Service*
- MSS SP-42, *Class 150 (PN 20) Corrosion Resistant Gate, Globe, Angle and Check Valves With Flanged and Butt Weld Ends*
- MSS SP-70, *Gray Iron Gate Valves, Flanged and Threaded Ends*
- MSS SP-71, *Gray Iron Swing Check Valves, Flanged and Threaded Ends*
- MSS SP-72, *Ball Valves with Flanged or Buttwelding Ends for General Service*
- MSS SP-78, *Gray Iron Plug Valves, Flanged and Threaded Ends*
- MSS SP-80, *Bronze Gate, Globe, Angle, and Check Valves*
- MSS SP-81, *Stainless-Steel or Stainless-Steel-Lined, Bonnetless, Knife Gate Valves with Flanged Ends*
- MSS SP-85, *Gray Iron Globe and Angle Valves, Flanged and Threaded Ends*
- MSS SP-88, *Diaphragm Valves*
- MSS SP-105, *Instrument Valves for Code Applications*

Listed valves are accepted for their specified pressure ratings. Some of these pressure ratings are tabulated in Appendix I. Valves that are not in accordance with one of the listed standards can be accepted as unlisted components in accordance with para. 302.2.3. Pressure ratings for unlisted valves have to be established in accordance with para. 304.7.2.

Two additional requirements are provided in para. 307. Bolted bonnet valves must be secured by at least four bolts, except in Category D Fluid Service. Bonnets secured with fewer than four bolts or with U-bolt bonnets are only permitted in Category D Fluid Service.

A requirement for positive stem retention was added to the 2006 edition. The requirement is that the valve stem cannot be removed from the valve, while it is under pressure, by simply removing the stem seal retainer (e.g., packing gland fasteners).

5.3 FLANGES

Most flanges in ASME B31.3 piping systems are in accordance with the following listed standards:

- ASME B16.1, *Gray Iron Pipe Flanges and Flanged Fittings*
- ASME B16.5, *Pipe Flanges and Flanged Fittings*
- ASME B 16.24, *Cast Copper Alloy Pipe Flanges and Flanged Fittings: Classes 150, 300, 600, 900, 1500, and 2500*
- ASME B 16.36, *Orifice Flanges, Classes 300, 600, 900, 1500, and 2500*
- ASME B 16.42, *Ductile Iron Pipe Flanges and Flanged Fittings, Classes 150 and 300*
- ASME B 16.47, *Large Diameter Steel Flanges, NPS 26 Through NPS 60*
- ASME B 16.48, *Steel Line Blanks*
- AWWA C115, *Flanged Ductile-Iron Pipe with Ductile-Iron or Gray-Iron Threaded Flanges*
- AWWA C207, *Steel Pipe Flanges for Water Works Service, Sizes 4 Inch Through 144 Inch (100 mm Through 3,600 mm)*
- MSS SP-44, *Steel Pipeline Flanges*
- MSS SP-51, *Class 150LW Corrosion Resistant Cast Flanges and Flanged Fittings*
- MSS SP-65, *High Pressure Chemical Industry Flanges and Threaded Stubs for Use with Lens Gaskets*
- MSS SP-106, *Cast Copper Alloy Flanges and Flanged Fittings Class 125, 150, and 300*

Listed flanges are accepted for their specified pressure ratings. Some of these pressure ratings are tabulated in Appendix I. Flanges that are not in accordance with one of the listed standards can be accepted as unlisted components in accordance with para. 302.2.3. If the Designer is satisfied that the composition, mechanical properties, method of manufacture, and quality control are comparable to the corresponding characteristics of a listed component, pressure design can be performed in accordance with ASME BPVC, Section VIII, Division 1, Appendix 2 or Appendix Y, as applicable; or ASME BPVC, Section VIII, Division 2, para. 4.16, Otherwise, qualification of the flange in accordance with para. 304.7.2 is required.

Paragraph 308 provides additional restrictions for flanges. Slip-on flanges are required to be double-welded when the service is (1) subject to severe erosion, crevice corrosion, or cyclic loading, (2) flammable, toxic, or damaging to human tissue, (3) under severe cyclic conditions, or (4) at temperatures below -101°C (-150°F). Further, the Designer is cautioned that the use of slip-on flanges should be avoided where many large temperature cycles are expected, particularly if the flange is not insulated.

A double-welded slip-on flange has a weld between the pipe and the flange hub and between the pipe and the bore of the flange. A single-welded slip-on flange only has the weld to the flange hub.

For double-welded slip-on flanges in hydrogen service, where hydrogen can diffuse into the annulus between the inner and outer welds, collect, and pressurize it, the flange should be drilled to vent this space. A precautionary consideration, F308.2, mentions this concern.

Slip-on flanges have been substituted for lap joint flanges (both ASME B16.5). However, the slip-on flange hub for some sizes is undersized for lap joint applications. As a result, Table 308.2.1 was added to the Code. It limits substitution of slip-on for lap joint flanges to DN 300 (NPS 12) or smaller for PN 20 (Class 150) flanges, and DN 200 (NPS 8) for PN 50 (Class 300) flanges, unless the pressure design has been qualified in accordance with the ASME BPVC, Section VIII, Division 1 or Division 2 rules. In addition, other requirements are provided relative to the lap joint in para. 308.2.1(c).

The limitations on joints discussed later in this chapter are applicable to the joint between the flange and the pipe and may affect flange type selection for a given service.

5.4 FITTINGS, BENDS, MITERS, AND BRANCH CONNECTIONS

Most fittings in ASME B31.3 piping systems are in accordance with the following listed standards:

- ASME B16.3, *Malleable Iron Threaded Fittings*
- ASME B16.4, *Gray Iron Threaded Fittings*
- ASME B16.9, *Factory-Made Wrought Steel Buttwelding Fittings*
- ASME B 16.11, *Forged Fittings, Socket-Welding and Threaded*
- ASME B16.14, *Ferrous Pipe Plugs, Bushings, and Locknuts with Pipe Threads*
- ASME B16.15, *Cast Bronze Threaded Fittings, Classes 125 and 250*
- ASME B16.18, *Cast Copper Alloy Solder Joint Pressure Fittings*
- ASME B 16.22, *Wrought Copper and Copper Alloy Solder Joint Pressure Fittings*
- ASME B 16.26, *Cast Copper Alloy Fittings for Flared Copper Tubes*
- ASME B 16.39, *Malleable Iron Threaded Pipe Unions, Classes 150, 250, and 300*
- ASTM F1476, *Performance of Gasketed Mechanical Couplings for Use in Piping Applications.*
- ASTM F1548, *Performance of Fittings for Use with Gasketed Mechanical Couplings Used in Piping Applications*
- AWWA C110, *Ductile-Iron and Gray-Iron Fittings, 3 Inch Through 48 Inch (75 mm Through 1200 mm), for Water and Other Liquids*
- AWWA C208, *Dimensions for Fabricated Steel Water Pipe Fittings*
- MSS SP-43, *Wrought Stainless Steel Buttwelding Fittings Including Reference to Other Corrosion Resistant Materials*

- MSS SP-75, *Specifications for High Test Wrought Buttwelding Fittings*
- MSS SP-79, *Socket-Welding Reducer Inserts*
- MSS SP-83, *Class 3000 Steel Pipe Unions, Socket-Welding and Threaded*
- MSS SP-95, *Swage(d) Nipples and Bull Plugs*
- MSS SP-97, *Integrally Reinforced Forged Branch Outlet Fittings—Socket Welding, Threaded,and Buttwelding Ends*
- MSS SP-119, *Factory-Made Wrought Belled End Socket Welding Fittings* (class MP only)
- SAE J513, *Refrigeration Tube Fittings—General Specifications*
- SAE J514, *Hydraulic Tube Fittings*
- SAE J518-1, *Hydraulic Flanged Tube, Pipe, and Hose Connections, 4-Screw Flange Connection Part 1: 3.5 MPa to 35 MPa (Code 61)*
- SAE J518-2, *Hydraulic Flanged Tube, Pipe, and Hose Connections, 4-Screw Flange Connection Part 2: 42 MPa (Code 62)*

Listed fittings are accepted for their specified pressure ratings. Note that some fittings are simply specified to have equivalent pressure ratings to matching straight seamless pipe. As discussed in Section 4.1, ASME B31.3 bases this on matching straight seamless pipe. Some ratings are tabulated in Appendix I. Fittings that are not in accordance with one of the listed standards can be accepted as unlisted components in accordance with para. 302.2.3. The relevant option of this paragraph for fittings is typically para. 304.7.2, because the pressure design equations in para. 304 usually are not relevant.

Branch connections are required to be listed components (para. 306.1.1); fabricated branch connections that are designed per para. 304.3 and welded per para. 311.1 (para. 306.5.1); fittings that are qualified by proof testing per ASME B16.9, MSS SP-97, or BPVC, Section VIII, Division 1, UG-101 [para. 306.1.3(a)]; or unlisted components that are qualified per para. 304.7.2 (para. 306.1.2).

The following fittings are prohibited from use under severe cyclic conditions.

- MSS SP-43 and MSS SP-119 fittings (para. 306.1.4)
- Proprietary "Type C" stub-end welding fittings (para. 306.1.4)
- Fittings (para. 306.1.4) other than forged, wrought, with 100% radiography of any welds, or cast, with $E_c \geq 0.90$.
- Corrugated or creased bends (para. 306.2.3)
- Miter bends with angle α >22.5° (para. 306.3.3)
- Fabricated branch connections with the branch set on (rather than set in) the run pipe [i.e., Fig. 328.5.4D, sketches (2) and (4)] are permitted (para. 306.5.2)

The following fittings are only permitted for Category D Fluid Service:

- Miter bends with a single joint angle $\alpha > 45°$.

Limitations for fabricated and flared laps for use with lapped flanges are provided in para. 306.4. Note that these limitations do not apply to stub ends that are in accordance with listed standards, such as ASME B16.9. Fabricated laps are constructed by forging or welding; flared laps are constructed by flaring a pipe section. The outside diameter of the lap is required to comply with the dimension specified in ASME B16.9. The lap thickness is required to be at least equal to the nominal wall thickness of the pipe for fabricated laps, and, for flared laps, 95% of the minimum pipe wall thickness (measured or minimum per purchase specification, considering mill tolerance) multiplied by the ratio of the pipe outside radius to the radius at which the lap thickness is measured (this essentially permits thinning due to the flaring process).

For a fabricated lap, the material for the lap must have an allowable stress at least as great as the pipe material. Welding must be per ASME B31.3 and typical details are provided in Fig. 328.5.5. Full penetration welds are required (by reference to the applicable requirements of para. 328.5.4, which covers branch connections).

Flared laps are required to have their pressure design qualified in accordance with para. 304.7.2. Thus, they are essentially treated as unlisted components in this regard. In addition, the radius of the fillet of the flared lap is not permitted to exceed 3 mm (1/8 in.), and the mating flange is required to be beveled or rounded approximately 3 mm (1/8 in.) (by reference to para. 308.2.5). This prevents interference between the lap and the flange at the pipe/lap junction.

5.5 BOLTING

Requirements for bolting are provided in para. 309. Listed bolting is acceptable. The only relevant component standards are ASME Bl.l, *Unified Inch Screw Threads (UN and UNR Thread Form).* ASME B 18.2.1, *Square and Hex Bolts and Screws (inch series)*, ASME B18.2.2, *Square and Hex Nuts (inch series)*, and ASME B18.31.2, *Continuous Thread Stud, Double-End Stud, and Flange Bolting Stud (Stud Bolt) (Inch Series)*. Tables A-2 and A-2M in ASME B31.3, Appendix A provides listed bolting material specifications. Unlisted bolting (i.e., not listed in Tables A-2 and A-2M) are subject to the requirements of para. 302.3.3 (per para. 309.1.2).

Requirements for flanged joints relative to bolt strength are provided as follows. Low strength is considered to refer to bolts with a specified minimum yield strength of 207 MPa (30 ksi) or less:

(a) Low-strength bolts are not permitted for flanges with ASME B16.5 Class 400 or higher ratings or with flanged joints using metallic gaskets, unless design calculations have been made to show the bolts have sufficient strength to maintain joint tightness. Although not specifically stated in ASME B31.3, this should be done per the bolting design requirements in ASME BPVC, Section VIII, Division 1, Appendix 2 or ASME BPVC, Section VIII, Division 2, para. 4.16,

(b) Certain flanges are considered to be weaker due to the material of construction or the flange design.

These include flanges per ASME B16.1 (gray iron), ASME B 16.24 (manufactured from an ASTM B61 or an ASTM B62 bronze casting), MSS SP-42, and MSS SP-51. These are required to use low-strength bolting to avoid overloading the flange, unless one of the following is satisfied: (i) The flanges are flat face and use a full face gasket. This essentially prevents excessive flange rotation due to bolt load; or (ii) the sequence and torque limits for the boltup are specified, giving appropriate consideration to the loads and the strength of the flange.

(c) Low-strength bolting is prohibited from flange joints under severe cyclic conditions.

The depth of tapped holes is required to be sufficient to provide a length of thread engagement at least seven-eighths times the nominal thread diameter.

5.6 WELDED JOINTS

Welded joints are covered by para. 311 and are required to follow the ASME B31.3 rules for fabrication and examination. The following specific rules are also provided.

Weld backing rings are generally permitted to be left in. However, if the resulting crevice would be detrimental (e.g., subject to corrosion, vibration, or severe cyclic conditions), para. 311.2.3 recommends removing the ring and grinding the internal joint face smooth. Split-backing rings are prohibited under severe cyclic conditions.

Socket-welded joints are generally permitted, with the following exceptions. They should not be used in services where crevice corrosion (due to the crevice between the pipe and socket) or severe erosion (which can cause local erosion at the gap between the pipe end and socket) may occur. Socket welds larger than DN 50 (NPS 2) are not permitted under severe cyclic conditions. Note that socket welds are generally limited in practice to sizes of DN 50 (NPS 2) or less in any case because the installed cost of butt welded joints is almost always lower.

The dimensions of the socket joint are required to conform to either ASME B16.5 for flanges or ASME B16.11 for other socket welding components (except for socket welding drain or bypass attachments to components that conform to Fig. 4 of ASME B16.5).

Weld dimensions are required to comply with the fabrication rules of ASME B31.3 (Figs. 328.5.2B and 328.5.2C). Socket joints are generally part of listed components, so their pressure design is satisfied by the component standard and compliance with the fillet weld size requirements of the fabrication rules of ASME B31.3.

Paragraph 311.2.5 permits fillet welds to be used for the primary welds to attach socket welding components and slip-on flanges. It also permits fillet welds to be used to attach reinforcement and structural attachments and to supplement the strength or reduce the stress concentration of other welds. Note, however, that the wording does not prohibit the use of fillet welds for other applications, and it has been the interpretation of the Committee that they are not prohibited from use in other joints, such as lined pipe sections connected by a metallic sleeve with fillet welds. This is reflected in Interpretation 10-04:

"Question: Does ASME/ANSI B31.3-1987 Edition and its Addenda permit the use of fillet welds as the primary pressure retaining welds for installation of a full encirclement tee type branch connections?"

"Reply: ASME/ANSI B31.3-1987 Edition and its Addenda do not specifically address rules for fillet welding of full encirclement tee type branch connections."

Seal welds are permitted to be used to prevent leakage of threaded joints. However, the seal weld is not permitted to be considered as contributing to the strength of the joint.

5.7 FLANGED JOINTS

Flanges are discussed in Section 5.3. Additional requirements for flanged joints are provided in para. 312 of ASME B31.3. These discuss conditions where flanges with different strengths are bolted together. This can involve two flanges of the same or different materials or with different ratings, or a nonmetallic and metallic flange bolted together. In addition, the pressure rating of the joint is limited by the flange with the lower rating.

Precaustions are generally required to avoid overtightening the bolts, which could damage the weaker flange. For metallic-to-nonmetallic flanges, use of full face gaskets is preferred, again, to prevent excessive distortion of the nonmetallic flange due to bolt loads. If full face gaskets are not used with metallic/nonmetallic flange joints, the bolt torque is required to be limited, to avoid overloading the nonmetallic flange.

5.8 THREADED JOINTS

Most threaded joints in ASME B31.3 piping systems are made with taper threads in accordance with ASME B 1.20.1. Figure 5.1 shows a taper thread. These threaded joints are generally acceptable for Normal Fluid Service and Category D Fluid Service. Limitations with respect to Category M Fluid Service are discussed in Chapter 16. Paragraph 314.1(a) states that threaded joints should be avoided in any service where crevice corrosion, severe erosion, or cyclic load may occur. Cyclic loading is not defined and this limitation is rather unclear, because all systems are subject to some cycles and threaded joints are commonly used in process piping.

As discussed in the Chapter 12, threaded joints that are to be seal-welded are prohibited from using thread sealing compounds. However, when seal welding is not an intent of the design, but is intended simply to seal leaks that are detected on hydrotest, threaded joints that use thread sealing compound may be seal-welded, provided the sealing compound is removed from the exposed thread. The joint does not have to be undone and remade without the compound.

Threaded joints are prohibited from use in severe cyclic service, except for two conditions. One is the use of threaded components of a specialty nature, which are not subject to external moment loading, such as thermowells. The other condition where threaded joints are permitted in severe cyclic conditions is provided in para. 314.2.2, which permits straight thread joints in which the tightness of the joint is provided by a seating surface other than the threads, provided the joint is safeguarded.

ASME B31.3 provides minimum thicknesses for male and female portions of threaded joints. For the female portions, it requires that they have equivalent strength to listed components (in Table 326.1, e.g., ASME B16.11). For the male portion, the minimum thickness is provided in Table 314.2.1, provided here as Table 5.1.

TABLE 5.1
MINIMUM SCHEDULE OF COMPONENTS WITH EXTERNAL THREADS (ASME B31.3 2020, Table 314.2.1)

Fluid Service	Notch-Sensitive Material [Note (1)]	Size Range [Note (2)] DN	Size Range [Note (2)] NPS	Minimum Schedule [Note (3)]
Normal	Yes	≤40	≤1½	80
		50–150	2–6	40
		>150	>6	None
	No	≤150	≤6	40
		>150	>6	None
Category D	Either	All	All	None

NOTES:
(1) Carbon steel is generally considered to be notch sensitive, whereas stainless steel is generally considered to be not notch sensitive.
(2) For sizes over DN 50 (NPS 2), the joint shall be safeguarded (see Appendix G) for a fluid service that is flammable, toxic, or damaging to human tissue.
(3) Minimum schedules 40 and 80 are listed in ASME B36.10M, which are identical to schedules 40S and 80S listed in ASME B36.19M for DN 150 (NPS 6) and smaller.

5.9 TUBING JOINTS

Tubing joints are covered by para. 315, and include flared, flareless, and compression-type tube fittings. ASME B31.3 provides some listed tubing joints; however, many tubing joints used in process piping are

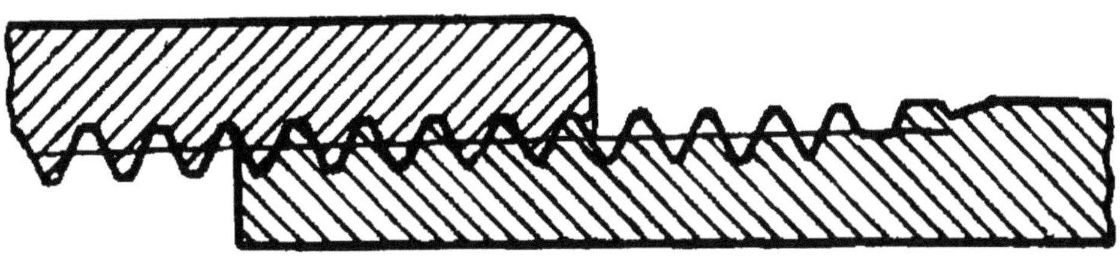

FIG. 5.1
TAPER THREAD

proprietary fittings that are qualified as unlisted components. The following are the listed standards covering tubing joints:

- SAE J513, *Refrigeration Tube Fittings—General Specifications*
- SAE J514, *Hydraulic Tube Fittings*

These listed components are permitted for use in Normal Fluid Service, provided that they are used within the pressure–temperature limitations of the fittings and joint. They are required to be safeguarded when used under severe cyclic conditions; Unlisted tube fittings must be qualified in accordance with para. 304.7.2.

5.10 MISCELLANEOUS JOINTS

Caulked and soldered joints are only permitted in Category D Fluid Service. Caulked joints are further limited to a maximum temperature of 93°C (200°F), and provisions are required to prevent disengagement of the joints and resist the effects of longitudinal forces due to internal pressure. With soldered joints, the low melting point of solder is to be considered where possible exposure to fire or elevated temperature is involved.

Brazed and braze-welded joints are permitted to be used in Normal Fluid Service. However, they are required to be safeguarded in fluid services that are flammable, toxic, or damaging to human tissue (see Section 2.6). They are prohibited from use under severe cyclic conditions, and the melting point of the brazing alloys is to be considered where possible exposure to fire is involved.

For other joints, such as bell-type and packed joints, the separation of the joint must be prevented by a means that has sufficient strength to withstand the anticipated conditions of service. Pressure tends to pull these joints apart. Furthermore, if the fluid service is flammable, toxic, or damaging to human tissues, or exposed to temperatures in the creep range, joint separation must be prevented by mechanical or welded interlocks. Friction within the joint or use of external anchors is not sufficient. However, joints in which the mechanical strength is developed by crimping a Female part onto a pipe or tube (e.g., Victaulic® Pressfit) do qualify as having a mechanical interlock (see Interpretation 19-46).

There are provisions that restrict mechanical joints and bell-and-gland-type joints in severe cyclic service.

CHAPTER

6

DESIGN FOR SUSTAINED AND OCCASIONAL LOADS

6.1 PRIMARY LONGITUDINAL STRESS

The wall thickness of pipe is nearly always selected based on the thickness required for internal pressure and allowances. The piping is then supported sufficiently such that the longitudinal stress (the stress in the axial direction of the pipe) is within Code limits and deflection is within acceptable limits.

Deflection limits are not Code requirements, but are generally accepted practices; a 13-mm (1/2-in.) deflection is a generally accepted guideline for general process plant piping. More stringent limits may be required for lines that must avoid pockets caused by sagging of the line; greater deflection is generally acceptable from a mechanical integrity standpoint, if not an operator confidence standpoint.

It is fortunate that the longitudinal pressure stress is one-half of the hoop stress in a cylinder. What this means is that if the pipe is designed for pressure, at least one-half of the strength in the longitudinal direction remains available for weight and other sustained loads.

While it is true that stresses caused by sustained loads are largely longitudinal for straight pipe, the sustained loads cause local bending stresses in fittings that are often much higher than the longitudinal stresses in the pipe. For that reason, the word "longitudinal" was dropped from the descriptions of sustained stresses starting in the 2018 edition.

The same basic allowable stress as is used for pressure design is used as the limit for sustained stresses in piping systems. These sustained and the occasional load stresses should encompass all of the load-controlled, primary-type stresses to which the pipe is subjected. They will fall either in the sustained or the occasional category, depending on duration. Whereas the weight of the pipe and contents are sustained, forces and moments due to wind, earthquake, and phenomena such as dynamic loads due to water hammer would be considered to be occasional loads.

6.2 SUSTAINED STRESS

Prior to the 2010 edition, the calculation of stress due to sustained loads was covered by descriptive text. Various providers of computer software to calculate pipe stress interpreted these words different in slight ways as some aspects were not specifically specified. The committee decided to provide equations primarily to provide consistency in calculated results. Because for most piping systems sustained stresses are well below the allowables, the addition of equations was not driven by piping systems failing due to sustained loads.

ASME B31.3, para. 302.3.5(c) requires that the thickness of the pipe used in calculating S_L be the nominal thickness minus the mechanical, corrosion, and erosion allowance. Unlike pressure design, it is not required to account for mill tolerance of the piping. Hence the forces, moments, and deflections in a piping system are calculated using nominal dimensions. This provides appropriate support loads for design. Section moduli used to compute the bending and torsion stresses are based on nominal pipe dimensions less allowances, i.e., the sum of mechanical (thread or groove depth), internal and external corrosion, and erosion allowances. Areas used to compute the axial stresses are based on nominal pipe dimensions less allowances affecting the inside diameter of the pipe, i.e., the sum of mechanical and internal corrosion and erosion allowances. Inclusion of the metal to be used as corrosion allowance in the weight but not the strength may be viewed as unduly conservative; however, corrosion could be local and at the highest loaded point, such as at a support.

Quality factors, E_j and E_c, are not used in the evaluation of stresses caused by sustained loads.

Insert 6.1 Span Limits for Elevated Temperature Piping

When a piping system is at a temperature sufficiently high for the material to creep, it is possible that significant sagging (deflection due to creep) can occur over time. Methods are readily available for calculation of elastic deflection of the pipe, and there are commonly used pipe span tables. However, such methods are not available for calculating the long-term deflection due to creep, which can be many times the initial elastic deflection for pipe operating in the creep regime of the material. Span tables, piping stress analysis programs, and generally available design methods only consider the elastic deflection of the pipe.

Becht and Chen (2000) developed closed form equations for calculating creep deflection for simple spans, in order to develop span tables for 980°C (1800°F) pipe for the Marble Hill Nuclear Reactor Pressure Vessel Annealing Demonstration Project. Closed form integrals to predict creep deflection of simply supported, cantilever and fixed end beams were developed. These equations follow.

The Norton creep equation is assumed,

$$\epsilon_c = B\sigma^n t$$

I_c and K, in the following equations, are defined as follows.

$$I_c = I \cdot \frac{8}{\left(3+\frac{1}{n}\right)\cdot\sqrt{\pi}} \cdot \frac{1-\left(\frac{r_i}{r_o}\right)^{3+\frac{1}{n}}}{1-\left(\frac{r_i}{r_o}\right)^4} \cdot \frac{\Gamma\left(1+\frac{1}{2\cdot n}\right)}{\Gamma\left(1.5+\frac{1}{2\cdot n}\right)}$$

$$K = \frac{2\cdot B\cdot t}{h}\cdot\left(\frac{h}{2.I_c}\right)^n\cdot\left(\frac{w}{2}\right)^n$$

Slope and deflection for simplysupported beam

$$y_p = K\cdot\int_0^{\frac{L}{2}} \left|\left(L\cdot x - x^2\right)^n\right| dx$$

$$y = \int_0^{\frac{L}{2}} y_p + K\cdot\int_0^{x} \left|\left(x\cdot L - x^2\right)^n\right| dx\, dx$$

Slope and deflection for cantilever beam

$$y_p = K \cdot \int_o^{\frac{L}{2}} \left(x^2\right)^n dx$$

closed form solution

$$y = \frac{2 \cdot B \cdot t}{h} \cdot \left(\frac{h}{2 \cdot I_c}\right)^n \cdot \left[\frac{w \cdot (L)^2}{2}\right]^n \cdot \frac{L^2}{2 \cdot (n+1)}$$

Slope and deflection for fixed ended beam

Slope at inflection point

$$y_{ps} = K \cdot \int_o^{s} - \left| \left[6 \cdot L \cdot x - (L)^2 - 6 \cdot x^2 \right]^n \right| dx$$

$$s = 0.2113L$$

deflection at inflection point

$$ys = \int_o^{s} K \cdot \left[\left| 16 \cdot L \cdot x - (L)^2 - 6x^2 \right| \right]^n \cdot (x - s)\, dx$$

deflection at mid span

$$y = y_s - \int_s^{\frac{L}{2}} K \cdot \left[\left| 16 \cdot Lx - (L)^2 - 6 \cdot x^2 \right| \right]^n \cdot (x - s)\, dx$$

where

- B constant in creep equation
- h outside diameter of pipe
- I moment of inertia of pipe
- I_c fictitious moment of inertia used to calculate outer fiber stress for a creeping beam
- L beam length
- n stress exponent in creep equation
- r_o outside radius of pipe
- r_i inside radius of pipe
- s distance from end of fixed-fixed beam to inflection point
- t time
- w weight of pipe and contents per unit length
- x dimension along beam length
- y beam deflection
- y_1 extreme fiber distance from the neutral axis
- y_p slope of deflected beam
- y_{ps} slope of fixed-fixed beam at inflection point

0_c creep strain
Γ gamma function
σ stress

Evaluating creep deflection provided insights into the problem of establishing allowable spans for high-temperature piping. The allowable span was found to be relatively insensitive to the allowable

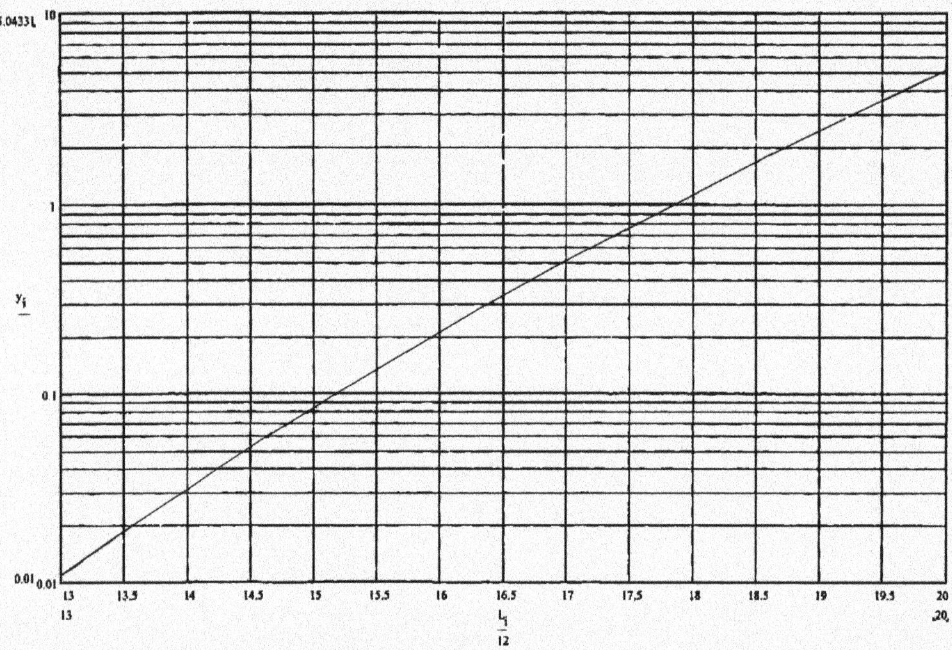

FIG. 6.1
CREEP DEFLECTION OF SIMPLY SUPPORTED BEAM AT 1000 HR VERSUS SPAN, 815°C (1500°F)

deflection and the constant B in the creep equation, within reasonable limits. It is the sensitivity of creep rate to stress, and, in turn, the sensitivity of the stress to span length, that dominates the creep deflection problem. Considering that stress is proportional to the span length to the second power, and assuming that creep strain rate is proportional to stress to, say, the sixth power, knowing that deflection is proportional to strain, we find that creep deflection rate is highly sensitive to span length.

Creep deflection goes up exponentially with span length, so that beyond a given length, at least at relatively high temperatures, deflection is unacceptable, and below that length, deflection is not particularly significant. Figure 6.1 shows the creep deflection of a simply supported Schedule 10S 304L stainless steel line, assuming 44 kg/m (30 lb/ft) steel and insulation weight with vapor as the contents, as a function of length at 815°C (1500°F) and for a duration of 1000 hours. One can see that, between span lengths of 5.2 and 5.5 m (17 and 18 ft), the deflection increases from 13 mm (½ in.) to over 25 mm (1 in.). At 4.9 m (16 ft), the deflection is about 5 mm (0.2 in.).

Figure 6.2 shows deflection as a function of span length, temperature and end restraint condition. These curves are based on 304 L material. The duration for the chart is 1000 hours and the temperature is 870°C (1600°F). They illustrate the effect of span length and support condition on deflection. The project for which this work was done used dual stamped material, so that the L-grade material properties were appropriate to evaluate the elevated temperature behavior of the material.

Figure 6.3 shows a chart for 304 L Schedule 20S pipe with 29 kg/m (20 lb/ft) insulation and a duration of 100,000 hours, about 10 years, for elastic plus creep deflection as well as for elastic deflection only. Included on the chart are results based on simply supported and fixed supports. Note that the

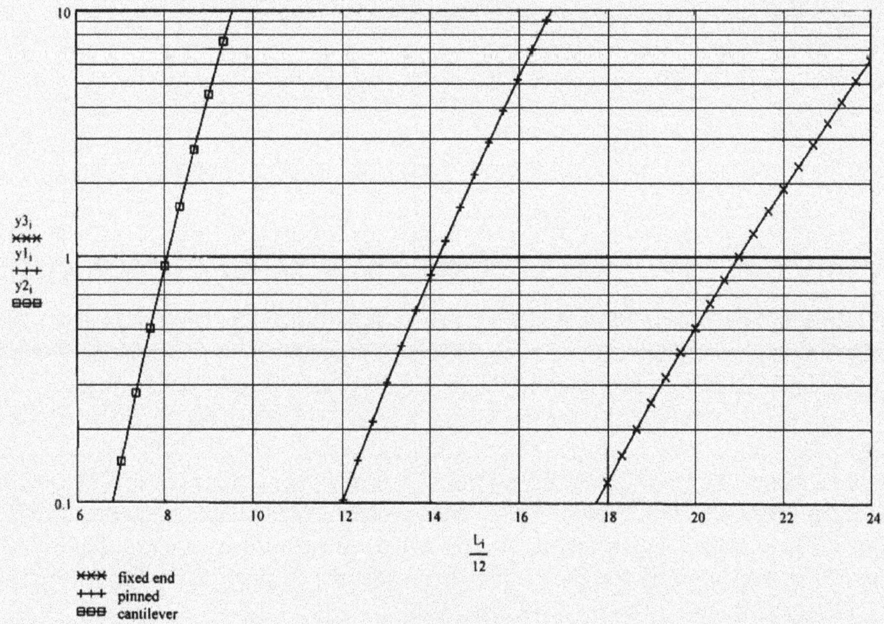

FIG. 6.2
CREEP DEFLECTION OF VERSUS SPAN LENGTH AT 1000 HR FOR DIFFERENT RESTRAINT CONDITIONS, 870°C (1600°F)

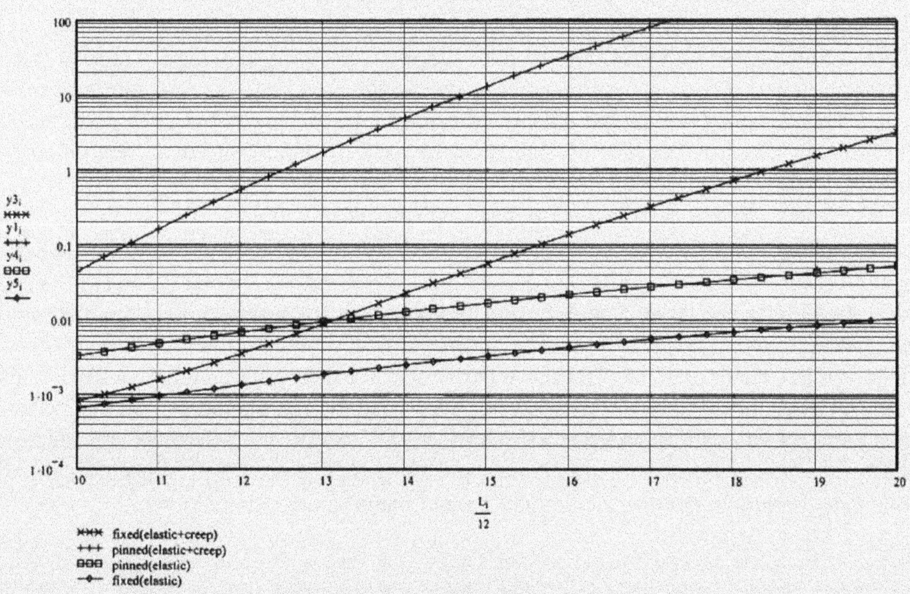

FIG. 6.3
COMPARISON OF CREEP AND ELASTIC DEFLECTION OF BEAMS AT 100,000 HR VERSUS SPAN LENGTH FOR PINNED AND FIXED RESTRAINT, 815°C (1500°F)

allowable span length based on allowable stress consideration only, per ASME B31-3, is about 4.9 m (16 ft) for simply supported and 7 m (23 ft) for fixed supports. The allowable span length, based on 13 mm (0.5 in.) permissible elastic deflection and a simply supported condition would be 9.4 m (31 ft) It is obvious that these span lengths based on these conventional criteria would result in excessive long-term deflection for this elevated temperature pipe.

Creep deflection should be considered in determining allowable span lengths for elevated temperature piping. The equation provided herein can be used to develop simplified span tables for specific applications.

6.3 LIMITS OF CALCULATED STRESSES DUE TO OCCASIONAL LOADS

The ASME B31.3 Code is consistent with the structural codes in the treatment of occasional loads. The sum of all stress, including both sustained and occasional loads, is limited to 1.33 times the basic-allowable stress. This differs from the ASME BPVC, Section VIII, Division 1 Pressure Vessel Code, which uses a factor of 1.2. The treatment of earthquake forces in this manner is generally considered to be extremely conservative.

The increase in allowable stress is only permitted as long as the yield strength of the material is not exceeded. This is one of the areas of the Code where use of the higher allowable stress relative to yield for austenitic stainless steels and nickel alloys with similar stress-strain properties can result in additional limits.

Although quality factors are not generally used in the evaluation of stresses due to sustained loads, for occasional loads there is an exception, in that the Code requires that the casting quality factor, E_c be multiplied by the basic allowable stress when evaluating occasional loads.

In the 2004 edition, alternative rules were added for occasional loads for piping at elevated temperatures, in the creep regime. In Elevated Temperature Fluid Service, the allowable stress is governed by long term creep rupture/rate properties, which are not particularly relevant to whether the piping can withstand short term loading. In the alternative rules for Elevated Temperature Fluid Service, the basic allowable is 90% of yield strength, reduced by a strength reduction factor which takes into account that some metals have reductions in yield strength over time when in elevated temperature service. This factor is set at 0.8 (which is a lower bound for relevant materials) or 1.0 for austenitic stainless steel which does not lose strength with time at elevated temperature. In a study documented in and ASME Research Report STP-PT-024, it was recommended that an additional limit of four times the basic allowable stress be used to avoid more than 20% creep damage if the duration of the occasional loads is ten hours or less, which is also in the rules. If the duration is greater than ten hours, the designer is required to calculate the limit using Appendix V.

In the drafting of the rules for later editions including 2018, a couple errors were introduced which may be corrected in a future edition of ASME B31.3. First, the word "weld" was added in (a) (2) (-a) in front of "strength reduction factor" which can lead the Code user to misinterpret this as "weld joint strength reduction factor" which is something entirely different. The "strength reduction factor" referred to in (-a) is defined further down in the text as either 1.0 of 0.8. The second error is that paragraph that requires casting quality factors be included and to also limit the allowable stress to yield strength at temperature was inadvertently incorporated into the alternative rules for elevated temperature in 302.3.6 (a)(2). These rules preceded the incorporation of the alternative rules for elevated temperatures and apply to all evaluations of occasional loads. It should be a standalone paragraph (e.g. (b) apart from the elevated temperature service rules.

CHAPTER

7

DESIGN FOR DISPLACEMENT STRAINS

7.1 ALLOWABLE STRESS FOR DISPLACEMENT STRAINS

The allowable stress for thermal expansion and other deformation-induced stresses is substantially higher than for sustained loads. This is due to the difference between load-controlled conditions, such as weight and pressure, and deformation-controlled conditions, such as thermal expansion or end displacements (e.g., due to thermal expansion of attached equipment).

When a load-controlled stress is calculated, it is an actual stress value. It is governed by equilibrium. For example, the stress in a bar when a tensile force is applied to it is the force divided by the area of the bar. This is not the case for thermal stresses. In the case of thermal stresses, it is the value of strain that is known. The elastically calculated stress is simply the strain value times the elastic modulus. This makes essentially no difference until the stress exceeds the yield strength of the material. In that case, the location on the stress-strain curve for the material is determined based on the calculated stress for load-controlled, or sustained, loads. The location on the stress-strain curve for the material is determined based on the calculated strain (or elastically calculated stress divided by elastic modulus) for deformation-controlled (e.g., thermal expansion) loads. This is illustrated in Fig. 7.1. Because the stress analyses are based on the assumption of elastic behavior, it is necessary to discriminate between deformation-controlled and load-controlled conditions in order to properly understand the post-yield behavior.

It is considered desirable for the piping system to behave in a substantially elastic manner so that the elastic stress analysis is valid. Furthermore, having plastic deformation every cycle carries with it uncertainties with respect to strain concentration and can be potentially far more damaging than calculated to be in the elastic analysis. One way to accomplish this would be to limit the total stress range to yield stress. However, this would be overly conservative and result in unnecessary expansion loops and joints. Instead, the concept of shakedown to elastic behavior is used in the Code. The basis for the Code equations is described by Markl (1960d). Rossheim and Markl (1960) also provide an interesting discussion on some of the thinking behind the rules.

The allowable displacement stress range in the Code is designed to result in shakedown to elastic behavior after a few operating cycles. The equation provided in the Code is[11]

$$S_A = f(1.25S_c + 0.25S_h) \tag{7.1}$$

[11]ASME B31.3, Eq. (1a).

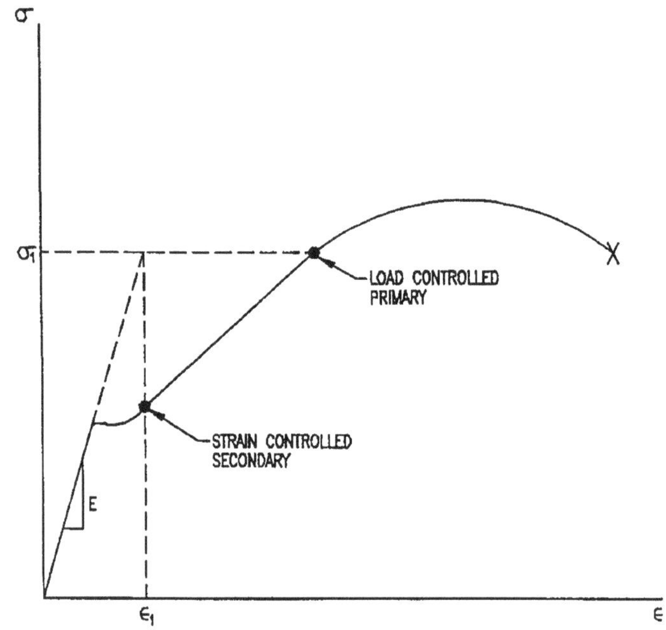

FIG. 7.1
LOAD-CONTROLLED VERSUS DEFORMATION-CONTROLLED BEHAVIOR σ= STRESS ε = STRAIN E = ELASTIC MODULUS

where

S_A = allowable displacement stress range
S_c = basic allowable stress at the minimum metal temperature expected during the displacement cycle under analysis
S_h = basic allowable stress at the maximum metal temperature expected during the displacement cycle under analysis
f = stress range reduction factor

The maximum values of S_c and S_h are limited to 138 MPa (20 ksi). There is a concern that the equation would overestimate the fatigue life for high strength steel components, so the impostion of the limitations avoids reducing the conservatism of the equation.

This equation assumes that the sustained stress consumes the entire allowable sustained stress, and it is simplified, in that it is not necessary to know the sustained stress in order to determine the allowable displacement stress range.

Note that the values of S_c and S_h do not include weld joint quality factors or strength reduction factors. However, casting quality factors, E_c, must be included.

Equation (lb) of the Code is a more detailed equation, and considers the magnitude of the sustained stress,

$$S_A = f[1.25(S_c + S_h) - S_L] \tag{7.2}$$

where
S_L = stress due to sustained loadings

The allowable displacement stress range can exceed the yield strength for the material, because both S_c and S_h may be as high as two-thirds of the yield strength. However, it is anticipated that the piping system will shake down to elastic behavior if the stress range is within this limit.

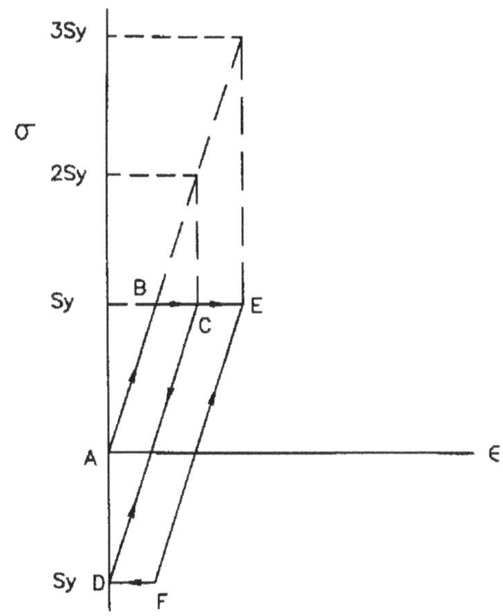

FIG. 7.2
STRESS-STRAIN BEHAVIOR ILLUSTRATING SHAKEDOWN

This behavior is illustrated in Fig. 7.2, which is based on the assumption of elastic, perfectly plastic material behavior. Consider, for example, a case where the elastically calculated displacement stress range is two times the yield strength of the material. Because it is a deformation-controlled condition, one must actually move along the strain axis to a value of stress divided by elastic modulus. In the material, assuming elastic, perfectly plastic behavior, the initial start-up cycle goes from point *A* to *B* (yield) to *C* (strain value of twice yield). When the system returns to ambient temperature, the system returns to zero strain and the piping system will unload elastically until it reaches yield stress in the reverse direction. If the stress range is less than twice yield, there is no yielding on the return to ambient temperature. On returning to the operating condition, the system returns from point *D* to point *C* elastically. Thus, the cycling will be between points *D* and C, which is elastic. The system has essentially self-sprung and is under stress due to displacement conditions in both the ambient and the operating conditions.

If twice the yield is exceeded, shakedown to elastic cycling does not occur. An example is if the elastically calculated stress range is three times the yield strength of the material. In this case, again referring to Fig. 7.2, the startup goes from point *A* to point *B* (yield) to point *E*. Shutdown results in yielding in the reverse direction, going from point *E* to *F* to *D*. Returning to the operating condition again results in yielding, from point *D* to *C* to *E*. Thus, each operating cycle results in plastic deformation and the system has not shaken down to elastic behavior.

This twice-yield condition was the original consideration. Since the yield strengths in the operating and the ambient conditions are different, the criterion becomes that the stress range must be less than the hot yield strength plus the cold yield strength, which, due to the allowable stress criterion, must be less than 1.5 times the sum of S_c and S_h (Note that the original ASME B31 criterion limited the allowable stress to 62.5% of yield, so the original factor that was considered was 1.6.) This 1.5 (1.6 originally) factor was conservatively reduced to 1.25. This total permissible stress range is then reduced by the magnitude of sustained stress in order to calculate the permissible thermal expansion stress range. The resulting equation is Eq. (lb) of the Code [Eq. (7.2) here]. Equation (la) of the Code [Eq. (7.1) here] simply assumes that $S_L = S_h$, the maximum permitted value, and assigns the remainder of the allowable stress range to displacement strains such as those caused by thermal expansion.

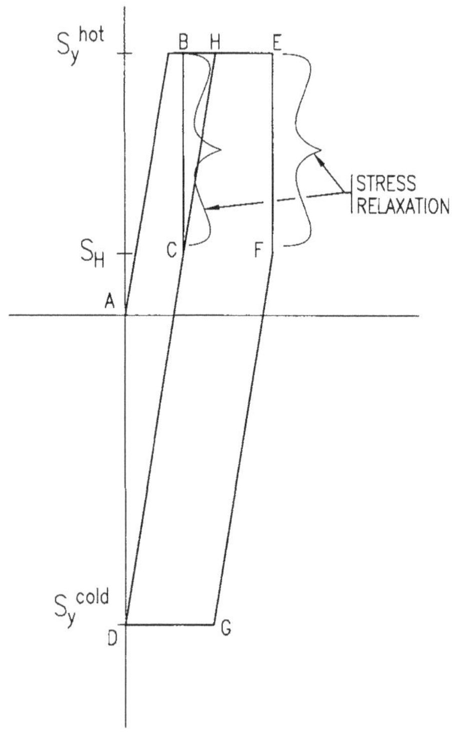

FIG. 7.3
STRESS-STRAIN BEHAVIOR ILLUSTRATING ELEVATED TEMPERATURE SHAKEDOWN

The same equation (7.2) also works in the creep regime. Deformation controlled stresses relax to a stress value sufficiently low that no further creep occurs. This stress value is the hot relaxation strength, S_H. Stress-strain behavior under the condition of creep, is illustrated in Figure 7.3. The initial start-up cycle, which can include some yielding, goes from point *A* to point *B*. During operation, the stresses relax to the hot relaxation strength, S_H, at which point no further relaxation occurs, point C. When the system returns to ambient temperature, the system returns to zero strain and the piping system will unload elastically until it reaches yield stress in the reverse direction. If the stress range is less than S_H plus to cold yield strength, there is no yielding on the return to ambient temperature. This is illustrated by going from point C to point D. On returning to the operating condition, the system returns from point D to point *C* elastically. Thus, if the stress range is less than the cold yield strength plus the hot relaxation strength, shakedown to elastic behavior also occurs at elevated temperature. If S_H is considered to be 1.25 S_H, then elevated temperature shakedown also is achieved with the Code allowable for displacement stresses, S_A. The anticipated behavior over time, with multiple shut downs, and a gradual relaxation process, is illustrated in Figure 7.4.

Figure 7.3 also shows the behavior when the allowable stresses are exceeded at elevated temperatures. In this case, the startup goes from A to E. Stresses relax to point F. When the system returns to ambient temperature, yielding in the reverse direction occurs, going from point F to G to D. Returning to operating condition again results in yielding, from point D to H to E. Since high stresses are re-established, another relaxation cycle then must occur. The behavior of this system over time is illustrated in Figure 7.5.

Even though the stress range is limited so as to result in shakedown to elastic behavior, there remains the potential for fatigue failure if there is a sufficient number of cycles. Therefore, the *f* factor is used to reduce the allowable stress range when the number of cycles exceeds about 7,000, which is about a once a day cycle for 20 years. As described later, *f* factors greater than 1.0 are permitted under some circumstances to increase the allowable stress range.

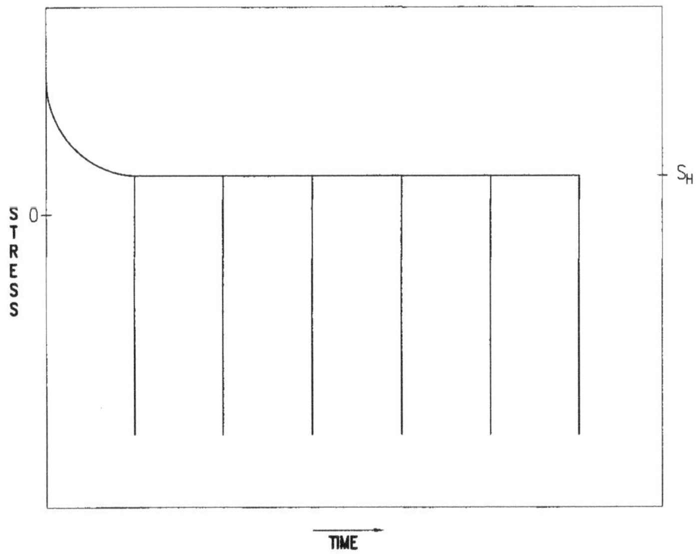

FIG. 7.4
CYCLIC STRESS HISTORY WITH SHAKEDOWN

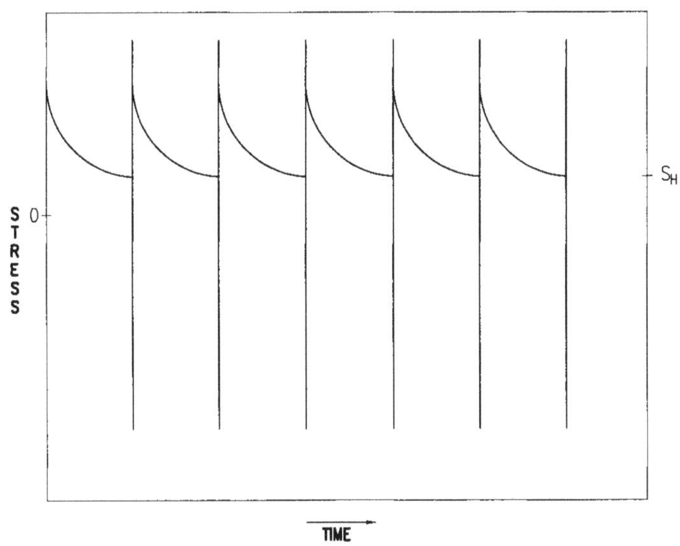

FIG. 7.5
CYCLIC STRESS HISTORY WITHOUT SHAKEDOWN

Figure 7.6 provides the basic fatigue curve for butt-welded pipe, developed by Markl (1960a) for carbon steel pipe. A safety factor of two on stress was applied to this curve, giving a design fatigue curve. It can be observed that the allowable thermal expansion stress range, prior to application of an f factor, intercepts the design fatigue curve at about 7000 cycles. For higher numbers of cycles, the allowable stress is reduced by the /factor to follow the fatigue curve, per the equation[12]

$$f = 6N^{-0.2} < f_m \qquad (7.3)$$

[12]ASMEB31.3, Eq. (lc).

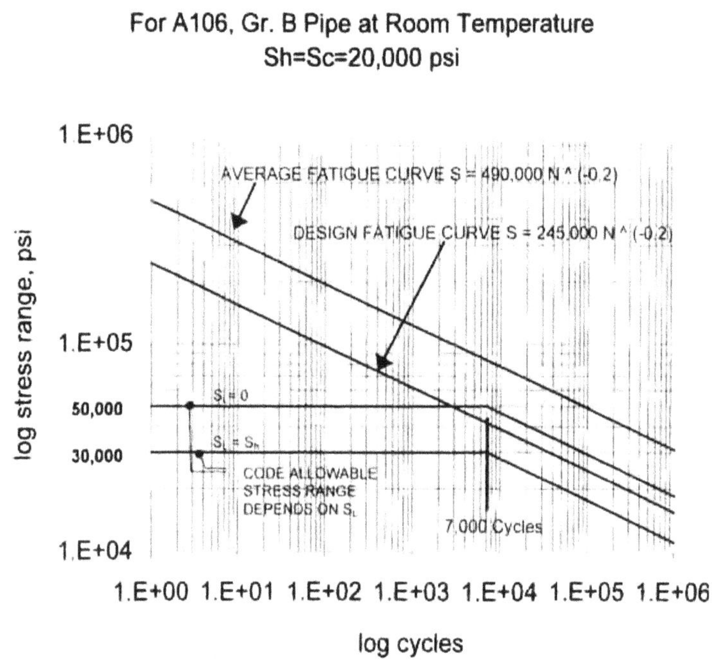

FIG. 7.6
MARKL FATIGUE CURVE FOR BUTT-WELDED STEEL PIPE

where

N = equivalent number of full displacement cycles during the expected service life of the piping system (see the next section for how to combine different cycles into an equivalent number of cycles)

f_m = maximum value of stress range factor: 1.2 for ferrous materials with a specified minimum tensile strength ≤ 517 MPa (75 ksi), and at metal temperatures ≤ 371°C (700°F); otherwise f_m = 1.0.

In the 2004 edition, the maximum permissible value of f was increased from 1.0 to 1.2, with certain limitations. A value of 1.2 corresponds to 3125 cycles. The rationale for allowing a factor as high as 1.2 is that stresses are permitted to be as high as two times yield when f = 1.2. Thus, the desired shakedown behavior is maintained. This change reduces the conservatism introduced when the original criteria were developed. The limitations are that

1. the specified minimum tensile strength of the material must be less than 517 MPa (75 ksi);
2. the material must be ferrous, and
3. the metal temperature must be less than or equal to 371°C (700°F).

The first limitation addresses a concern regarding the conservatism of the present f factors for high strength steel. There is a concern that the rules would overestimate the fatigue life for high strength steel components, so the limitation avoids further reducing the conservatism. The second limitation addresses a similar concern, but for non-ferrous alloys. The f factors were originally developed based on fatigue testing of carbon steel and austenitic stainless steel piping components; their application to other alloys such as aluminum and copper are not necessarily conservative. The third limitation is included because the rationale for increasing f to 1.2 does not apply to components operating in the creep regime.

In the 2004 edition, the equation for *f* was also extended from a maximum of 2,000,000 cycles to an unlimited number of cycles. The minimum value is $f = 0.15$, which results in an allowable displacement stress range for an indefinitely large number of cycles. The term endurance limit was not used, as it is associated with stress amplitude, rather than stress range. The background on the derivation of the value of 0.15 is provided in Insert 7.1. See Section 7.3 for more discussion on high cycle fatigue.

Insert 7.1 What About Vibration?

ASME B31.3 is a code for design and construction of new piping. Typically, piping systems have not been analyzed for vibrating conditions during design. While the Code does require that piping be designed to eliminate excessive and harmful effects of vibration (para. 301.5.4), this is typically done by attempting to design systems to not vibrate, rather than by performing detailed vibration analysis. Guidance on designing systems that are unlikely to vibrate is provided in *Guidelines for Avoidance of Vibration Induced Fatigue Failure in Process Pipework*, published by the Energy Institute. It describes ways to reduce the probability of harmful vibration from simple rule-of-thumb type approaches to detailed vibration assessments performed for piping systems such as is done for certain reciprocating compressor installations. The Energy Institute guidelines also describe how to evaluate piping systems that are found to be vibrating during initial operation.

Excellent guidance for the evaluation of vibrating piping systems can be found in Part 3 of ASME Standard OM-S/G, ASME OM-3, *Requirements for Preoperational and Initial Stan-Up Vibration Testing of Nuclear Power Plant Piping Systems.* The procedures contained therein were developed based on experience with reciprocating compressor systems for gas pipelines. Vibration can be evaluated via the procedures contained therein based on peak measured velocity of the vibrating pipe or calculated stress. A screening velocity criterion that is generally very conservative. 0.5 in./sec peak velocity, is provided. Endurance limit stress ranges are also provided.

Based on the endurance limit stress ranges, an *f* factor for an unlimited number of cycles may be derived. The "endurance limit"[1] stress ranges are provided for carbon steel and stainless steel, Including all the factors provided in the document gives the "endurance limit"[1] stress range of 106 MPa (15.4 ksi) for carbon steel. Assuming a typical S_A of 345 MPa (50 ksi) ($1.25\,S_c + 1.25\,S_h$), and considering that the Code flexibility analysis equations calculate about one-half of the actual peak stress, we find an "endurance limit" *f* factor of 0.15 (at about 10^s cycles). The "endurance limit" *f* factor for stainless steel would be higher, if calculated by the same procedure. However, the same "endurance ltmi" *f* factor is applied to all materials, consistent with existing rules.

7.2 HOW TO COMBINE DIFFERENT DISPLACEMENT CYCLE CONDITIONS

The Designer may have more than one thermal expansion or other displacement cycle conditions to consider. Cycles at lower stress ranges are substantially less damaging than cycles at higher stress ranges. ASME B31.3 uses the cycle with the highest stress range (full displacement cycle), S_E, to compare to the allowable stress range, S_A. However, cycles with lower stress ranges are converted into equivalent numbers of full displacement cycles to determine the *f* factor. This procedure, described in para. 302.3.5(d), uses the equation[13]

$$N = N_E + \sum \left(r_i^5 N_i\right) \quad \text{for } i = 1,2,\ldots,n \tag{7.4}$$

[1]The term "endurance limit" stress range is used tor convenience although endurance limit is typically stress amplitude.

[13]ASME B31.3, Eq. (Id).

TABLE 7.1
COMBINATION OF DIFFERENT DISPLACEMENT CYCLES[1]

Load Case	Design Cycles	Calc. Stress Range	S_i/S_E	N_{equiv}
1	1,000	30,000	1	1000 .
2	7,000	10,000	0.33	27
3	20,000	6,000	0.20	6

NOTE:

(1) Total equivalent number of cycles with stress range of 30,000 is 1033 < 7000. The stress range reduction factor *f*= 1.

where

N_E = number of cycles of maximum computed displacement stress range, S_E
$r_i = S_i/S_E$
S_i = any computed displacement stress range smaller than S_E
N_i = number of cycles associated with displacement stress range S_i
n = total number of displacement stress conditions to be considered

Based on this procedure, the maximum stress range is limited to S_a, which satisfies the shakedown limit. Lesser cycles are converted into equivalent, with respect to fatigue damage, numbers of cycles at S_E to determine if an *f* factor less than one is required to protect against fatigue failure.

Table 7.1 provides an example of combining several displacement cycle conditions with different stress ranges and numbers of cycles. Note that due to the sensitivity of fatigue damage to stress (the fifth power in the equation), displacement cycles at significantly lower stress ranges than S_E produce very little damage or a significantly reduced number of equivalent cycles.

7.3 HIGH CYCLE FATIGUE

It has been shown (Hinnant and Paulin, 2008) that the slope of the fatigue curve developed by Markl, from which the *f* factors were developed, is significantly flatter than has been observed in more recent fatigue tests of welded structures. The newer tests provided much more data in the higher cycle regime, and the difference is apparent. A new fatigue curve has been developed in the cited reference, and the two fatigue curves cross at about 850,000 cycles. At higher cycles, fatigue design using the Markl equation becomes less conservative and potentially unconservative. A further, more comprehensive treatment of high cycle fatigue was developed and resulted in the inclusion of Appendix W in the 2018 edition. Changing the slope of the *f* curve in the base Code is under consideration by the committee.

The lower bound *f* factor is 0.15 for which the piping is considered to be suitable for an unlimited number of cycles. The development of this is described in Insert 7.1. Note that the unconservatism of the slope of the fatigue curve has no impact on this endurance limit *f* factor. Changing the slope would simply put the number of cycles where the *f* factor becomes 0.15 at a lower number of cycles.

The primary focus for the development of Appendix W was offshore platforms where displacement cycles due to wave motion are quite numerous. Use of the Appendix is optional and subject to Owner's approval. Its use is limited to situations where there are more than 100,000 cycles.

For cycles other than wave motion, the method for evaluating the displacement strain cycles follows the ASME B&PV Code Section VIII, Division 2 structural stress method. This method is for welded joints. The slope of the curve in log-log space is linear with no endurance limit. However, most of the standards dealing with fatigue of welded joints have bilinear curves with the slope changing at about 10^7 cycles (with exceptions such as being in saltwater service) and generally recognize endurance limits (with an exception being if there are occasional much larger stress cycles). On recognition of the generally accepted (other than Division 2) bilinear nature of the fatigue curve for welded joints, optional fatigue material coefficients are provided in para. W302.1.3-3 when the number of cycles exceeds 10^7. The initial draft of the rules were consistent with Division 2 and did not recognize an endurance limit. One was added, but it is in the opinion of this author, exceedingly conservative. Stress cycles with a range less than 3,000 psi can be ignored in the fatigue assessment. Note this is much lower than the stress range that is permitted with an f factor of 0.15, as permitted in the base Code.

The fatigue analysis approach for wave loading is not stress range-cycle pairs. Instead it is loading represented by a Weibull distribution which represents the variable amplitude random loadings based on wave height. The fatigue analysis procedure for cyclic stress due to wave motion does not provide a bilinear fatigue curve, but para. W302.3 permits the designer, with the Owner's approval, to use a more rigorous analysis which can include a bilinear (on log-log scale) fatigue curve.

CHAPTER

8

FLEXIBILITY ANALYSIS

8.1 FLEXIBILITY ANALYSIS

Piping flexibility analysis is also known as piping stress analysis. Neither phrase satisfactorily describes all that is involved in this piping design discipline. In flexibility analysis, the response of the system to loads is calculated. The objectives of flexibility analysis are to calculate stress in the pipe; loads on supports, restraints, and equipment; and displacement of the pipe. It is essentially a beam analysis model on pipe centerlines. The fundamental principles include the following:

1. The analysis is based on nominal dimensions of the pipe.
2. The effect of components such as elbows and tees on piping flexibility and stress is considered by inclusion of flexibility factors and stress intensification factors.
3. Stresses due to bending, torsion and direct axial loads are included.
4. The modulus of elasticity at 21°C (70°F) is normally used in the analysis. As a result, no elastic modulus adjustment is required. For a more detailed discussion, see below, Section 8.8.

What temperatures should be considered in the flexibility analysis? The temperature range to considered for the maximum computed stress range is between the extreme displacement condition and the original (as-installed) condition or any anticipated condition with a greater differential effect. The range is usually considered to be the range from the coldest temperature the piping system can see to the warmest temperature the piping system can see. The temperature of concern is the pipe wall temperature.

For the warmest temperature, the design temperature is often used, but it is not always the correct approach. Design temperature is only used in the Code for pressure design. Interpretation 19-41 addresses this.

> Question: In accordance with ASME B31.3, 2002 Edition, does the phrase, "maximum metal temperature... for the thermal cycle under analysis" in para. 319.3.1(a) require the use of the design temperature (defined in accordance with para. 301.3) in "determining total displacement strains for computing the stress range", S_E?
>
> Reply: No.

It is common to evaluate piping flexibility for the design temperature, but it is not a Code requirement. It is certainly permitted, but in all cases, the worst case temperature conditions must be evaluated. For example, the temperature may be higher than the design temperature during a low pressure steam out. Consider another example with two operating conditions with a carbon steel line, 100 kPa (15 psi) at 315°C (600°F) and 1400 kPa (200 psi) at 205°C (400°F). The design conditions would be 1400 kPa (200 psi) at 205°C (400°F) since those are the coincident conditions that govern the pressure design. However, the 315°C (600°F) condition would govern in the flexibility analysis. Alternatively, there may be additional margin put on the operating temperature to calculate the design temperature; again, it is not a Code requirement that this higher temperature be used in the flexibility analysis.

For the coldest temperature, the design minimum temperature is often used, but again it is not always the correct approach. The design minimum temperature is defined as "the lowest component temperature expected in service." For an outdoor water line in a cold climate, the lowest temperature in service could be no lower that 0°C (32°F). If it was anticipated that the piping would be drained during the winter and subject to ambient conditions, the coldest temperature would be much lower than the design minimum temperature.

Flexibility factors for typical components are included in ASME B31J, *Stress Intensification Factors (i-Factors), Flexibility Factors (k-Factors), and Their Determination for Metallic Piping Components.* The flexibility factor is the length of straight pipe having the same flexibility as the component divided by the centerline length of the component. It can be used in hand calculations of piping flexibility, and is included in all modern piping stress analysis programs.

Additional flexibility is introduced in the system by elbows. Elbows derive their flexibility from the fact that the cross section ovalizes when the elbow is bent. This ovalization reduces the moment of inertia of the pipe cross section, reducing its stiffness and increasing its flexibility. Note that the presence of a flange at the end of an elbow will reduce the ability of the elbow to ovalize, and thus ASME B31J provides reduced flexibilities for elbows with flanges welded to one or both ends.

Appendix S was added in the 2004 edition. It provides piping stress analysis examples. The 2004 edition only includes one sample problem. Examples illustrating effects of lift-off at supports and moment reversal were added in the 2006 edition. Significant updating for revised thermal expantion properties and revised stress intensification factors were made for the 2020 edition.

8.2 WHEN IS FORMAL FLEXIBILITY ANALYSIS REQUIRED?

8.2.1 Sustained and Occasional Loads

ASME B31.3 requires a formal flexibility analysis for sustained and occasional loads. These loads may be evaluated by detailed analysis, approximate methods, or simplified means such as pipe support span tables. When detailed analysis is performed, the Code prescribes how to calculate stresses due to sustained and occasional loads, which usually requires the use of a computer analysis.

8.2.2 Displacement Strains

ASME B31.3 requires a formal flexibility analysis for displacement strains unless certain exemption criteria are met, as described in para. 319.4.1.

A formal flexibility analysis does not mean a computer stress analysis; it can be by simplified, approximate, or comprehensive methods. These vary from simple charts and methods such as Kellog's guided cantilever method to detailed computer stress analysis of the piping system.

There are three exemptions from formal flexibility analysis:

- A system that duplicates or replaces, without significant change, a system operating with a successful service record
- A system that can be readily judged adequate by comparison with previously analyzed systems
- A system that satisfies the simplified equation (16) of ASME B31.3, reproduced below as Eq. (8.1)

The first exemption is basically a means of grandfathering a successful design. The difficulty comes when trying to determine how long a system must operate successfully in order to demonstrate that the design is acceptable. Considering that some systems may cycle less than one time per year and the design criteria are based on fatigue considerations, the fact that a piping system has not experienced a failure for some period of time provides little indication that it actually complies with the Code or that it will not fail sooner or later. Interpretation 13-14 on this subject reads as follows:

> "Question: In accordance with para. 319.4.1, how many years of operation or number of operating cycles are required to qualify a piping system as having a successful service record?"
>
> "Reply: Such determination is the responsibility of the designer. See para. 300(b)(2)."

This matter is left as a judgment call, but should not be blindly used to accept a design without careful consideration.

The second exemption relies on the judgment of an engineer or designer who, based on experience, can determine that a system has adequate flexibility.

The third exemption uses a simplified condition that has limited applicability. The requirements for its use are that the system is of uniform size, has no more than two points of fixation, and has no intermediate restraints. The equation reads[14]

$$\frac{Dy}{(L-U)^2} \leq K_1 \tag{8.1}$$

where

D = outside diameter of pipe, mm (in.)
K_1 = $208{,}000 S_A/E_a$, (mm/m)^2, for SI units
= $30\ S_A/E_a$, (in./ft)^2, for U.S. customary units
E_a = reference modulus of elasticity at 21°C (70°F), MPa (ksi)
S_A = allowable displacement stress range per Eq. (la) of ASME B31.3 [Eq. (7.1) of this book], MPa (ksi)
L = developed length of piping between anchors, m (ft)
U = anchor distance, straight line between anchors, m (ft)
y = resultant of total displacement strains, mm (in.), to be absorbed by the piping system; displacements of equipment to which the pipe is attached (end displacements of the piping system) should be included

This equation tends to be very conservative. However, there are a number of warning statements within the Code regarding limits to its applicability. One such warning is for near-straight sawtooth runs, which are described below in Insert 8.1. In addition, the equation provides no indication of the end reactions, which would need to be considered in any case for load-sensitive equipment.

8.3 WHEN IS COMPUTER STRESS ANALYSIS TYPICALLY USED?

The Code does not indicate when computer stress analysis is required. It is difficult to generalize when a particular piping flexibility problem should be analyzed by computer methods, because this depends on the type of service, actual piping layout and size, and severity of temperature. However, there are quite a few guidelines in use by various organizations that indicate which types of lines should be evaluated by computer in a project. These tend to require that lines at higher combinations of size and temperature or larger lines that are attached to load-sensitive equipment be computer-analyzed. One set of recommended criteria is provided below.

(a) In the case of general piping systems; according to the following line size/flexibility temperature criteria:
- All DN 50 (NPS 2) and larger lines with a temperature range over 260°C (500°F)
- All DN 100 (NPS 4) and larger lines with a temperature range exceeding 205°C (400°F)

[14]ASME B31.3, Eq. (16).

- All DN 200 (NPS 8) and larger lines with a temperature range exceeding 150°C (300°F)
- All DN 300 (NPS 12) and larger lines with a temperature range exceeding 90°C (200°F)
- All DN 500 (NPS 20) and larger lines at any temperature

(b) All DN 75 (NPS 3) and larger lines connected to rotating equipment
(c) All DN 100 (NPS 4) and larger lines connected to air fin heat exchangers
(d) All DN 150 (NPS 6) and larger lines connected to tankage
(e) Double-wall piping with a design temperature differential between the inner and the outer pipe greater than 20°C (40°F)

Again it is emphasized that the intent of the above criteria is to identify in principle only typical lines that should be considered at least initially for detailed stress analysis.Such criteria are used as a basis for estimating engineering hours for a project.Obviously, the final decision as to whether or not a computer analysis should be performed should depend on the complexity of the specific piping layout under investigation and the sensitivity of equipment to piping loads.

Just because a line may pass some exemption criteria from computer stress analysis does not mean that it is exempt from other forms of formal analysis, nor that it will always pass the Code criteria if analyzed in detail. The procedure is intended to separate the more trouble-free types of systems from those that are more subject to overload or overstress. The lines exempted from computer stress analysis are less sensitive to potential problems when properly laid out with sufficient flexibility by the designer.

8.4 STRESS INTENSIFICATION FACTORS

Stress intensification factors are used to relate the stress in a component to the stress in nominal-thickness straight pipe. As discussed in the previous section, the analysis is based on nominal pipe dimensions, so the calculated stress is the stress in straight pipe unless some adjustment is made. The stress can be higher in components such as branch connections.

Stress intensification factors that relate the stress in components to that in butt-welded pipe have been developed from so-called Markl fatigue testing of piping components, which generally follow the procedures developed by Markl (1960a-d).

As mentioned previously, Markl developed a fatigue curve for butt-welded pipe. It was based on displacement-controlled fatigue testing, bending the pipe in a cantilever bending mode. Figure 8.1 shows a Markl-type fatigue test machine. Use of a butt-welded pipe fatigue curve has several practical advantages. One is that the method was developed for butt-welded pipe, and the stress analyst typically does not know where the welds will be in the as-constructed system. Using a butt-welded pipe fatigue curve as the baseline fatigue curve allows for the possibility that butt welds could be anywhere in the system. Furthermore, from a testing standpoint, appropriate fatigue curves could not readily be developed for straight pipe without welds in a cantilever bending mode, because the failure will occur at the point of fixity, where effects of the method of anchoring the pipe could significantly affect results.

The stress intensification factors were developed from component fatigue testing. The stress intensification factor is the nominal stress from the butt-welded pipe fatigue curve at the number of cycles to failure in the component test, divided by the nominal stress in the component. The nominal stress in the component is the range of bending moment at the point of failure divided by the section modulus of matching pipe with nominal wall thickness. In a flexibility analysis, it is precisely this nominal stress that is calculated. When the nominal stress is multiplied by the stress intensification factor and then compared to the fatigue curve for butt-welded pipe, one can determine the appropriate number of cycles to failure of the component.

Different stress intensification factors are provided in ASME B31J for in-plane and out-plane loads. The direction of these moments are illustrated in Fig. 8.2.

One of the less well known aspects of piping flexibility analysis per the ASME B31 Codes is that in piping stress analysis, the calculated stress range due to bending loads is about one-half of the peak stress range. This

FIG. 8.1
MARKL-TYPE FATIGUE TEST MACHINE (COURTESY OF WFI)

is because the stress concentration factor for typical as-welded pipe butt welds is two. Since the stresses are compared to a butt-welded pipe fatigue curve, one-half of the actual peak stresses is calculated. Thus, the theoretical stress, for example, in an elbow due to bending loads is two times what is calculated in a piping flexibility analysis following Code procedures. This is not significant when performing standard design calculations, because the Code procedures are self-consistent. However, it can be very significant when trying to do a more detailed analysis, for example, in a fitness-for-service assessment.

One example occurs in the design of high-pressure piping in ASME B31.3, Chapter IX. This chapter requires a detailed fatigue analysis, using polished bar-type fatigue curves (rather than butt-welded pipe fatigue curves). When calculating stresses due to piping thermal expansion via a flexibility analysis, these calculated stresses must be multiplied by a factor of two in order to arrive at the actual stress range. This should not be confused with the difference between stress range and stress amplitude, which is an additional consideration.

Another commonly misunderstood item relates to where the peak stresses are in an elbow, as this is counterintuitive. With in-plane bending (see Fig. 8.2 for in-plane versus out-plane directions) of an elbow, the highest stresses are not at the intrados or extrados, they are in the elbow sidewalls (crown). They are through-wall bending stresses due to ovalization of the elbow. Again, this is not significant in design, but can be so in failure analysis or fitness-for-service evaluations.

Stress intensification factors were developed for several common components by Markl around 1950. The stress intensification factors presented in Appendix D of ASME B31.3 through the 2018 edition were based on Markl's work as modified over the years by further testing on selected components.

More recently, as a result of some findings of nonconservatism and the development of newer products, additional fatigue tests have been performed. Results of these tests were first presented in an ASME report, then published as B31J, *Stress Intensification Factors (i-Factors), Flexibility Factors (k-Factors), and Their*

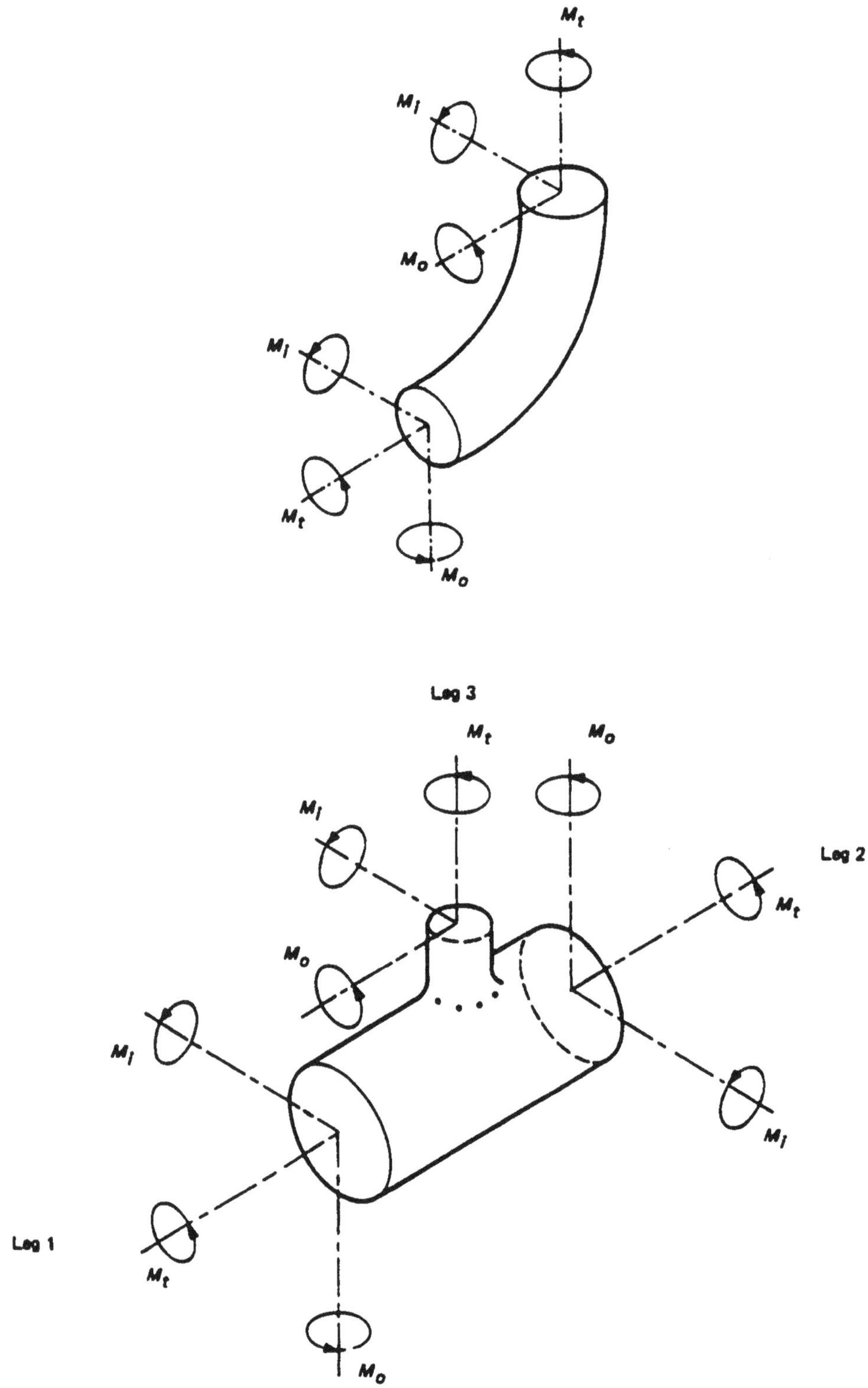

FIG. 8.2
IN-PLANE AND OUT-PLANE BENDING MOMENTS IN BENDS AND BRANCH CONNECTIONS
(ASME B31.3, FIGS. 319.4.4A AND 319.4.4B)

Determination for Metallic Piping Components in 2018. In response to this publication, the B31.3 Section Committee added a note to Appendix D in the 2018 edition that said:

The designer may use the more applicable stress intensification factors and flexibility factors from ASME B31J instead of the stress intensification factors and flexibility factors herein and is encouraged to do so when both $S_E > 0.5S_A$ and significant cycles are present. Alternatively. stress intensification factors and branch connection flexibility factors may be developed using ASME B31J, Nonmandatory Appendices A and B, respectively.

For the 2020 edition, the Section Committee removed Appendix D all together, instead referring to ASME B31J for stress intensification factors.

8.5 FLEXIBILITY ANALYSIS EQUATIONS

8.5.1 Sustained and Occasional Loads

ASME B31.3 provides the following set of equations for calculating the stress due to sustained and occasional loads in piping systems. The sustained stress, S_L, is calculated from the combination of stresses due to axial stresses, bending stresses, and shear stresses due to torsion. These are combined per Mohr's circle.

$$S_L = \sqrt{\left(\left|S_a\right| + S_b\right)^2 + \left(2S_t\right)^2} \quad \text{(8.2) [B31.3 equation 23a]}$$

$$S_a = \frac{I_a F_a}{A_p} \quad \text{(8.3) [B31.3 equation 23d]}$$

$$S_b = \frac{\sqrt{\left(I_i M_i\right)^2 + \left(I_o M_o\right)^2}}{Z} \quad \text{(8.4) [B31.3 equation 23b]}$$

$$S_t = \frac{I_t M_t}{2Z} \quad \text{(8.5) [B31.3 equation 23c]}$$

Where:

A_p = cross-sectional area of the pipe computed using nominal pipe dimensions less allowances
I_a = sustained longitudinal force index. In the absence of more-applicable data, I_a is taken as 1.00.
I_i = sustained in-plane moment index. In the absence of more-applicable data, I_i is taken as the greater of $0.75i_i$ or 1.00.
I_o = sustained out-plane moment index. In the absence of more-applicable data, I_o is taken as the greater of $0.75i_o$ or 1.00.
I_t = sustained torsional moment index. In the absence of more-applicable data, I_t is taken as 1.00.
i_i = in-plane stress intensification factor
i_o = out-plane stress intensification factor
F = longitudinal force due to sustained loads
M_i = In-plane moment due to sustained loads
M_o = out-plane moment due to sustained loads
M = torsional moment due to sustained loads
Z = section modulus computed using nominal pipe dimensions less allowances

The concern with respect to sustained loads is collapse. Stress indices have been developed by Paulin Research (Becht V, 2014) using the twice elastic slope method, as described in B31J Nonmandatory Appendix D. The stress index data was not readily available when the equations for calculating stress were introduced to the Code. As are result, when developing these rules 0.75i was specified to be used in the absence of more applicable data. This is consistent with ASME B31.1. However, it is generally not a good measure of collapse load, since it is based on fatigue failure data and not collapse load data. It was specified so that users of the Code would have consistent calculations. However, the analyst should use stress index data when possible. It is planned to add stress indices in a future edition of B31J.

8.5.2 Displacement Strains

ASME B31.3 provides the following set of equations for calculating the stress due to displacement strains in piping systems. The displacement stress range, S_E, is calculated from the combination of stresses due to axial stresses, bending stresses, and shear stresses due to torsion. These are combined per Mohr's circle.

$$S_E = \sqrt{\left(|S_a| + S_b\right)^2 + \left(2S_t\right)^2} \qquad \text{(8.6) [B31.3 equation 17]}$$

$$S_b = \frac{\sqrt{\left(i_i M_i\right)^2 + \left(i_o M_o\right)^2}}{Z} \qquad \text{(8.7) [B31.3 equation 18]}$$

Where

A_p = cross-sectional area of pipe computed using nominal pipe dimensions
F_a = axial force range between any two conditions being evaluated
i_a = axial stress intensification factor. In the absence of more-applicable data, i_a = 1.0 for elbows, pipe bends, and miter bends (single, closely spaced, and widely spaced), and $i_a = i_o$ for other components
i_i = in-plane stress intensification factor
i_o = out-plane stress intensification factor
i_t = torsional stress intensification factor. In the absence of more-applicable data, i_t = 1.0
M_i = in-plane bending moment range between any two conditions being evaluated
M_o = out-plane bending moment range between any two conditions being evaluated
M_t = torsional moment range between any two conditions being evaluated
S_q = axial stress range due to displacement strains = $i_a F_a / A_p$
S_b = bending stress range due to displacement strains
S_t = torsional stress range due to displacement strains = $i_t M_t / 2Z$
Z = section modulus of pipe computed using nominal pipe dimensions

8.6 COLD SPRING

Cold spring is the deliberate introduction of a cut short in the system to offset future thermal expansion. It is used to reduce loads on equipment; however, it does not affect the strain range. Thus, the cold spring does not affect the calculation of the displacement stress range, S_A, or the allowable stress range, S_A.

The range of loads on equipment due to thermal expansion is also unchanged by cold spring. However, the magnitude of the load at any given operating condition can be changed. Cold spring is typically used to reduce the load in the operating condition. It does this by shifting the load to the nonoperating, ambient condition.

The effectiveness of cold spring is generally considered to be questionable, but it is occasionally the only reasonable means of satisfying equipment load limits. It should not be used indiscriminately. Although it may provide an easy way for an analyst to solve an equipment load problem, there are a number of considerations.

Its implementation in the field is generally difficult to achieve accurately. After the plant has been operated, deliberately installed cold spring can be misunderstood to be piping misalignment and "corrected." Additionally, when evaluating an existing piping system that is designed to include cold spring, it is highly questionable what the actual condition is.

Because of difficulties in accomplishing desired cold spring, ASME B31.3 requires that the reactions be computed both with the assumption that only two-thirds of the design cold spring is present, and with four-thirds of the design cold spring present. Equations are provided in para. 319.5.1 for calculating the maximum reaction force or moment, including cold spring, at the operating condition and at the piping installation temperature. The equations are limited to a two-anchor system without intermediate restraints. Currently, any piping requiring cold spring would most likely be evaluated using computer flexibility analysis.

Insert 8.1 How to Increase Piping Flexibility

There are a number of means of increasing piping flexibility. The intent of this insert is to describe some general concepts in order to better enable the designer to develop increased flexibility in piping systems.

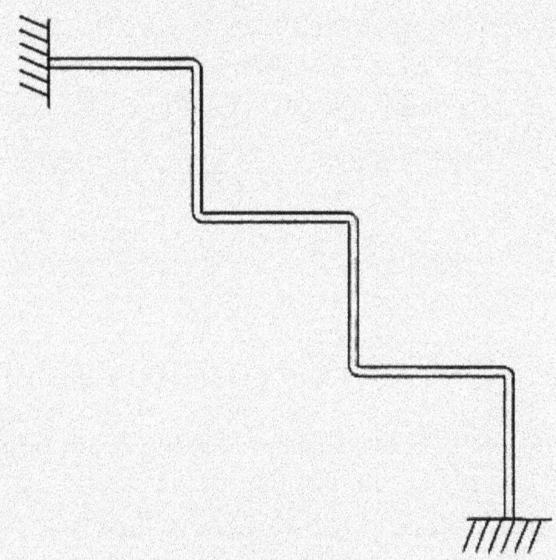

FIG. 8.3
PIPING LAYOUT 1

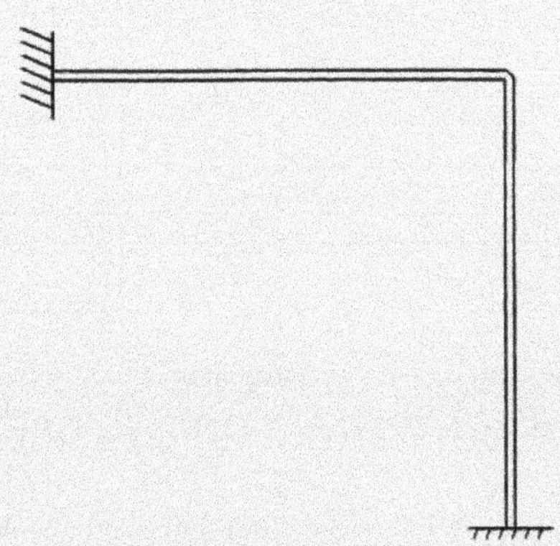

FIG. 8.4
PIPING LAYOUT 2

One alternative is to add expansion joints. However, this is usually undesirable. Although it may make the job easier for the analyst, it typically reduces the overall reliability of the piping system. Expansion joints should be the solution of last resort.

Here are a few fundamental concepts in flexibility:

(a) Flexibility of a pipe in bending is roughly proportional to the cube of the length of a straight run of pipe. Therefore, longer straight runs are much more flexible than a series of shorter runs chopped up into short lengths and separated by elbows.
(b) Flexibility can be provided by a length of piping perpendicular to the direction of thermal expansion that must be accommodated.
(c) In general, the further the center of gravity of the piping system is from a straight line between piping system end points (for a two-anchor system), the lower are the thermal expansion stresses.

These concepts are illustrated in the classic, near-sawtooth piping configuration. The question is whether the system in Fig. 8.3 or that in Fig, 8.4 is more flexible. The reader should look and decide before reading the next paragraph.

The thermal stresses in the second figure are significantly lower than those in the first figure. To resolve any doubt, run a flexibility analysis for both cases. The first figure is considered a near-sawtooth pipe configuration. The fact that it has additional elbows does not compensate for the fact that the pipe has been chopped into a number of smaller lengths. By either measure (a) or (c) above. Fig. 8.4 can be visually discerned to be more flexible.

The concepts related above are simple, but powerful when understood, in application to flexibility problems.

8.7 ELASTIC FOLLOW-UP/STRAIN CONCENTRATION

The analysis procedures in the Code essentially assume that the strain range in the system can be determined from an elastic analysis. That is, strains are proportional to elastically calculated stresses. The stress range is limited to less than two times the yield stress, in part to achieve this. However, in some systems, strain concentration or elastic follow-up occurs. A classic reference for elastic follow-up due to creep conditions is Robinson (1960).

As an example, consider a cantilevered pipe with a portion adjacent to the fixed end constructed with a reduced-diameter or - thickness pipe or lower-yield-strength material that has the free end laterally displaced.

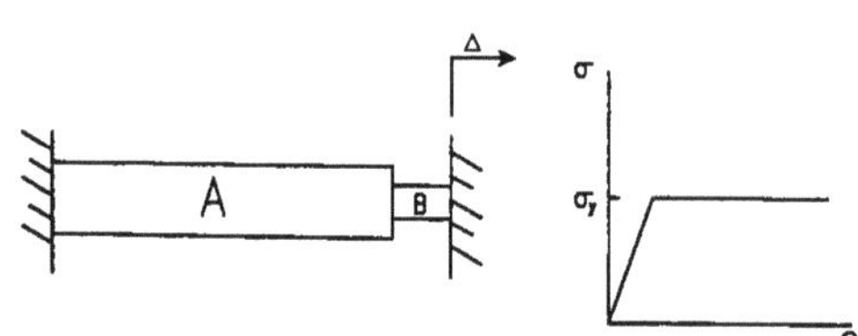

FIG. 8.5
STRAIN CONCENTRATION TWO-BAR MODEL

The elastic analysis assumes that strains will be distributed in the system in accordance with the elastic stiffnesses. However, consider what happens when the locally weak section yields. As the material yields, a greater proportion of additional strain due to displacement occurs in the local region, because its effective stiffness has been reduced by yielding of the material. Thus, there is plastic strain concentration in the local region. In typical systems, this strain concentration is generally not considered to be significant. However, it can be highly significant under specific conditions, such as unbalanced systems. Paragraph 319.2.2(b) provides warnings regarding these conditions.

An example of a two-bar system under axial compression is provided to illustrate elastic follow-up, although the concern in piping is generally bending. However, the axial compression case illustrates the problem.

The problem is illustrated in Fig. 8.5. The elastic distribution of the total displacement, *A*, between bar *A* and bar *B* is as follows.

$$\Delta_A = \frac{K_B}{K_A + K_B}\Delta \tag{8.6}$$

$$\Delta_B = \frac{K_A}{K_A + K_B}\Delta \tag{8.7}$$

where

Δ_A = displacement absorbed by bar *A*
Δ_B = displacement absorbed by bar *B*
K_A = $A_A E_A / L_A$, elastic stiffness of bar *A*
K_B = $A_B E_B / L_B$, elastic stiffness of bar *B*
A_A = area of bar *A*
A_B = area of bar *B*
E_A = elastic modulus of bar *A*
E_B = elastic modulus of bar *B*
L_A = length of bar *A*
L_B = length of bar *B*
Δ = total axial displacement imposed on two bars

For elastic, perfectly plastic behavior of the material with a yield stress σ_y, the stress in bar *B* cannot exceed yield, so the load in bar *A* cannot exceed $\sigma_y A_B$. Therefore, after bar *B* starts to yield, the displacement in each bar is given by

$$\Delta_A^{e-p} = \frac{\sigma_y A_A}{K_A} \tag{8.8}$$

$$\Delta_B^{e-p} = \Delta - \Delta_A^{e-p} \tag{8.9}$$

where

$\Delta_A^{e\text{-}p}$ = actual displacement absorbed by *A*, elastic plastic case
$\Delta_B^{e\text{-}p}$ = actual displacement absorbed by *B*, elastic plastic case

The strain concentration is the actual strain in bar *B,* considering elastic plastic behavior, divided by the elastically calculated strain in bar *B*. Since strain is displacement divided by length, the strain concentration is given as

$$\text{strain concentration} = \frac{\Delta_B^{e-p}}{\Delta_B} = \frac{\left[\Delta - \left(\sigma_y A_B / K_A\right)\right] / L_B}{\{[K_A / \left(K_A + K_B\right)\}\Delta / L_B} \tag{8.10}$$

This can be rearranged as

$$\text{strain concentration} = \left(1 + \frac{K_B}{K_A}\right)\left(1 - \frac{\sigma_y A_B}{\Delta K_A}\right) \tag{8.11}$$

If we consider a general region under lower stress or in a stronger condition coupled with a local region under higher stress or with weaker material (e.g., lower σ_y), then the more flexible is the general region (stiffer is the local region), the more severe is the elastic follow-up. This can be seen by considering Eq. (8.11). As the stiffness of bar B increases relative to bar A, the plastic strain concentration becomes more severe.

For a system subject to plastic strain concentration, the simplest solution in design is typically to limit the thermal expansion stress range to less than the yield strength of the material. This avoids plastic strain concentration by keeping the component elastic. Although the stresses due to sustained loads such as weight and pressure usually do not need to be added to those due to thermal expansion when satisfying this limit, proper consideration of this requires a very detailed understanding of the phenomenon. Thus, it is generally preferable to conservatively add the stresses due to sustained loads to the thermal expansion stresses in this type of evaluation.

Under creep conditions, elastic follow-up can have very severe effects. This is due to the implicit assumption in the code allowable stress basis that thermal expansion stresses will relax. Near-yield-level stresses cannot be sustained very long at the very high temperatures permitted in ASME B31.3 without rapid creep rupture failures. With elastic follow-up, creep strain in the local region does not result in a corresponding reduction in thermal expansion load/stress. In severe cases of elastic follow-up, rapid creep rupture failures can occur, even though the calculated stresses fall well within Code limits. The most straightforward solution to this condition is to design the system such that the level of thermal expansion stresses in the local region subject to elastic follow-up does not exceed S_h.

A well known circumstance where elastic followup can occur in creep conditions is in refractory lined piping systems with local sections that are hot walled. For example, in fluid catalytic cracking units, it is common to use carbon steel pipe with an insulating refractory lining to carry high temperature fluid solids and flue gas, at temperatures to 760°C (1400°F) and higher. The metal temperature is much lower, as a result of the insulating lining. However, in some circumstances, hot wall sections such as hot wall valves are included in the system. The wall temperature for these components is generally the hot process temperature, so they will creep over time. In this circumstance, the thermal strain in the carbon steel portion of the system gradually transfers to the creeping hot walled section. Further, the carbon steel portion of the system acts as a spring, keeping the load on the hot walled section, preventing it from relaxing. The elastic analysis that was performed in design, does not consider this behavior. Thus, piping systems that met the allowable thermal expansion stresses have cracked in service as a result of elastic follow-up. Becht (1988) describes an evaluation of such a system that had failed.

As an analogous situation, consider a cantilever beam. At the built in end, heat up the beam to creep temperatures. The elastic calculation predicts a distribution of strain over the entire beam. However, during service, the strain that was predicted to be in the low temperature portions of the beam will gradually shift to the creeping portion. Further, the low temperature portion of the beam will act as a spring, keeping a load on the high temperature portion, preventing it from relaxing.

8.8 EFFECT OF ELASTIC MODULUS VARIATIONS DUE TO TEMPERATURE

In a typical flexibility analysis, the elastic modulus at a reference temperature of 21°C (70°F) is used. It was changed to this from the elastic modulus at installation temperature in the 1999 Edition, to be simpler to apply and more theoretically correct.

Fatigue, in terms of strain range versus cycles to failure, is generally considered to be independent of temperature at temperatures where creep effects are not significant. Markl fatigue testing was done at ambient temperature, and thus an ambient temperature fatigue curve was developed. If one were to calculate the stress in a flexibility analysis using the elastic modulus at temperature, one would need to divide this by the elastic modulus at temperature and multiply this result by the ambient temperature elastic modulus to compare it to the Markl fatigue curve. Dividing the stress by hot elastic modulus yields the strain, which is then multiplied by ambient temperature elastic modulus to yield the stress range at ambient temperature. This is the procedure used in fatigue analysis per ASME BPVC, Section VIII, Division 2. Using the ambient temperature elastic modulus in the flexibility analysis accomplishes the same thing, directly.

It is recognized by the Code Committee that the analysis could be done with the hot modulus of elasticity, with the calculated stress multiplied by the ratio of the cold modulus to the hot modulus, and that in some cases it may be preferable to do this. The following text was recently added [para. 319.2.2(b)(4)]:

"When differences in the elastic modulus within a piping system will significantly affect the stress distribution, the resulting displacement stresses shall be computed based on the actual elastic moduli at the respective operating temperatures for each segment in the system and then multiplied by the ratio of the elastic modulus at ambient temperature to the modulus used in the analysis for each segment."

CHAPTER

9

SUPPORTS

9.1 OVERVIEW OF SUPPORTS

Requirements for supports and other devices to restrain the piping are provided in Chapter II of ASME B31.3, specifically in Section 321, Piping Support.

ASME B31.3 provides general requirements for piping supports as well as descriptions of conditions for which they must be designed. The support elements (e.g., springs, hanger rods, etc.) are within the scope of ASME B31.3, but the support structures to which they are attached are not. The supports must achieve the objectives in the design of the piping for sustained and occasional loads as well as thermal displacement.

9.2 MATERIALS AND ALLOWABLE STRESS

Pipe support elements may be constructed from a variety of materials, including the materials listed in the allowable stress tables, other metallic materials, steel of unknown specification, wood and other materials. Some specific requirements are provided in para. 321.1.4 and the allowable stress basis is provided in para. 321.1.3 [para. 321.1.4(c) for steel of unknown specification].

Material that is bonded or welded to the pipe must be compatible with the piping and service and must be of known specification. Material that is not welded directly to the pressure-containing piping component may be of unknown specification. Material of unknown specification may also be welded to a pad or sleeve of known specification that is welded in turn to the pressure boundary.

The Code does not presently provide any toughness requirements (i.e. impact testing) for welded attachments. In the opinion of the author, welded attachments (base material and weld) should comply with the same toughness requirements as the pipe material.

Steel of unknown specification is assigned a basic allowable stress in tension or compression of 82 MPa (12 ksi) and is limited to metal temperatures in the range of -29°C to 343°C (-20°F to 650°F).

The allowable stress for other materials, except for springs and materials for which the criteria are not applicable (e.g., wood) is required to be in accordance with para. 302.3.1. This makes the allowable stress for supports (via reference to para. 302.3.2 and Appendix A) the same as for other piping components. This includes the limits on shear and bearing, which are 1.6 times the basic allowable stress in tension. However, weld joint quality factors are not used for supports. Structural stability (i.e., buckling) must also be given due consideration for elements in compression. Although there is no specific margin specified for structural elements in compression, a design margin of two (about equal to the margin provided for elastic buckling in the American Institute of Steel Construction Structural Steel Code) to three (consistent with ASME B31.3 external pressure buckling margin) should be considered.

Recommendations are made for gray ductile, and malleable iron with respect to their use in pipe support elements. Gray iron is not recommended for elements that may be in tension (due to the brittle nature of the material). Gray iron is also not recommended for services where the piping may be subject to impact-type loading resulting from pulsation or vibration. These limitations are not applied to ductile or malleable iron. However, para. 321.1.4(c) specifically permits use of ductile and malleable iron only for pipe and beam clamps, hanger flanges, clips, brackets, and swivel rings.

9.3 DESIGN OF SUPPORTS

ASME B31.3, Section 321 discusses the types of loads and conditions that must be considered. Detailed criteria are generally not provided, with the exception of the allowable stress. The objectives to be satisfied are described in para. 321.1.1, reproduced below. Layout and design of piping and its supporting elements shall be directed toward preventing the following.

"(a) piping stresses in excess of those permitted in this Code [ASME B31.3];
(b) leakage at joints;
(c) excessive thrusts and moments on connected equipment (such as pumps and turbines);
(d) excessive stresses in the supporting (or restraining) elements;
(e) resonance with imposed or fluid-induced vibrations;
(f) excessive interference with thermal expansion and contraction in piping which is otherwise adequately flexible;
(g) unintentional disengagement of piping from its supports;
(h) excessive piping sag in piping requiring drainage slope;
(i) excessive distortion or sag of piping (e.g., thermoplastics) subject to creep under conditions of repeated thermal cycling;
(j) excessive heat flow, exposing supporting elements to temperature extremes outside their design limits."

These are generally requirements for sustained and thermal loads, stated elsewhere in the Code as well as in the section on supports.

The pipe support elements must be designed for all the loads to which they can be subjected, including surge, thermal expansion, and weight. In addition, for supports that can slide, the lateral load due to friction must be considered [paras. 321.2.1(b) and 321.2.2(c)].

The allowable load for all threaded parts (e.g., rods and bolts) is required to be based on the root diameter [para. 321.2.2(b)]. Screw threads are required to be per ASME B1.1 unless other threads are required for adjustment under heavy loads. Turnbuckles and adjusting nuts are required to have full thread engagement and all threaded adjustments are required to be provided with a means of locking (e.g., locknut) (para. 321.1.5).

General requirements for spring supports, constant-effort supports, counterweight supports, and hydraulic supports are provided in paras. 321.2.3 and 321.2.4.

Paragraph 321.3 requires consideration of stresses in the pipe caused by pipe attachments. This includes local stresses due to lugs, trunions, and supporting elements welded to the pipe or attached by other means. The stated criteria are performance-based (e.g., shall not cause undue flattening of the pipe, excessive localized bending stress, or harmful thermal gradients in the pipe wall). Specific criteria are not provided. However, the design-by-analysis rules of ASME BPVC, Section VIII, Division 2, Part 5, would be considered to be an acceptable approach to evaluating these stresses. Insert 9.2 below describes allowable stress criteria of Part 5. Approaches to calculating the stresses due to loads on attachments are provided in Bednar (1986), WRC 107 (Wichman et al, 1979), and WRC 297 (Mershon et al, 1984).

A specific reference to the shear stress limits in 302.3.1(b) was added in the 2008 edition. Further, it is specified that if the allowable stress values in the pipe and support differ, the lower value shall be used (for the weld).

Although it is not referred to in the pipe support section, there is a referenced component standard for pipe supports in Table 326.1. This standard is MSS SP-58, *Pipe Hangers and Supports—Materials, Design, and Manufacture.* Piping supports per this standard are acceptable for ASME B31.3 piping systems.

Insert 9.1 Spring Design

Springs are used to provide continued support, while permitting the pipe to move vertically. There are variable-effort springs, which vary supporting force when the pipe moves up or down, and constant-effort springs, which provide a nearly constant supporting force as the pipe moves up or down. Because it is normally desirable to limit the change in load in the spring caused by pipe movement (typically a 25% limit on load change), when the movement becomes large, constant-effort supports, which typically provide load changes of 6% or less, are considered.

The three primary considerations in selecting a variable-effort spring are for the spring to provide sufficient support (i.e., be strong enough), sufficient movement capability to accommodate the movement of the pipe, and sufficiently low stiffness to prevent excessive change in load as a result of the pipe movement.

Springs are typically selected from charts. An example is provided in Fig. 9.1. The first step is to find the design load and movement. The design load is used to make the initial selection of spring size. For example, if the load is 3600 lb, a size 13 or size 14 variable spring could be selected. The movement can then be considered by calculating the maximum permissible stiffness or by looking at the spring chart. The maximum stiffness can be determined from the equation

$$K_{max} = \frac{\text{force} \times \text{variation}}{\text{displacement}}$$

where

K_{max} = maximum spring stiffness
force = required supporting force from the spring
variation = allowable load variation, expressed as a fraction
displacement = vertical movement of piping

The allowable load variation is typically 25%, or a traction of 0.25.

Continuing with the same example, for a movement of 1 in. and an allowable variation of 0.25, the maximum allowable spring stiffness would be 900 lb/in. Springs that are more flexible would be acceptable, because their load variation caused by the piping movement would be less than 25%. Looking at the bottom of the chart in Fig. 9.1, we see that a size 14 figure 268 spring would be acceptable. The stiffness of this spring is 800 lb/in., which is less than 900 lp/in. A size 13 spring is also acceptable from a stiffness standpoint.

The spring should be installed at a load such that the supporting force in the operating condition will be the desired supporting force, 3600 lb in this example. If the movement from ambient to operating condition is up, the spring supporting force will be reduced. Thus, the load in the installed, nonoperating or cold condition would need to be higher. The required load would be the desired force in the operating condition plus the spring stiffness times the displacement. In this example, with a size 14, figure 268 spring, the initial force in the cold condition would be 4400 lb [3600 (lb) + 1 (in.) x 800 (lb/in.)]. Noting that this force is within the operating range of a size 14 spring, 2800 ib to 4800 lb, we see that the spring is an acceptable choice. A size 13 spring would not be acceptable, because the required cold load, 4200 Ib, exceeds the maximum design load of the spring, 3600 lb.

Another way to approach the problem is to look at the spring charts. Looking at a size 14, figure 268 spring, we can use the displacement scale at the left to determine what the cold setting of the spring will be and if it will be in the appropriate range. Noting that the spring will be moving up (lesser displacement) as it goes from cold to hot, and the movement is 1 in., one can look at the displacement scale and determine that 1 in, down from an operating load setting of 3600 lb (1 in. on the load scale).

the cold load would be 4400 lb (2 in. on the load scale), and it would remain in scale. The next step would be to check load variation, the difference between hot and cold settings divided by the hot setting, and make sure it is within acceptable limits (typically 25%).

SIZE AND SERIES SELECTION

HOW TO USE HANGER SELECTION TABLE:

In order to choose a proper size hanger, it is necessary to Know the actual load which the spring is to support and the amount and direction of the pipe line movement from the cold to the hot position:

Find the actual load of the pipe in the load table. As it is desirable to support the actual weight ot the pipe when the line is not, the actual load is the hot load.

To determine the cold load, read the spring scale, up or down, for the amount of expected movement. The chart must be read opposite from the direction of the pipe's movement. The load arrived at is the cold load.

If the cold load falls outside of the working load range of the hanger selected, relocate the actual or hot load in the adjacent column and find the cold load. When the hot and cold loads are both within the working range of a hanger, the size number of that hanger will be found at the top of the column.

Should it be impossible to select a hanger in a particular series such that both loads occur within the working range, consideration should be given to a variable spring hanger with a wider working range or a constant support hanger.

spring hangers

The cold load is calculated by adding (for up movement) or subtracting (for down movement) the product of spring rate times movement to or from the hot load Cold load = hot load (±) movement x spring rate.

A key criteria in selecting the size and series of a variable spring is a factor known as variability. This is a measurement of the percentage change in supporting force between the hot and cold positions of a spring and is calculated from the formula:

$$\text{Variability} = \frac{\text{Movement} \times \text{ Spring Rate}}{\text{Hot Load}}$$

If an allowable variability is not specified, good practice would be to use 25% as specified by MSS-SP58.

load table in pounds: for selection of hanger size

Quadruple	Triple	fig. 98	fig. 268	fig. 82	000*	00*	0	1	2	3	4	5	6	7	8	9	10	11	12	13	14	15	16	17	18	19	20	21	22	fig. 82	fig. 268	fig. 98
working range, in.					hanger size																									spring deflection, in.		
					7	19	43	63	81	105	141	189	252	336	450	600	780	1020	1350	1800	2400	3240	4500	6000	7990	10610	14100	18750	25005	0	0	0
					7	20	44	66	84	109	147	197	263	350	469	625	813	1063	1406	1875	2500	3375	4688	6250	8322	11053	14688	19531	26047			
					8	22	46	68	88	114	153	206	273	364	488	650	845	1105	1463	1950	2600	3510	4875	6500	8655	11495	15275	20313	27089			
					9	24	48	71	91	118	159	213	284	378	506	675	878	1148	1519	2025	2700	3645	5063	6750	8987	11938	15863	21094	28131			
0	0	0	0	0	10	26	50	74	95	123	165	221	294	392	525	700	910	1190	1575	2100	2800	3780	5250	7000	9320	12380	16450	21875	29173	¼	½	1
					11	28	52	76	98	127	170	228	305	406	544	725	943	1233	1631	2175	2900	3915	5438	7250	9652	12823	17038	22656	30215			
					12	30	54	79	101	131	176	236	315	420	563	750	975	1275	1688	2250	3000	4050	5625	7500	9985	13265	17625	23438	31256			
					12	31	56	81	105	136	182	244	326	434	581	775	1008	1318	1744	2325	3100	4185	5813	7750	10317	13708	18213	24219	32298			
2	1½	1	½	¼	14	34	58	84	108	140	188	252	336	448	600	800	1040	1360	1800	2400	3200	4320	6000	8000	10650	14150	18800	25000	33340	½	1	2
					14	35	59	87	111	144	194	260	347	462	619	825	1073	1403	1856	2475	3300	4455	6188	8250	10982	14592	19388	25781	34382			
					15	38	61	89	115	149	200	268	357	476	638	850	1105	1445	1913	2550	3400	4590	6375	8500	11315	15035	19975	26563	35424			
					16	40	63	92	118	153	206	276	368	490	656	875	1138	1488	1969	2625	3500	4725	6563	8750	11647	15477	20563	27344	36466			
4	3	2	1	½	17	41	65	95	122	158	212	284	378	504	675	900	1170	1530	2025	2700	3600	4860	6750	9000	11980	15920	21150	28125	37508	¾	1½	3
					18	43	67	97	125	162	217	291	389	518	694	925	1203	1573	2081	2775	3700	4995	6938	9250	12312	16362	21738	28906	38549			
					19	45	69	100	128	165	223	299	399	532	713	950	1235	1615	2138	2850	3800	5130	7125	9500	12645	16805	22325	29688	39591			
					20	47	71	102	132	171	229	307	410	546	731	975	1268	1658	2194	2925	3900	5265	7313	9750	12977	17247	22913	30469	40633			
6	4½	3	1½	¾	21	49	73	105	135	175	235	315	420	560	750	1000	1300	1700	2250	3000	4000	5400	7500	10000	13310	17690	23500	31250	41675	1	2	4
					21	50	74	108	138	179	241	323	431	574	769	1025	1333	1743	2306	3075	4100	5535	7688	10250	13642	18132	24088	32031	42717			
					22	53	76	110	142	184	247	331	441	588	788	1050	1365	1785	2363	3150	4200	5670	7875	10500	13975	18575	24675	32813	43759			
					23	55	78	113	145	188	253	339	452	602	806	1075	1398	1828	2419	3225	4300	5805	8063	10750	14307	19017	25263	33594	44801			
8	6	4	2	1	24	56	80	116	149	193	258	347	462	616	825	1100	1430	1870	2475	3300	4400	5940	8250	11000	14640	19460	25850	34375	45843	1¼	2½	5
					25	58	82	118	152	197	264	354	473	630	844	1125	1463	1913	2531	3375	4500	6075	8438	11250	14972	19902	26438	35156	46885			
					26	60	84	121	155	201	270	362	483	644	863	1150	1495	1955	2588	3450	4600	6210	8625	11500	15305	20345	27025	35938	47926			
					27	62	86	123	159	206	276	370	494	658	881	1175	1528	1998	2644	3525	4700	6345	8813	11750	15637	20787	27613	36719	48968			
10	7½	5	2½	1¼	28	64	88	126	162	210	282	378	504	672	900	1200	1560	2040	2700	3600	4800	6480	9000	12000	15970	21230	28200	37500	50010	1½	3	6
					28	66	89	129	165	214	288	386	515	686	919	1225	1593	2083	2756	3675	4900	6615	9188	12250	16302	21672	28788	38281	51052			
					29	68	91	131	169	219	294	394	525	700	938	1250	1625	2125	2813	3750	5000	6750	9375	12500	16635	22115	29375	39063	52094			
					30	70	93	134	172	223	300	402	536	714	956	1275	1658	2168	2869	3825	5100	6885	9563	12750	16967	22557	29963	39844	53136			
					31	72	95	137	176	228	306	410	546	728	975	1300	1690	2210	2925	3900	5200	7020	9750	13000	17300	23000	30550	40625	54178	1¾	3½	7

spring rate — lb. per in.

	000*	00*	0	1	2	3	4	5	6	7	8	9	10	11	12	13	14	15	16	17	18	19	20	21	22
82			30	42	54	70	94	126	168	224	300	400	520	680	900	1200	1600	2160	3000	4000	5320	7080	9400	12500	16670
268	7	15	15	21	27	35	47	63	84	112	150	200	260	340	450	600	800	1080	1500	2000	2660	3540	4700	6250	8335
98			7	10	13	17	23	31	42	56	75	100	130	170	225	300	400	540	750	1000	1330	1770	2350	3125	4167
Triple			5	7	9	12	16	21	28	37	50	67	87	113	150	200	267	360	500	667	887	1180	1567	2083	2778
Quadruple			4	5	7	9	12	16	21	28	38	50	65	85	113	150	200	270	375	500	665	885	1175	1563	2084

*Available in fig. B268 & C-B268 only

FIG. 9.1
VARIABLE-SPRING HANGER TABLE (COURTESY OF ANVIL INTERNATIONAL)

Most springs are figure 268. A figure 82 spring is twice as stiff (essentially a half-spring). A figure 98 spring is twice as flexible as a figure 268 spring (essentially two springs stacked). A triple is three times as flexible and a quadruple is four times as flexible. Figure 82 springs are typically used when there are space limitations and the stiffer spring is an acceptable choice. The more flexible springs are used to accommodate greater movement.

When the load variation for variable-support springs is too great, a constant-effort spring can be used. Figure 9.2 provides a typical selection table for constant-effort springs. A spring size is selected based on the desired movement and supporting force. Note that the chart requires a margin be placed on the travel. The greater of 1 in. or 20% is added to the calculated travel, and then rounded up to the nearest ½ in.

If, for example, a constant support with a supporting force of 2100 lb and a calculated movement of 6.2 in. was required, a size 40 spring would be picked from the table in Fig. 9.2. This comes from looking in the column under total travel of 7.5 in. (6.2 in. plus 1 in., rounded up to the nearest ½ in.) for a spring size with a load greater than or equal to 2100 lb.

figs. 80-V and 81-H

model R

load travel table

constant supports

load in pounds for total travel in inches

hanger size no.	total travel* in inches: 1½	2	2½	3	3½	4	4½	5	5½	6	6½	7	7½	8	8½
1	144	108	86	72	62	54	48	43	39	36	33	31	29	27	
	173	130	104	87	74	65	58	52	47	43	40	37	35	33	
2	204	153	122	102	87	77	68	61	56	51	47	44	41	38	
3	233	175	140	117	100	88	78	70	64	58	54	50	47	44	
4	280	210	168	140	120	105	93	84	76	70	65	60	56	53	
5	327	245	196	163	140	123	109	98	89	82	75	70	65	61	
6	373	280	224	187	160	140	124	112	102	93	86	80	75	70	
7	451	338	270	225	193	169	150	135	123	113	104	97	90	85	
8	527	395	316	263	226	198	176	158	144	132	122	113	106	99	
9	600	450	360	300	257	225	200	180	164	150	138	129	120	113	
10	727	545	436	363	311	273	242	218	198	182	168	156	145	136	
11	851	638	510	425	365	319	284	255	232	213	196	182	170	160	
12	977	733	586	489	419	367	326	293	267	244	226	209	195	183	
13	1177	883	706	589	505	442	392	353	321	294	272	252	235	221	
14	1373	1030	824	687	589	515	458	412	375	343	317	294	275	258	
15	1573	1180	944	787	674	590	524	472	429	393	363	337	315	295	
16	1893	1420	1136	947	811	710	631	568	516	473	437	406	379	355	
17	2217	1663	1330	1109	950	832	739	665	605	554	512	475	443	416	
18	2540	1905	1524	1270	1089	953	847	762	693	635	586	544	508	476	
19															448
		2025	1620	1350	1157	1013	900	810	736	675	623	579	540	506	476
20		2145	1716	1430	1226	1073	953	858	780	715	660	613	572	536	505
21		2335	1868	1557	1334	1168	1038	934	849	778	718	667	623	584	549
22		2525	2020	1683	1443	1263	1122	1010	918	842	777	721	673	631	594
23		2710	2168	1807	1549	1355	1204	1084	985	903	834	775	723	678	638
24		2910	2328	1940	1663	1455	1293	1164	1058	970	895	831	776	728	685
25		3110	2488	2073	1777	1555	1382	1244	1131	1037	957	889	829	778	732
26		3310	2648	2207	1891	1655	1471	1324	1204	1103	1018	946	883	828	779
27		3630	2904	2420	2074	1815	1613	1452	1320	1210	1117	1037	968	908	854
28		3950	3160	2633	2257	1975	1756	1580	1436	1317	1215	1129	1053	988	929
29		4270	3416	2847	2440	2135	1898	1708	1553	1423	1314	1220	1139	1068	1005
30		4535	3628	3023	2591	2268	2016	1814	1649	1512	1396	1296	1209	1134	1067
31		4795	3836	3197	2740	2398	2131	1918	1744	1598	1475	1370	1279	1199	1128
32		5060	4048	3373	2891	2530	2249	2024	1840	1687	1557	1446	1349	1265	1191
33		5295	4236	3530	3026	2648	2353	2118	1925	1765	1629	1513	1412	1324	1246
34		5525	4420	3683	3157	2763	2456	2210	2009	1842	1700	1579	1473	1381	1300
35			4696	3913	3354	2935	2609	2348	2135	1957	1806	1677	1565	1468	1381
36			4968	4140	3549	3105	2760	2484	2258	2070	1911	1774	1656	1553	1461
37			5240	4367	3743	3275	2911	2620	2382	2183	2015	1871	1747	1638	1541
38			5616	4680	4011	3510	3120	2808	2553	2340	2160	2006	1872	1755	1652
39			5988	4990	4277	3743	3327	2994	2722	2495	2303	2139	1996	1871	1761
40			6360	5300	4543	3975	3533	3180	2891	2650	2446	2271	2120	1988	1871
41			6976	5813	4983	4360	3876	3488	3171	2907	2683	2491	2325	2180	2052
42			7588	6323	5420	4743	4216	3794	3449	3162	2919	2710	2529	2371	2232
43			8200	6833	5857	5125	4556	4100	3727	3417	3154	2929	2733	2563	2412
44			8724	7270	6231	5453	4847	4362	3965	3635	3355	3116	2908	2726	2566
45			9284	7737	6631	5803	5158	4642	4220	3868	3571	3316	3095	2901	2731
46			9760	8133	6971	6100	5422	4880	4436	4067	3754	3486	3253	3050	2871
47			10376	8647	7411	6485	5764	5188	4716	4323	3991	3706	3459	3243	3052
48			10988	9157	7848	6868	6104	5494	4995	4578	4226	3924	3663	3434	3232
49			11600	9667	8286	7250	6444	5800	5273	4833	4462	4143	3867	3625	3412
50				10367	8886	7775	6911	6220	5655	5183	4785	4443	4147	3888	3659
51				11067	9486	8300	7378	6640	6036	5533	5108	4743	4427	4150	3906
52				11847	10154	8885	7898	7108	6462	5923	5468	5077	4739	4443	4181
53				12623	10820	9468	8415	7574	6886	6311	5826	5410	5049	4734	4455
54				13400	11486	10050	8933	8040	7309	6700	6185	5743	5360	5025	4730
55				14713	12611	11035	9809	8828	8026	7356	6791	6306	5886	5518	5193
56				16023	13734	12018	10682	9614	8740	8011	7396	6867	6409	6008	5655
57				17333	14857	13000	11556	10400	9455	8666	8000	7429	6933	6500	6118
58				18423	15791	13818	12282	11054	10049	9211	8503	7896	7369	6909	6503
59				19510	16723	14633	13007	11708	10642	9755	9006	8362	7804	7316	6886
60				20600	17657	15450	13733	12360	11236	10300	9508	8829	8240	7725	7271
61				21890	18763	16418	14593	13134	11940	10945	10103	9382	8756	8209	7726
62				23176	19865	17383	15451	13906	12642	11588	10697	9933	9270	8691	8180
63				24463	20968	18348	16309	14678	13344	12231	11291	10484	9785	9174	8634
"B" average inches	1¾	1⅞	2¼	2⅝	3⅛	3⅝	4⅛	4⅝	5⅛	5½	6	6½	6⅞	7⅜	7⅞

table continued

table continued

FIG. 9.2

CONSTANT-EFFORT-SPRING HANGER TABLE (COURTESY OF ANVIL INTERNATIONAL)

*NOTE Total Travel equals Actual Travel plus 1 inch or 20% (whichever is greater), rounded up to the nearest ½ inch as applicable. Constant supports are readily available for travel and loads not listed in this table. Dimensions and lug locations may vary from those shown on the following pages.

constant supports

load travel table *(continued from opposite page) hanger size Nos. 64 to 110 on next page* ▶

load in pounds for total travel in inches

hanger size no.	total travel* in inches														
	9	9½	10	10½	11	11½	12	12½	13	13½	14	14½	15	15½	16
1															
2															
3															
4															
5															
6															
7															
8															
9															
10															
11															
12															
13															
14															
15															
16															
17															
18															
19	423 450	401 426	381 405												
20	477	452	429												
21	519	492	467												
22	561	532	505												
23	602	571	542												
24	647	613	582												
25	691	655	622												
26	736	697	662												
27	807	764	726												
28	878	832	790												
29	949	899	854												
30	1008	956	907												
31	1066	1009	959												
32	1124	1065	1012												
33	1177	1115	1059												
34	1228	1163	1105												
35	1304	1236	1174	1053 1118	1005 1067	962 1021	922 978	885 939	851 903	819 870	790 838				
36	1380	1307	1242	1183	1129	1080	1035	994	955	920	887				
37	1456	1379	1310	1248	1191	1139	1092	1048	1008	970	936				
38	1560	1478	1404	1337	1276	1221	1170	1123	1080	1040	1003				
39	1663	1576	1497	1426	1361	1302	1247	1198	1151	1109	1069				
40	1767	1674	1590	1514	1445	1383	1325	1272	1223	1178	1136				
41	1938	1836	1744	1661	1585	1516	1453	1395	1341	1292	1246				
42	2108	1997	1897	1807	1724	1649	1581	1518	1459	1405	1355				
43	2278	2158	2050	1952	1863	1782	1708	1640	1577	1518	1464				
44	2423	2296	2181	2077	1983	1896	1817	1745	1678	1615	1558				
45	2579	2443	2321	2210	2110	2018	1934	1857	1785	1719	1658				
46	2711	2568	2440	2324	2218	2122	2033	1952	1877	1807	1743				
47	2882	2730	2594	2470	2358	2255	2162	2075	1995	1921	1853				
48	3052	2891	2747	2616	2497	2389	2289	2198	2113	2035	1962				
49	3222	3053	2900	2762	2636	2522	2417	2320	2231	2148	2071				
50	3458	3274	3110	2962	2827	2704	2592	2488	2392	2304	2221	2001 2145	1934 2073	1871 2006	1813 1944
51	3689	3495	3320	3162	3018	2887	2767	2656	2554	2459	2371	2289	2213	2142	2075
52	3949	3741	3554	3384	3231	3090	2962	2843	2734	2632	2538	2451	2369	2293	2221
53	4208	3986	3787	3606	3442	3293	3156	3030	2913	2805	2705	2612	2524	2443	2367
54	4467	4231	4020	3828	3654	3495	3350	3216	3092	2978	2871	2772	2680	2593	2513
55	4904	4646	4414	4203	4012	3838	3678	3531	3395	3269	3152	3044	2942	2847	2759
56	5341	5060	4807	4518	4370	4180	4006	3846	3698	3561	3433	3315	3204	3101	3004
57	5778	5474	5200	4952	4727	4521	4333	4160	4000	3852	3714	3586	3466	3355	3250
58	6141	5816	5527	5263	5024	4806	4606	4422	4251	4094	3947	3811	3684	3565	3454
59	6603	6161	5853	5574	5320	5089	4877	4682	4502	4335	4180	4036	3902	3776	3658
60	6867	6505	6180	5885	5618	5374	5150	4944	4754	4578	4414	4262	4120	3987	3863
61	7297	6912	6567	6254	5969	5710	5472	5254	5051	4864	4690	4529	4378	4236	4104
62	7725	7319	6953	6621	6320	6046	5794	5562	5348	5150	4965	4795	4635	4485	4346
63	8164	7725	7339	6989	6671	6381	6116	5871	5645	5436	5242	5061	4892	4734	4587
"B" average inches	8¼	8¾	9¼	9¾	10⅛	10⅝	11	11½	12	12⅜	12⅞	13⅜	13⅞	14¼	14¾

FIG. 9.2
CONTINUED

*NOTE Total Travel equals Actual Travel plus 1 inch or 20% (whichever is greater), rounded up to the nearest ½ inch as applicable. Constant supports are readily available for travel and loads not listed in this table. Dimensions and lug locations may vary from those shown on the following pages.

fig. 80-V and 81-H
model R

constant supports

See pages ph-120, 121, for sizes 1 to 63

load travel table

load in pounds for total travel in inches

hanger size no.	total travel* in inches																
	4	4½	5	5½	6	6½	7	7½	8	8½	9	9½	10	10½	11	11½	12
64	19225	17089	15380	13982	12816	11831	10986	10253	9613	9047	8544	8094	7690	7323	6990	6686	6408
65	20100	17866	16080	14618	13400	12370	11486	10720	10050	9459	8933	8463	8040	7657	7308	6991	6700
66	22068	19615	17654	16049	14711	13580	12610	11769	11034	10385	9808	9291	8827	8406	8024	7675	7356
67	24033	21362	19226	17478	16021	14790	13733	12817	12016	11310	10681	10119	9613	9154	8738	8359	8011
68	26000	23111	20800	18909	17333	16000	14857	13866	13000	12238	11555	10947	10400	9904	9454	9043	8666
69	27635	24564	22108	20098	18423	17007	15792	14738	13818	13005	12282	11635	11054	10527	10048	9611	9211
70	29268	26015	23414	21266	19511	18011	16725	15609	14634	13773	13008	12323	11707	11149	10642	10179	9755
71	30900	27468	24720	22473	20599	19016	17657	16480	15450	14542	13733	13010	12360	11770	11235	10747	10300
72	32835	29186	26268	23880	21889	20207	18763	17512	16418	15452	14593	13825	13134	12508	11939	11420	10945
73	34768	30904	27814	25286	23177	21396	19868	18542	17384	16362	15452	14639	13907	13244	12641	12092	11589
74	36700	32622	29360	26691	24466	22585	20972	19573	18350	17271	16311	15452	14680	13980	13344	12764	12233
75	38800	34489	31040	28218	25866	23878	22172	20693	19400	18259	17244	16336	15520	14780	14108	13495	12933
76	40900	36355	32720	29746	27266	25170	23372	21813	20450	19248	18178	17221	16360	15580	14871	14225	13633
77	43000	38222	34400	31273	28666	26462	24572	22933	21500	20236	19111	18105	17200	16380	15635	14955	14333
78	45335	40297	36268	32971	30222	27899	25906	24178	22668	21335	20149	19088	18134	17269	16484	15768	15111
79	47668	42371	38134	34666	31779	29335	27239	25422	23834	22432	21185	20070	19067	18158	17332	16579	15889
80	50000	44444	40000	36364	33332	30770	28572	26666	25000	23530	22222	21052	20000	19046	18180	17390	16666
81	52500	46666	42000	38182	35000	32309	30000	27999	26250	24707	23333	22105	21000	19998	19089	18260	17500
82	55000	48888	44000	40000	36665	33847	31429	29333	27500	25883	24444	23157	22000	20951	20000	19129	18333
83	57500	51111	46000	41819	38332	35386	32858	30666	28750	27060	25555	24210	23000	21903	20907	20000	19166
84			49200	44726	40998	37847	35144	32799	30750	28942	27333	25894	24600	23427	22361	21390	20500
85			52400	47637	43665	40309	37429	34932	32750	30824	29111	27578	26200	24950	23816	22781	21832
86			55400	50364	46165	42616	39572	36932	34625	32589	30777	29157	27700	26379	25179	24085	23082
87			58400	53091	48665	44924	41715	38932	36500	34354	32444	30736	29200	27807	26543	25389	24332
88			61400	55819	51165	47232	43858	40932	38375	36119	34111	32315	30700	29236	27906	26694	25582
89			66000	60000	54998	50771	47144	43999	41250	38825	36666	34736	33000	31426	29997	28694	27500
90					61331	56617	52572	49065	46000	43295	40888	38736	36800	35045	33451	31998	30665
91					67164	62002	57573	53732	50375	47413	44777	42420	40300	38378	36633	35041	33582
92					73500	67848	63001	58799	55125	51884	49000	46420	44100	41996	40087	38345	36749
93					80830	74617	69287	64665	60625	57060	53888	51051	48500	46187	44087	42171	40415
94					87500	81540	75716	70665	66250	62355	58888	55788	53000	50472	48177	46084	44165
95							78930	73665	69063	65002	61388	58156	55250	52615	50222	48040	46040
96							82145	76665	71875	67649	63888	60525	57500	54757	52268	50000	47915
97							85360	79665	74688	70296	66388	62893	59750	56900	54313	51953	49790
98							87500	82665	77500	72943	68888	65261	62000	59043	56358	53909	51665
99								85998	80625	75884	71666	67893	64500	61423	58631	56083	53748
100								87500	83750	78826	74444	70524	67000	63804	60903	58257	55831
101									86875	81767	77221	73156	69500	66185	63176	60430	57914
102									87500	84706	80000	75787	72000	68566	65448	62604	60000
103										87500	83610	79210	75250	71661	68402	65430	62706
104											87221	82629	78500	74756	71357	68256	65414
105											87500	86050	81750	77851	74311	71082	68122
106												87500	85000	80946	77265	73908	70831
107													87500	84469	80628	77125	73914
108														87500	83992	80342	77000
109															87446	83646	80163
110															87500	86950	83330
"B" dim sizes 64-83	3⅝	4⅛	4⅝	5⅛	5½	6	6½	6⅞	7⅜	7⅞	8¼	8⅝	9¼	9⅝	10⅛	10⅝	11
"B" dim sizes 84-110	....	....	4 3/16	4 9/16	5	5⅜	5 13/16	6¼	6⅝	7 1/16	7½	7⅞	8 5/16	8¾	9⅛	9 9/16	10

table continued

FIG. 9.2
CONTINUED

*NOTE Total Travel equals Actual Travel plus 1 inch or 20% (whichever is greater), rounded up to the nearest ½ inch as applicable. Constant supports are readily available for travel and loads not listed in this table. Dimensions and lug locations may vary from those shown on the following pages.

constant supports

load travel table *(continued from opposite page)*

load in pounds for total travel in inches

hanger size no.	total travel' in inches 12½	13	13½	14	14½	15	15½	16	16½	17	17½	18	18½	19	19½	20
64	6152	5915	5696	5492	5303	5126	4961	4806								
65	6432	6184	5955	5742	5544	5359	5187	5025								
66	7062	6790	6538	6304	6087	5884	5694	5517								
67	7690	7394	7120	6966	6629	6408	6201	6008								
68	8320	8000	7703	7428	7172	6933	6709	6500								
69	8843	8503	8188	7895	7623	7369	7131	6909								
70	9366	9005	8671	8361	8073	7804	7552	7317								
71	9888	9507	9155	8828	8523	8239	7973	7725								
72	10507	10103	9728	9380	9057	8755	8473	8209								
73	11126	10697	10301	9932	9590	9270	8971	8692								
74	11744	11292	10873	10484	10123	9786	9470	9175								
75	12416	11938	11496	11084	10703	10346	10012	9700								
76	13088	12584	12118	11684	11282	10906	10554	10225								
77	13760	13230	12740	12284	11861	11466	11096	10750								
78	14507	13949	13432	12951	12505	12088	11698	11334								
79	15254	14666	14123	13618	13149	12710	12300	11917								
80	16000	15384	14814	14284	13792	13332	12902	12500								
81	16800	16153	15555	14998	14482	14000	13547	13125								
82	17600	16922	16295	15712	15171	14665	14192	13750								
83	18400	17692	17036	16427	15861	15332	14837	14375								
84	19680	18922	18221	17569	16964	16398	15869	15375								
85	20960	20153	19406	18712	18068	17465	16902	16375								
86	22160	21307	20517	19783	19102	18465	17869	17313								
87	23360	22461	21628	20855	20136	19465	18837	18250								
88	24560	23614	22739	21926	21171	20465	19805	19188								
89	26400	25384	24443	23569	22757	21998	21288	20625								
90	29440	28307	27258	26283	25377	24531	23740	23000								
91	32240	31000	29850	28782	27791	26864	25998	25188								
92	35280	33922	32665	31496	30411	29397	28449	27563								
93	38800	37306	35824	34639	33446	32330	31287	30313								
94	42400	40768	39257	37853	36549	35330	34190	33125								
95	44200	42498	40924	39460	38100	36830	35642	34531	32119 33482	31175 32498	30285 31570	29442 30691	28647 29863	27894 29078	27179 28332	26500 27625
96	46000	44230	42590	41067	39652	38330	37093	35938	34845	33822	32856	31941	31080	30262	29486	28750
97	47600	45960	44257	42673	41204	39829	39545	37344	36209	35145	34141	33191	32295	31446	30640	29875
98	49600	47690	45923	44280	42755	41329	40000	38750	37572	36468	35427	34441	33511	32631	31794	31000
99	51600	49613	47775	46066	44479	42996	41609	40313	39067	37939	36855	35830	34862	33946	33076	32250
100	53600	51536	49627	47851	46203	44662	43221	41875	40602	39409	38264	37219	36214	35282	34358	33500
101	55600	53459	51479	49637	47927	46329	44834	43438	42117	40880	39712	38607	37565	36578	35640	34750
102	57600	56382	53330	51422	49651	47995	46447	45000	43632	42350	41141	39996	38916	37894	36922	36000
103	60200	57882	55738	53744	51892	50162	48544	47031	45602	44262	42998	41801	40673	39604	38588	37625
104	62800	60382	58145	56065	54134	52328	50640	49063	47571	46174	44855	43807	42429	41315	40255	39260
105	65400	62882	60552	58386	56375	54495	52737	51094	49541	48085	46712	45412	44166	43025	41921	40875
106	68000	65382	62960	60707	58616	56661	54834	53125	51510	50000	48569	47218	45943	44738	43686	42500
107	70960	68228	65700	63350	61168	59127	57220	55438	53752	52173	50683	49273	47942	46683	45485	44350
108	73920	71074	68441	65992	63719	61594	59607	57750	55994	54350	52797	51328	49942	48630	47383	46200
109	76960	74000	71255	68706	66340	64127	62069	60125	58297	58585	54969	53439	52000	50630	49331	48100
110	80000	76920	74070	71420	68960	66660	64510	62500	60600	58820	57140	55550	54050	52630	51280	50000
"B" dim sizes 64-93	11½	12	12⅜	12⅞	13⅜	13⅛	14¼	14⅝	....	....	....	....	....	....	....	....
"B" dim sizes 64-110	10⅜	10 13/16	11 3/16	11⅝	12 1/16	12½	12⅞	13 5/16	13 11/16	14⅛	14 9/16	14 15/16	15⅜	15¾	16 3/16	16⅝

FIG. 9.2 CONTINUED

Constant supports are readily available for travel and loads not listed in this table. Dimensions and lug locations may vary from those shown on the following pages.

Insert 9.2 Stress Classification

The design-by-analysis rules of ASME BVPC Section VIII, Division 2, Part 5 are used to assess the results of detailed stress analysis of nozzles and attachments to pressure vessels. These rules include consideration of primary, secondary, and peak stresses. Primary stresses are load-controlled stresses and have essentially the same allowable limit for membrane stress as the basic allowable stress provided in Appendix A. A primary stress is one that is necessary for satisfying the laws of equilibrium of external and internal forces and moments.

Primary stresses include primary membrane, primary bending, and local primary stress. Primary membrane stress is limited to the basic allowable stress; it is the average of the stress through the wall thickness. Primary bending stress is rarely a consideration in these evaluations. Although through-wall bending is often present, it is a secondary stress. A typical example of a primary bending stress in a pressure vessel is in a flat head.

Local primary membrane stress is a primary membrane stress that is in a limited region. To be classified local, the meridional (longitudinal in a pipe) extent over which the stress intensity exceeds 1.1 times the basic allowable stress cannot exceed $(Rt)^{½}$, where R is the midsurface radius of curvature measured normal to the surface from the axis of rotation and t is the minimum thickness in the region considered.

The limit for local primary membrane stress is 1.5 times the basic allowable stress. This higher stress can be used when the extent of the local stress region can be determined. This typically requires a detailed (e.g.. finite-element) analysis.

Secondary stresses are deformation-controlled. The throughwall bending stresses in the vessel wall, calculated using WRC 107 (Wichman et al, 1979), WRC 297 (Mershon et al, 1984), Bednar (1986), finite-element analysis, or similar methods, are considered to be secondary stresses. Local yielding is considered to accommodate the imposed deformation, rather than result in gross failure. The combination of primary plus secondary stress is limited to three times the basic allowable stress (where the allowable stress is the average of the allowable stress at the temperature extremes of the condition being evaluated). At temperatures below the creep regime, this essentially limits the stress range to twice yield, or a shakedown limit (discussed in Section 7.1).

Peak stresses are highly localized and of concern in a fatigue analysis and limited in the ASME BVPC Section VIII, Division 2, by the Appendix 5 fatigue curves. However, these are not calculated in shell analysis, WRC 107 (Wichman etal, 1979), WRC 297 (Mershon et al. 1984), or Bednar (1986), although a stress concentration factor at the welds can be included. However, peak stresses are not generally considered in these evaluations and fatigue life calculations are generally not performed.

9.4 FABRICATION OF SUPPORTS

ASME does not provide much in the way of requirements for the fabrication of support elements. Supports and restraints that are cold-formed to a centerline radius less than twice the support or restraint thickness are required to be annealed or normalized after forming [para. 321.1.4(a)].

Welded attachments to the pressure boundary are required to comply with the Code requirements for preheating, welding, and heat treatment. These are covered in Chapter V of ASME B31.3.

CHAPTER

10

LOAD LIMITS FOR ATTACHED EQUIPMENT

10.1 OVERVIEW OF EQUIPMENT LOAD LIMITS

It is an ASME B31.3 requirement that piping systems be designed such that the reactions are not detrimental to the equipment to which they are connected. This equipment can be machinery, pressure vessels, tankage, or air fin heat exchangers, as well as a myriad of other possibilities. The Designer needs to be aware of the applicable load limits. Some are described herein as examples; however, this is not a complete list. The remainder of this chapter does not provide specific Code requirements; it provides general information on load limits for equipment.

10.2 PRESSURE VESSELS

There are several approaches to dealing with load limits for pressure vessels.

- Specify the loads the nozzles are required to be able to accept in the vessel specifications and design the piping to comply with those limits.
- Calculate loads on nozzles, and when they appear high by some rule of thumb, send them to the vessel manufacturer to confirm the nozzles are acceptable for the imposed loads.
- Perform nozzle stress calculations to confirm the nozzles are acceptable for the loads, again typically when they exceed some rule-of-thumb value.

No rule of thumb works for every vessel, and a variety may well be in use. One of these is to be concerned when the total stress in the piping, including sustained and displacement loads at the vessel nozzle, exceeds S_h.With a little planning ahead, target allowable loads can be placed in the vessel specifications, which simplifies the entire process for new installations.

Another screening tool was developed by Koves et. al in a paper *"Establishing Allowable Nozzle Loads"* ASME PVP2011-57100. They developed a set of tables giving allowable nozzle loads for Class 150 and 300 nozzles with worst-case geometries. The tables are applicable for

- Vessels constructed to ASME BPVC Section VIII, Division 1
- Carbon steels (SA-516 Gr. 70, SA-106 Gr. B, SA-105)
- Design pressure not greater than the flange rating
- Design temperature not greater than 650°F
- Vessel wall thickness at the minimum, but not less than ¼ and 3/8 in. remaining after corrosion

Tables 10.1, 10.2 and 10.3 show the allowable loads as presented by Koves et.al.

Table 10.1
Class 150 Nozzle Loads
Vessels with Thickness >=0.25 in.

	Load (Lbs)			Moments (Ft-Lbs)		
NPS	Axial Fx	Long Fy	Cire Fz	Long Mz	Cire My	Torsion Mx
2	670	670	500	340	220	420
3	1,030	1,030	770	890	590	1,110
4	1,160	1,160	870	1,370	910	1,720
6	1,430	1,430	1,070	2,140	1,430	2,680
8	1,510	1,510	1,130	3,290	2,190	4,110
10	1,600	1,600	1,200	4,320	2,880	5,400
12	1,680	1,680	1,260	5,040	3,360	6,300
14	1,790	1,790	1,350	6,240	4,160	7,800
16	1,910	1,910	1,430	7,340	4,900	9,180
18	2,020	2,020	1,520	8,270	5,510	10,340
20	1,930	1,930	1,450	9,010	6,010	11,260
22	1,840	1,840	1,380	9,440	6,300	11,800
24	1,740	1,740	1,310	9,500	6,340	11,880

Table 10.2
Class 150 Nozzle Loads
Vessels with Thickness >=0.375 in.

	Load (Lbs)			Moments (Ft-Lbs)		
NPS	Axial Fx	Long Fy	Cire Fz	Long Mz	Cire My	Torsion Mx
2	980	980	740	490	330	610
3	1,130	1,130	850	980	650	1,220
4	1,290	1,290	970	1,510	1,010	1,890
6	1,600	1,600	1,200	2,390	1,600	2,990
8	1,740	1,740	1,300	3,730	2,490	4,670
10	1,880	1,880	1,410	5,020	3,340	6,270
12	2,020	2,020	1,510	6,050	4,030	7,560
14	2,240	2,240	1,680	7,710	5,140	9,640
16	2,470	2,470	1,850	9,390	6,260	11,740
18	2,700	2,700	2,020	11,030	7,350	13,790
20	3,190	3,190	2,390	14,360	9,570	17,950
22	3,690	3,690	2,770	18,270	12,180	22,840
24	4,180	4,180	3,140	22,810	15,210	28,510

Table 10.3
Class 300 Nozzle Loads
Vessels with Thickness >=0.375 in.

NPS	Load (Lbs)			Moments (Ft-Lbs)		
	Axial Fx	Long Fy	Cire FY	Long Mz	Cire My	Torsion Mx
2	1,030	1,030	770	500	370	620
3	1,320	1,320	900	1,040	780	1,300
4	1,600	1,600	1,200	1,700	1,280	1,130
6	2,180	2,180	1,630	3,170	2,380	3,960
8	2,270	2,270	1,700	5,210	3,900	6,510
10	2,360	2,360	1,770	7,470	5,600	9,330
12	2,450	2,450	1,840	9,790	7,340	12,240
14	2,980	2,980	2,240	12,150	9,110	15,190
16	3,150	3,510	2,640	14,330	10,750	17,910
18	4,050	4,050	3,030	16,190	12,140	20,230
20	4,900	4,900	3,670	19,430	14,570	24,290
22	5,750	5,750	4,310	22,850	17,140	28,560
24	6,600	6,600	4,950	26,400	19,800	33,000

When a detailed stress calculation of the nozzle, with the imposed loads, is required, it is typical to turn to a design-by-analysis approach, per ASME BPVC, Section VIII, Division 2, Appendix 4. This is done whether the vessel is Division 1 or Division 2. Division 1 does not contain any rules for this evaluation. Stresses are calculated by hand calculations or the finite-element method.

A variety of hand calculation methods are available, based on work by Bijlaard (1954-1959). WRC 107 (Wichman et al, 1979) and WRC 297 (Mershon et al, 1984) contain procedures for analysis of stresses in nozzles due to external loads in cylinders and spheres. These are somewhat tedious to apply; but the procedure is available in commercial programs, including most piping flexibility analysis programs. Bednar (1986) contains a more simplified approach, which is quicker to apply and is based on the same original theory. When calculating the stresses due to external loads one needs to keep in mind that these must be added to the membrane stresses in the vessel wall due to internal pressure. This is commonly forgotten.

The pressure thrust load (pressure times inside area of the nozzle, in the absence of expansion joints) must be included in the nozzle load. As some flexibility analysis programs do not include this in reaction loads, it is sometimes missed.

Finite-element analysis of nozzles has become more common, because a variety of finite-element programs have features that make it relatively easy to automatically create such models.

Once the stresses are calculated, they are typically compared to the primary, local primary, and secondary stress limits described in Appendix 4 of ASME BPVC, Section VIII, Division 2. These limits are described in Insert 9.2 of Chapter 9.

10.3 OTHER EQUIPMENT LOAD LIMITS

The load limits for other equipment must come from the equipment manufacturer. For equipment in accordance with certain standards, the manufacturer is required by the standard to design the equipment for certain minimum loads. In that case, the manufacturer often refers the designer to the applicable standard for allowable loads. The following lists some that are more commonly used:

- Centrifugal pumps, API 610
- Centrifugal compressors, API 617
- Air-cooled heat exchangers, API 661
- Steam turbines, NEMA SM 23 (withdrawn, but still in use by some manufacturers)
- Steam turbines, API 611
- Steam turbines, API 612
- Rotary-type positive-displacement compressors, API 619

An alternative would be for the purchaser to specify minimum design loads in the purchase specification.

10.4 CALCULATING REACTIONS

The accuracy of a piping flexibility analysis computer program is almost always required to give sufficiently accurate calculated reactions to compare to allowable loads.

Because installation practices like that described in API 686, *Machinery Installation and Installation Design*, the loads from the piping are assumed to be minimal at the time the piping is attached to the equipment. So rather than calculating the displacement strains from the coldest condition to the warmest condition, for the purposes of calculating reactions on equipment, the displacement strains from the installation condition to the extreme condition (cold or warm) is used.

When the temperature at the extreme condition is remarkably different than ambient temperature, using the elastic modulus of the piping material at temperature improves the accuracy of the calculation. For example, when the calculated reactions for steel piping, a change from ambient temperature of 150°C (270°F) will cause about a 5% increase for colder temperatures and about a 5% decrease for warmer temperatures because of the change in elastic modulus.

Including the rotational and axial flexibility of nozzles in the analysis can significantly affect the calculations. For most machinery, enough information to calculate the nozzle flexibilities is generally unobtainable. But for fabricated equipment like pressure vessels and heat exchangers, enough information is frequently available. Nozzle flexibilities can be calculated using

- *Spherical Shells* – BS PD 5500
- *Cylindrical Shells* – WRC Bulletin 297
- *Tanks* – "Stiffness Coefficients and Allowable Loads for Nozzles in Flat-Bottom Storage Tanks", HD Billimoria & J Hagstrom, ASME Journal of Pressure Vessel Technology, 1978.
- *Tanks* – "Stiffness Coefficients for Nozzles in API 650 Tanks", M Lengsfeld, et. al., 2002 ASME Pressure Vessel and Piping Conference
- *All Cases* – Finite element analysis

10.5 MEANS OF REDUCING LOADS ON EQUIPMENT

When faced with a situation where excessive loads are calculated to be exerted on equipment, typically the equipment has already been purchased, since the piping stress work is often done after equipment procurement. In such a case, there are a number of alternatives. Some of these are listed as follows in general order of preference:

- If the loads are due to weight, add appropriate support.
- Rerun the analysis using the hot modulus of elasticity to calculate the equipment loads.
- If the loads are due to friction, consider using low-friction slide plates.
- Add restraints to direct thermal expansion and resulting loads away from load-sensitive equipment.
- Add flexibility to the piping system.
- Consider using cold spring or expansion joints.

CHAPTER

11

MATERIALS OF CONSTRUCTION

11.1 OVERVIEW OF MATERIAL REQUIREMENTS

The requirements for materials are covered in Chapter III of ASME B31.3. Selection of the appropriate material for a given service is not covered by the Code. As stated in para. 323.5, "Selection of materials to resist deterioration in service is not within the scope of this Code." Practical considerations make it impossible for the Code to provide guidance relevant to the essentially unlimited process applications covered by the Code. Some precautionary considerations are included in Appendix F based on specific experience or interests of Code Committee members, but this is a very limited list of considerations.

ASME B31.3 contains listed materials, which are simply those that are included in the allowable stress tables and those that are permitted in component standards listed in Table 326.1. However, the Designer is not prevented from using other materials. ASME B31.3 only lists materials in more common use. In addition, the Designer is not generally prevented from using a material at temperatures for which allowable stresses are not provided. Rules are provided in Chapter III for qualification of unlisted materials and for use of materials outside of the temperature limits for which allowable stresses are provided.

Note that the Code typically identifies material by the appropriate American Society for Testing and Materials (ASTM) designation (e.g., A106) rather than the ASME designation (e.g., SA106). An "SA" designation identifies the material as being ASME BPVC material in accordance with specifications listed in Section II, Part A, Ferrous Material Specifications. An "SB" designation identifies the material as being ASME BPVC material in accordance with specifications list in Section II, Part B, Nonferrous Material Specifications. The SA and SB materials typically correspond to ASTM specification designations, and can be used interchangeably in piping systems.

The editions listed for the material specifications in Appendix E cannot always be the latest edition, if only because of publication schedules. A component ordinarily is not marked to indicate the edition date of the standard to which it is manufactured. It is therefore possible that an item taken from inventory was produced in accordance with an earlier or later edition than the one shown in Appendix E.If compliance with a specific edition is a requirement for the intended service, it usually will be necessary to state the specific requirement in the purchase specification and to maintain identification of the component until it is put in service.

Detailed rules covering notch toughness requirements for materials and exemptions from impact testing are provided in ASME B31.3, Chapter III along with some specific limitations for some materials, such as gray iron.

11.2 QUALIFICATION OF UNLISTED MATERIALS

A listed material is one that is listed in one of the allowable stress tables and/or one that is permitted in a component standard listed in Table 326.1 of ASME B31.3. Use of unlisted materials is covered by para. 323.4, along with use of materials at temperatures above the maximum temperature for which allowable stresses are provided in the Code. In these cases, it is the Designer's responsibility to determine the allowable stresses and determine the suitability of the material for the service.

Per para. 323.1.2, unlisted materials may be used if

"they conform to a published specification covering chemistry, physical and mechanical properties, method and process of manufacture, heat treatment, and quality control,"

and otherwise meet any applicable requirements of ASME B31.3, and allowable stresses are determined in accordance with the ASME B31.3 rules or on a more conservative basis.

The first requirement can be satisfied if the material is in accordance with some published standard, such as an ASTM, EN or JIS standard.

The allowable stress basis is provided in ASME B31.3, para. 302.3. Using these criteria, allowable stresses can be developed using data obtained from a sound scientific program carried out in accordance with recognized technology [para. 323.2.4(b)]. However, an alternative, simpler approach can be used for many materials. If the material is listed in ASME BPVC, Section II, Part D the allowable stress from that document may be used as follows.

The allowable stresses provided for use with ASME BPVC, Section VIII, Division 1 are slightly more conservative for some materials and temperatures and can be used because of this. Essentially, the only difference is that the margin on tensile strength is 1/3.5 versus 1/3.

Below the creep range, the yield and ultimate strengths described in ASME BPVC Section II, Part D may be used in the future to directly determine the allowable stresses for unlisted materials in accordance with para. 302.3.

In addition to determining the allowable stress, the Code requires consideration of:

- Applicability and reliability of the data, especially for extremes of the temperature range
- Resistance of the material to deleterious effects of the fluid service and of the environment throughout the temperature range

The latter is actually a requirement for listed materials as well.

11.3 USE OF MATERIALS ABOVE THE HIGHEST TEMPERATURE FOR WHICH ALLOWABLE STRESSES ARE PROVIDED

ASME B31.3 permits the use of materials above the highest temperature for which allowable stresses are provided in Appendix A. Furthermore, it permits the use of materials above the highest temperature permitted in listed standards. For example, ASME B16.5 limits the use of Type 304L stainless steel flanges to temperatures at or below 427°C (800°F); however, ASME B31.3 does not prohibit their use at higher temperatures. In fact, Type 304L flanges can be used satisfactorily at temperatures higher than 815°C (1500°F) if properly designed/rated. Note, however, per para. 302.2.1, the extension of pressure-temperature ratings of a component beyond the ratings of the listed standard requires the Owner's approval.

In circumstances such as these, it is the Designer's responsibility to determine the serviceability of the material and the allowable stresses. The allowable stresses are determined the same way as for unlisted materials (described in the previous section).

11.4 LOW-TEMPERATURE SERVICE

The tables of Appendix A of ASME B31.3 provide an allowable stress in a column headed "Min. Temp. to 40 [°C]" in the MPa tables and "Min. Temp. to 100 [°F]" in the psi tables. This does not mean the material is limited to temperatures above the minimum temperature, also provided in Appendix A, either directly or by reference to Fig. 323.3.2A.

The minimum temperature simply provides a break point in impact test requirements for most materials. If a material is used below the minimum temperature, the same allowable stresses are used as are provided in the column "Min. Temp. to xxx." Although materials typically become increasingly stronger as the temperature is further decreased, the Designer is not permitted to take credit for that.

11.5 WHEN IS IMPACT TESTING REQUIRED?

The Code requires impact testing when, due to various considerations, there is a greater risk of brittle fracture. This risk is due to the combination of material condition, constraint (thickness), toughness, and stress.

Table 323.2.2 of the Code, reproduced here as Table 11.1, sets forth the requirements for impact testing for various materials. Column A provides requirements when the design minimum temperature is at or above the minimum temperature listed in Tables A-1 and A-1M (Appendix A) or Fig. 323.2.2A of ASME B31.3. These are temperatures above which the probability of brittle fracture is considered to be lower. Column B provides requirements for when the design minimum temperature is below the minimum temperature in Tables A-1 and A-1M or Fig. 323.2.2A of ASME B31.3.

Carbon steel materials were formerly assumed to be acceptable down to either -29°C (-20°F) or -46°C (-50°F) without impact testing. However, it was recognized that these materials in greater wall thicknesses could fail in a brittle manner. As a result, a thickness dependence of minimum temperature was provided in ASME BPVC, Section VIII, Division 1 in Fig. UCS-66. A similar chart was provided in ASME B31.3 in Fig. 323.2.2A.

Figure 323.2.2A of ASME B31.3, provided here as Fig. 11.1, is similar to the curve of ASME BPVC, Section VIII, Division 1. The primary differences are that it provides more detailed treatment of carbon steel piping materials, and curve B, which covers many common carbon steel piping materials, was shifted so that the minimum temperature was -29°C (-20°F) through 13-mm (½-in.) wall thickness.

The design minimum temperature is the minimum temperature that the metal can experience during service, including conditions such as autorefrigeration (see para. 301.3.1). Generally, when the design minimum temperature is below the minimum temperature specified in the Code, impact testing is required. This consists of using impact-test-qualified weld procedures and impact tested materials, as set forth in Table 323.3.1.

When the temperature is at or above the minimum temperature specified in the Code, there are different requirements listed for the base material [column A(a) of Table 11.1] and the weld metal and heat-affected zone [column A(b)]. Again, when impact testing is required for the weld metal, this is satisfied by impact testing done as part of the weld procedure qualification See Note (2) of Table 11.1. When impact testing of the base material is required, impact-tested material must be used.

The low-stress [41 MPa (6 ksi)] exemption from impact testing under certain conditions that was previously provided was replaced in the 2000 Addenda with a curve similar to the curve that was in ASME BPVC, Section VIII, Division 2 at the time. This provides a more comprehensive treatment. Note that the Division 2 curve was used rather than the Division 1 curve, because the Division 2 curve in 2000 was consistent with the ASME B31.3 allowable stress basis (one-third of tensile strength). This curve is provided as Figure 323.2.2B of ASME B31.3, reproduced as Fig. 11.2 here.

TABLE 11.1
REQUIREMENTS FOR LOW TEMPERATURE TOUGHNESS TESTS FOR METALS
(ASME B31.3-2020, Table 323.2.2)

	Type of Material	Column A Design Minimum Temperature at or Warmer Than Minimum Temperature in Table A-1, Table A-1M, or Figure 323.2.2A or as Described in Para. 323.2.2(h)		Column B Design Minimum Temperature Colder Than Minimum Temperature in Table A-1, Table A-1M, or Figure 323.2.2A or as Described in Para. 323.2.2(h)
Listed Materials	1 Gray iron	A-1 No additional requirements		B-1 No additional requirements
	2 Malleable and ductile iron; carbon steel in accordance with Note (1)	A-2 No additional requirements		B-2 Materials designated in Box 2 shall not be used.
		(a) Base Metal	(b) Weld Metal [Note (2)]	
	3 Other carbon steels; low and intermediate alloy steels; and ferritic, martensitic, and duplex stainless steels	A-3 (a) No additional requirements	A-3 (b) The weld metal of the welding procedure qualification test coupon shall be impact tested at or below the design minimum temperature when that temperature is less than −18°C (0°F), except when Note (3) or Note (4) is satisfied; impact testing of the weld metal is not required when the classification of the welding consumables according to the filler metal specification requires impact testing at or below the design minimum temperature.	B-3 Each heat, test unit, or lot of base material as identified in the applicable base metal specification shall be impact tested except when Note (3) or Note (4) is satisfied. When materials are subjected to heat treatment during fabrication, including hot bending or forming and postweld heat treatment such as that required by Table 331.1.1, impact testing shall be performed on material that has been subjected to heat treatment simulating the anticipated heat treatment, including, for postweld heat treatment, at least 80% of the time at temperature. Each welding procedure specification shall be qualified in accordance with Table 323.3.1.
	4 Austenitic stainless steels	A-4 (a) If (1) carbon content by analysis >0.1% or (2) material is not in solution heat treated condition, then impact test in accordance with para. 323.3 for design minimum temperature <−29°C (−20°F) except as provided in Notes (5) and (6)	A-4 (b) Weld metal deposits shall be impact tested in accordance with para. 323.3 if design minimum temperature <−29°C (−20°F) except as provided in para. 323.2.2 and in Notes (5) and (6)	B-4 Base metal and weld metal deposits shall be impact tested in accordance with para. 323.3. See Notes (2), (5), and (6).
	5 Austenitic ductile iron, ASTM A571	A-5 (a) No additional requirements	A-5 (b) Welding is not permitted.	B-5 Base metal shall be impact tested in accordance with para 323.3. Do not use <−196°C (−320°F). Welding is not permitted.
	6 Aluminum, copper, nickel, and their alloys; unalloyed titanium	A-6 (a) No additional requirements	A-6 (b) No additional requirements unless filler metal composition is outside the range for base metal composition; then test in accordance with item B-6	B-6 Designer shall be assured by suitable tests [see Note (7)] that base metal, weld deposits, and HAZ are suitable at the design minimum temperature
Unlisted Materials	7 An unlisted material shall conform to a published specification. Where composition, heat treatment, and product form are comparable to those of a listed material, requirements for the corresponding listed material shall be met. Other unlisted materials shall be qualified as required in the applicable section of column B.			

NOTES:

(1) Carbon steels conforming to the following are subject to the limitations in Box B-2: plates in accordance with ASTM A36, A283, and A570; pipe in accordance with ASTM A134 when made from these plates; structural shapes in accordance with ASTM A992; and pipe in accordance with ASTM A53 Type F and API 5L Gr. A25 butt weld.

(2) Impact tests that meet the requirements of Table 323.3.1, which are performed as part of the weld procedure qualification, will satisfy all requirements of para. 323.2.2, and need not be repeated for production welds.

(3) See paras. 323.2.2(g) through (i).

(4) Impact tests are not required when the maximum obtainable Charpy specimen has a width along the notch of less than 2.5 mm (0.098 in.). Under these conditions, and where the stress ratio defined in para. 323.2.2(b) is greater than 0.3, the design minimum temperature shall not be colder than the lower of −48°C (−55°F) or the minimum temperature for the material in Table A-1 or Table A-1M. See also para. 323.2.2(g).

(5) Impact tests are not required when the maximum obtainable Charpy specimen has a width along the notch of less than 2.5 mm (0.098 in.).

(6) For austenitic stainless steels, impact testing is not required if the design minimum temperature is warmer than or equal to −104°C (−155°F), and the stress ratio as defined in para. 323.2.2(b) is 0.3 or less. See also para. 323.2.2(g).

(7) Tests may include tensile elongation, sharp-notch tensile strength (to be compared with unnotched tensile strength), or other tests, conducted at or colder than design minimum temperature. See also para. 323.3.4.

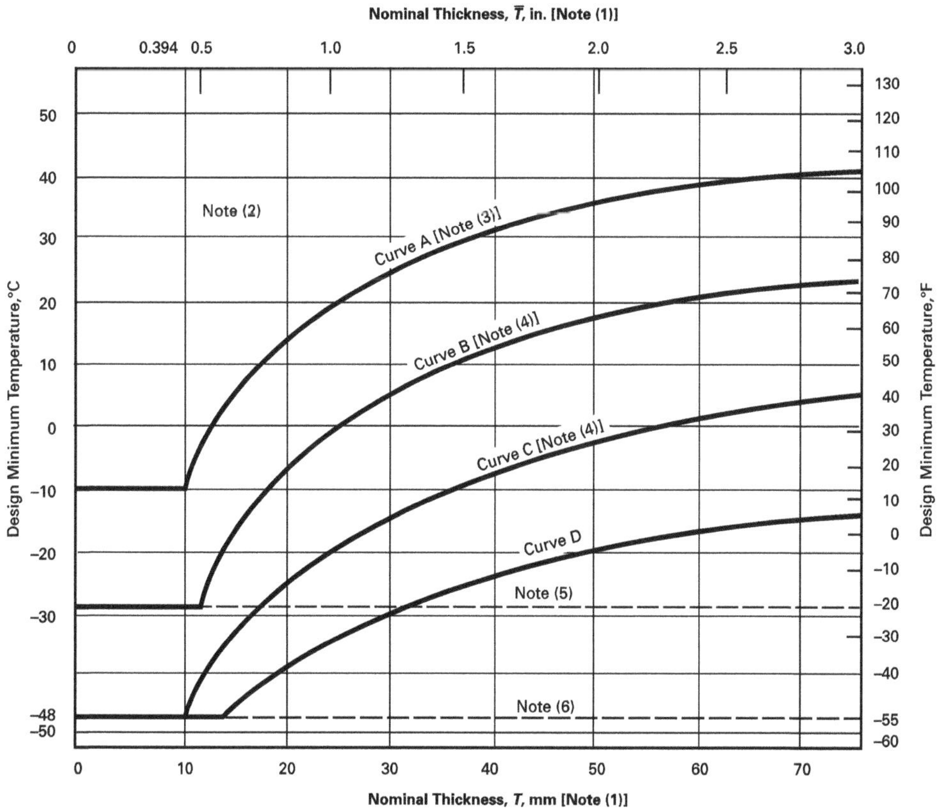

NOTES:

(1) For blind flanges and blanks made from materials with a letter designation in the Min. Temp. column of Table A-1 or Table A-1M, $\bar{T}$ shall be ¼ of the total thickness, where the total thickness is the thickness of the blind flange or blank including the thickness of the facing(s), if applicable.

(2) Any carbon steel material may be used to a minimum temperature of −29°C (−20°F) for Category D Fluid Service.

(3) X Grades of API 5L, and ASTM A381 materials, may be used in accordance with Curve B if normalized or quenched and tempered.

(4) The following materials may be used in accordance with Curve D if normalized:
   (a) ASTM A516 plate, all grades
   (b) ASTM A671 pipe made from A516 plate, all grades
   (c) ASTM A672 pipe made from A516 plate, all grades

(5) A welding procedure for the manufacture of pipe or components shall include impact testing of welds and HAZ for any design minimum temperature below −29°C (−20°F), except as provided in Table 323.2.2, A-3(b).

(6) Impact testing in accordance with para. 323.3 is required for any design minimum temperature below −48°C (−55°F), except as permitted by Note (3) in Table 323.2.2.

FIG. 11.1
MINIMUM TEMPERATURE WITHOUT IMPACT TESTING FOR CARBON STEEL MATERIALS (ASME B31.3, FIG. 323.2.2A). See Table A-1 or Table A-1M for Designated Curve for a Listed Material; see Table 323.2.2A for Tabular Values)

Figure 11.2 provides a reduction in the minimum temperature, without impact testing, for carbon steel material when the stress in the piping is less than the maximum allowable stress. For minimum design metal temperatures of −48°C (−55°F) and warmer, the temperature reduction provided in Fig. 11.2 may be applied to the minimum permissible temperature. For minimum design metal temperatures colder than −48°C (−55°F) and at or warmer than −104°C (−155°F), carbon steel is permitted if the stress ratio as defined in para. 322.2.2(b) is 0.3 or less. Note that this last exemption also applies from −29°C (−20°F) down to −104°C (−155°F), per para. 322.2.2(g) for austenitic stainless steel, intermediate-alloy steels, high-alloy ferritic steels, and duplex stainless steels.

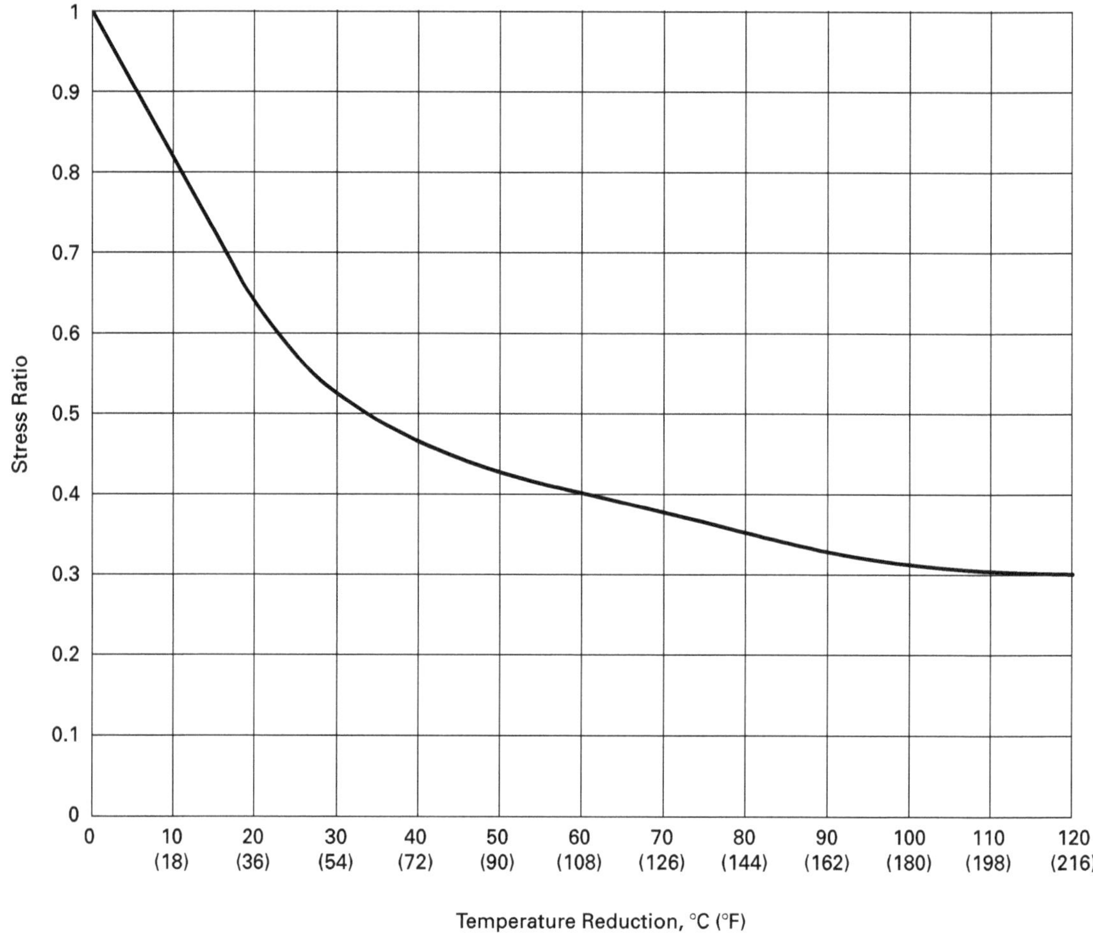

GENERAL NOTE: See para. 323.2.2(b) to determine stress ratio.

FIG. 11.2
REDUCTION IN MINIMUM DESIGN METAL TEMPERATURE WITHOUT IMPACT TESTING (ASME B31.3-2020, FIG. 323.2.2B)

The stress ratio must consider both hoop-type and longitudinal stresses. The stress ratio is the maximum of the following:

a. circumferential pressure stress for the condition under consideration (based on minimum pipe wall thickness less allowances) divided by the basic allowable stress at the condition under consideration.
b. for piping components with pressure ratings, the pressure for the condition under consideration divided by the pressure rating at the condition under consideration.
c. combined stress due to pressure, dead loads, live loads, and displacement strain for the condition under consideration divided by the basic allowable stress at the condition under consideration. In calculating this combined stress, the forces and moments in the piping system for these combined sustained loads and displacements trains shall be calculated using nominal dimensions,and the stresses shall be calculated using eqs.(23a) through (23d) with all of the stress indices taken as 1.0 and using section properties based on the nominal dimensions less corrosion, erosion, and mechanical allowances.

The calculated stress, which is compared to the allowable stress at the design minimum temperature, is calculated for the loads that are present at the condition being considered. For example, if a system has a temperature range from –80°C to 500°C, it is the thermal stresses at –80°C that would be considered in the stress ratio, not the thermal expansion stress range nor the thermal stress at 500°C.

Stress intensification factors are not required to be included when calculating the stress in the piping system. The stress is considered to be sufficiently low that even if a crack were to be initiated at some local high-stress region, it would be arrested (stop advancing) when it entered the general low-membrane-stress regions of the piping system.

Additional requirements are provided for systems to be eligible for the low-stress exemption.

The low-stress exemption is permitted only when all the following apply:

1. The piping is not in Elevated Temperature Fluid Service.
2. Local stresses caused by shock loading, thermal bowing, and differential expansion between dissimilar metals (e.g., austenitic welded to ferritic) are less than10% of the basic allowable stresses at the condition under consideration.
3. The piping is safeguarded from maintenance loads, e.g., using a valve wheel wrench on a small-bore valve.

The first requirement recognizes the potential introduction of residual stresses that could add to the calculated stresses. The second addresses other loads that could cause the actual stresses to be much higher than those calculated. The third requirement is a concern regarding unanticipated applied loads that might occur while the piping is at the cold condition.

Another exemption deals with the situation when the material is sufficiently thin. Thinner sections are less susceptible to brittle fracture, because of the reduced constraint. When the maximum obtainable Charpy specimen has a width along the notch of less than 2.5 mm (0.098 in.), impact testing is not required. However, the exemption is limited to design minimum temperatures above −48°C (−55°F) or the minimum temperature listed in ASME B31.3, Table A-l, whichever is colder for carbon steel and other alloys listed in the third row of Table 11.1.

The temperatures of −46°C (−50°F) and −101°C (−150°F) were generally changed to −48°C (−55°F) and –104°C (−155°F) with the 2000 Addenda, consistent with the changes to ASME BPVC, Section VIII. There is essentially no difference in the material properties between the prior and revised temperatures, but the change permits inclusion of propylene for the higher threshold and ethylene for the lower threshold.

There are no impact testing requirements for welded attachments to the pipe. The only requirements are in 321.3.2 (...material shall be good weldable quality... preheating, welding and heat treatment shall be in accordance with Chapter V...) and 321.1.4. These do not prescribe impact test requirements. In the view of the author, impact test requirements should apply to attachment welds. Section VIII could be used for guidance.

11.6 IMPACT TEST REQUIREMENTS

Impact testing is required to be performed in accordance with ASTM A370. This is Charpy V-notch impact testing. Each set of impact test specimens consists of three standard bars with the standard 10-mm (0.394-in.) square-cross-section Charpy V-notch configuration. In a standard test, a weight on a pendulum swings through the sample, fracturing it. The initial stored potential energy of the pendulum is consumed partly in fracturing the specimen, and the remaining energy can be measured by how high the pendulum swings after fracturing the specimen (see Fig. 11.3). The initial stored energy minus the energy remaining after breaking the specimen is the energy consumed in fracture, which is a measure of the notch toughness of the material. The specimen must be cooled so that it is at or below the design minimum temperature during the impact test.

Another measure of notch toughness used in ASME B31.3 for some materials is the lateral expansion of the specimen (after the fracture) opposite the notch. The greater the lateral expansion, the greater is the ductility of the fracture. Figure 11.3 shows a Charpy impact test specimen before and after the test.

Reduced-size specimens may be used, if required, because the material shape or thickness does not permit machining a full-size specimen. However, smaller specimens exhibit improved toughness properties.

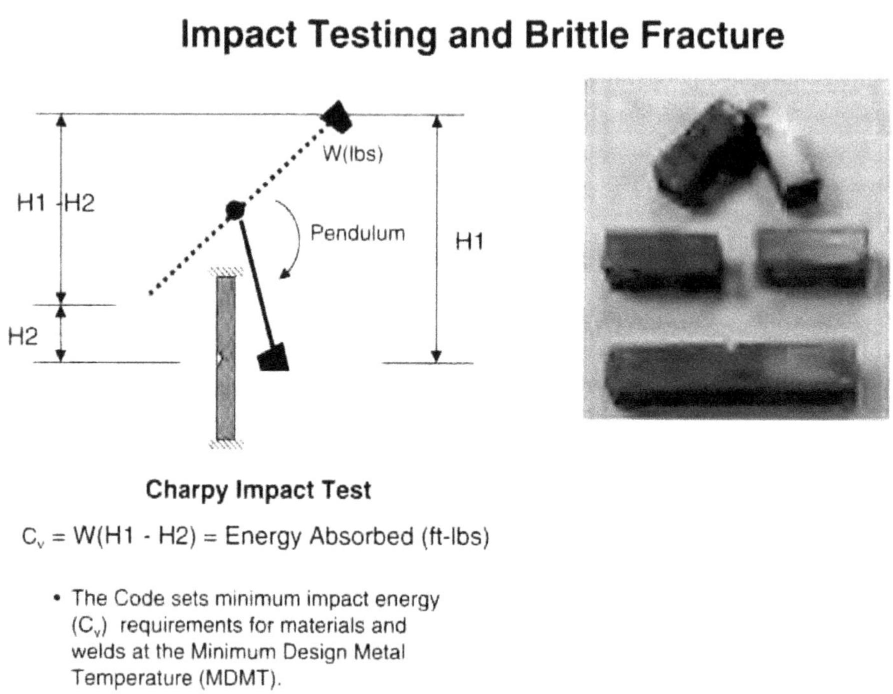

FIG.11.3
SCHEMATIC OF CHARPY IMPACT TEST AND TEST SPECIMENS BEFORE AND AFTER FRACTURE

To compensate for this, the test temperature is reduced for subsize specimens in accordance with Table 323.3.4. For example, if the Charpy impact specimen width along the notch is 2.5 mm (0.098 in.), which is one-fourth of the full size, the test must be performed at a temperature that is 27.8°C (50°F) below the design minimum temperature.

The acceptance criteria for the tests are provided in Table 323.3.5. For carbon and low-alloy steels with specified minimum tensile strengths below 656 MPa (95 ksi), the acceptance criteria are based on energy absorption. A minimum acceptable average for the three specimens and a minimum value for any of the three specimens are provided. For higher strength carbon and low-alloy steels and other materials, the criteria are based on lateral expansion opposite the notch. The minimum acceptable lateral expansion value is 0.38 mm (0.015 in.).

Under limited circumstances, described in para. 323.3.5, retesting is permitted if the initial set of Charpy specimens fails to meet the acceptance criteria.

11.7 FLUID SERVICE LIMITATIONS

Limitations on specific materials with respect to their use for pressure-containing parts are provided in para. 323.4. These include the following for iron materials:

- Temperature limitations with respect to use of ductile iron
- Prohibition of the use of welding for fabrication, repair, or assembly of ductile iron components
- Prohibition of the use of gray, malleable, or high-silicon iron in severe cyclic service
- Pressure-temperature limitations with respect to use of gray or malleable iron in flammable fluid service
- Minimum temperature limit for malleable iron
- Prohibition of the use of high-silicon iron in flammable fluid service.

In all cases, the Code requires that gray iron, malleable iron, and high-silicon iron be safeguarded against excessive heat and thermal shock and mechanical shock and abuse. Lead, tin, and their alloys are prohibited from use in flammable fluid services, because of their low melting points.

Determination of the allowable stresses for cast aluminum alloys that are subject to the heat of weldin or thermal cutting is the Designer's responsibility. These are hardened materials, and the welding process results in a reduction of the material strength. Generally, annealed aluminum material properties must be used under these conditions.

11.8 HOW TO USE THE ALLOWABLE STRESS TABLES IN APPENDIX A OF THE CODE

The allowable stresses for use with the metallic materials in accordance with the base Code are listed in Appendix A, Tables A-1 and A-1M for metals other than bolting and Tables A-2 and A-2M for bolting. An illustration of typical pages of allowable stress values taken from Table A-1 is shown in Table 11.2.

Materials in Tables A-1 and A-1M are first grouped by general alloy content. They are separated into iron, carbon steel, low and intermediate alloys, stainless steel, copper and copper alloys, nickel and nickel alloys, titanium and titanium alloys, zirconium and zirconium alloys, and aluminum alloys. Within each alloy grouping they are separated into pipes and tubes, plates and sheets, forgings and fittings, bar, and castings, as applicable.

In addition to the designation of the material by alloy content and specification number, additional information is provided. This includes the P-Number as well as the specified minimum tensile and yield strengths. The specified minimum strengths are from the material specifications.

TABLE 11.2

BASIC ALLOWABLE STRESSES IN TENSION METALS (ASME B31.3-2020, Table A-1)

Numbers in Parentheses Refer to Notes for Appendix A Tables; Specifications Are ASTM Unless Otherwise Indicated

BE SURE TO READ THE REFERENCED NOTES

									Specified Min. Strength, ksi		Basic Allowable Stress, *S*, ksi, at Metal Temperature, °F [Notes (1), (4a)]	
Material	**Spec. No.**	**Type/ Grade**	**UNS No.**	**Class/ Condition/ Temper**	**Size, in.**	**P-No. (5)**	**Notes**	**Min. Temp., °F (6)**	**Tensile**	**Yield**	**Min. Temp. to 100**	**200**
Carbon Steel — Pipes and Tubes												
A285 Gr. A	A134	A285A	K01700	...	...	1	(8b) (57)	B	45	24	15.0	14.7
A285 Gr. A	A672	A45	K01700	...	...	1	(57) (59) (67)	B	45	24	15.0	14.7
Butt weld	API 5L	A25	...	...	...	1	(8a) (77)	−20	45	25	15.0	15.0
Smls & ERW	API 5L	A25	...	...	...	1	(57) (59) (77)	B	45	25	15.0	15.0
...	A179	...	K01200	...	...	1	(57) (59)	−20	47	26	15.7	15.7
Type F	A53	A	K02504	...	...	1	(8a)	20	48	30	16.0	16.0
...	A139	A	...	...	...	1	(8b)	A	48	30	16.0	16.0
...	A587	...	K11500	...	...	1	(57) (59)	−20	48	30	16.0	16.0
...	A53	A	K02504	...	...	1	(57) (59)	B	48	30	16.0	16.0
...	A106	A	K02501	...	...	1	(57)	B	48	30	16.0	16.0
...	A135	A	...	...	...	1	(57) (59)	B	48	30	16.0	16.0
...	A369	FPA	K02501	...	...	1	(57)	B	48	30	16.0	16.0
...	API 5L	A	...	...	...	1	(57) (59)	B	48	30	16.0	16.0
A285 Gr. B	A134	A285B	K02200	...	...	1	(8b) (57)	B	50	27	16.7	16.5
A285 Gr. B	A672	A50	K02200	...	...	1	(57) (59) (67)	B	50	27	16.7	16.5
A285 Gr. C	A134	A285C	K02801	...	...	1	(8b) (57)	A	55	30	18.3	18.3
...	A524	II	K02104	...	...	1	(57)	−20	55	30	18.3	18.3
...	A333	1	K03008	...	...	1	(57) (59)	−50	55	30	18.3	18.3
...	A334	1	K03008	...	...	1	(57) (59)	−50	55	30	18.3	18.3
A285 Gr. C	A671	CA55	K02801	...	...	1	(59) (67)	A	55	30	18.3	18.3
A285 Gr. C	A672	A55	K02801	...	...	1	(57) (59) (67)	A	55	30	18.3	18.3
A516 Gr. 55	A672	C55	K01800	...	...	1	(57) (67)	C	55	30	18.3	18.3
A516 Gr. 60	A671	CC60	K02100	...	...	1	(57) (67)	C	60	32	20.0	19.5
A515 Gr. 60	A671	CB60	K02401	...	...	1	(57) (67)	B	60	32	20.0	19.5
A515 Gr. 60	A672	B60	K02401	...	...	1	(57) (67)	B	60	32	20.0	19.5
A516 Gr. 60	A672	C60	K02100	...	...	1	(57) (67)	C	60	32	20.0	19.5
...	A139	B	K03003	...	...	1	(8b)	A	60	35	20.0	20.0
...	A135	B	K03018	...	...	1	(57) (59)	B	60	35	20.0	20.0
...	A524	I	K02104	...	...	1	(57)	−20	60	35	20.0	20.0

LETTER CORRESPONDS TO CURVE IN FIG. 323.2.2 WHICH DESCRIBES DESIGN MIN. TEMP NORMALLY OK W/O IMPACT TESTING

TABLE 11.2

BASIC ALLOWABLE STRESSES IN TENSION METALS (Cont'd)

Numbers in Parentheses Refer to Notes for Appendix A Tables; Specifications Are ASTM Unless Otherwise Indicated

Basic Allowable Stress, *S*, ksi, at Metal Temperature, °F [Notes (1), (4a)]															
300	400	500	600	650	700	750	800	850	900	950	1,000	1,050	1,100	Type/ Grade	Spec. No.
														Carbon Steel — Pipes and Tubes	
14.2	13.7	13.0	12.3	11.9	11.5	10.7	9.2	7.9	5.9	...	...	...	...	...	A134
14.2	13.7	13.0	12.3	11.9	11.5	10.7	9.2	7.9	5.9	4.0	2.5	1.6	1.0	A45	A672
14.7	14.2	...	...	...	...	...	...	...	...	...	...	...	...	A25	API 5L
14.7	14.2	...	...	...	...	...	...	...	...	...	...	...	...	A25	API 5L
15.3	14.8	14.1	13.3	12.8	12.4	10.7	9.2	7.9	5.9	4.0	2.5	1.6	1.0	...	A179
16.0	16.0	...	...	...	...	...	...	...	...	...	...	...	...	A	A53
16.0	...	...	...	...	...	...	...	...	...	...	...	...	...	A	A139
16.0	16.0	16.0	15.3	14.6	12.5	10.7	9.2	7.9	...	...	...	...	...	...	A587
16.0	16.0	16.0	15.3	14.6	12.5	10.7	9.2	7.9	5.9	4.0	2.5	1.6	1.0	A	A53
16.0	16.0	16.0	15.3	14.6	12.5	10.7	9.2	7.9	5.9	4.0	2.5	1.6	1.0	A	A106
16.0	16.0	16.0	15.3	14.6	12.5	10.7	9.2	7.9	5.9	4.0	2.5	1.6	1.0	A	A135
16.0	16.0	16.0	15.3	14.6	12.5	10.7	9.2	7.9	5.9	4.0	2.5	1.6	1.0	FPA	A369
16.0	16.0	16.0	15.3	14.6	12.5	10.7	9.2	7.9	5.9	4.0	2.5	1.6	1.0	A	API 5L
15.9	15.4	14.7	13.8	13.3	12.5	10.7	9.2	7.9	5.9	...	...	...	...	...	A134
15.9	15.4	14.7	13.8	13.3	12.5	10.7	9.2	7.9	5.9	4.0	2.5	1.6	1.0	A50	A672
17.7	17.1	16.3	15.3	14.8	14.3	13.0	10.8	8.7	5.9	...	...	...	...	...	A134
17.7	17.1	16.3	15.3	14.8	14.3	13.0	10.8	8.7	5.9	4.0	2.5	...	...	II	A524
17.7	17.1	16.3	15.3	14.8	14.3	13.0	10.8	8.7	5.9	4.0	2.5	1.6	1.0	1	A333
17.7	17.1	16.3	15.3	14.8	14.3	13.0	10.8	8.7	5.9	4.0	2.5	1.6	1.0	1	A334
17.7	17.1	16.3	15.3	14.8	14.3	13.0	10.8	8.7	5.9	4.0	2.5	1.6	1.0	CA55	A671
17.7	17.1	16.3	15.3	14.8	14.3	13.0	10.8	8.7	5.9	4.0	2.5	1.6	1.0	A55	A672
17.7	17.1	16.3	15.3	14.8	14.3	13.0	10.8	8.7	5.9	4.0	2.5	1.6	1.0	C55	A672
18.9	18.2	17.4	16.4	15.8	15.3	13.0	10.8	8.7	5.9	4.0	2.5	...	...	CC60	A671
18.9	18.2	17.4	16.4	15.8	15.3	13.0	10.8	8.7	5.9	4.0	2.5	1.6	1.0	CB60	A671
18.9	18.2	17.4	16.4	15.8	15.3	13.0	10.8	8.7	5.9	4.0	2.5	1.6	1.0	B60	A672
18.9	18.2	17.4	16.4	15.8	15.3	13.0	10.8	8.7	5.9	4.0	2.5	1.6	1.0	C60	A672
20.0	...	...	...	...	...	...	...	...	...	...	...	...	...	B	A139
20.0	19.9	19.0	17.9	17.3	16.7	13.9	11.4	8.7	5.9	4.0	2.5	...	...	B	A135
20.0	19.9	19.0	17.9	17.3	16.7	13.9	11.4	8.7	5.9	4.0	2.5	...	...	I	A524

MAXIMUM TEMPERATURE FOR WHICH STRESS VALUES ARE TABULATED. WARMER TEMPERATURES ARE PERMITTED UNLESS PROHIBITED IN REFERENCED NOTES.

TABLE 11.2

BASIC ALLOWABLE STRESSES IN TENSION METALS (Cont'd)

Numbers in Parentheses Refer to Notes for Appendix A Tables; Specifications Are ASTM Unless Otherwise Indicated

Nominal Composition	Product Form	Spec. No.	Type/ Grade	UNS No.	Class/ Condition/ Temper	Size, in.	P-No. (5)	Notes	Min. Temp., °F (6)	Specified Min. Strength, ksi: Tensile	Yield	Basic Allowable Stress, *S*, ksi, at Metal Temperature, °F [Notes (1), (4a)]: Min. Temp. to 100	200	300	400
Stainless Steel — Pipes and Tubes (3)(4a)															
18Cr-10Ni-Ti	Smls. pipe	A312	TP321	S32100	...	$>\frac{3}{8}$ thk.	8	(28)	-425	70	25	16.7	*16.7*	*16.7*	*16.7*
18Cr-10Ni-Ti	Smls. pipe	A376	TP321	S32100	...	$>\frac{3}{8}$ thk.	8	(28) (36)	-425	70	25	16.7	*16.7*	*16.7*	*16.7*
18Cr-8Ni	Tube	A213	TP304L	S30403	...	...	8	(14) (36)	-425	70	25	16.7	*16.7*	*16.7*	**15.8**
18Cr-8Ni	Tube	A269	TP304L	S30403	...	...	8	(14) (36)	-425	70	25	16.7	*16.7*	*16.7*	**15.8**
18Cr-8Ni	Tube	A270	TP304L	S30403	...	...	8	(14)	-425	70	25	16.7	*16.7*	*16.7*	**15.8**
18Cr-8Ni	Pipe	A312	TP304L	S30403	...	...	8	...	-425	70	25	16.7	*16.7*	*16.7*	**15.8**
18Cr-8Ni	Pipe	A358	304L	S30403	...	...	8	(36)	-425	70	25	16.7	*16.7*	*16.7*	**15.8**
16Cr-12Ni-2Mo	Tube	A213	TP316L	S31603	...	...	8	(14) (36)	-425	70	25	16.7	*16.7*	*16.7*	**15.7**
16Cr-12Ni-2Mo	Tube	A269	TP316L	S31603	...	...	8	(14) (36)	-425	70	25	16.7	*16.7*	*16.7*	**15.7**
16Cr-12Ni-2Mo	Tube	A270	TP316L	S31603	...	...	8	(14)	-425	70	25	16.7	*16.7*	*16.7*	**15.7**
16Cr-12Ni-2Mo	Pipe	A312	TP316L	S31603	...	...	8	...	-425	70	25	16.7	*16.7*	*16.7*	**15.7**
16Cr-12Ni-2Mo	Pipe	A358	316L	S31603	...	...	8	(36)	-425	70	25	16.7	*16.7*	*16.7*	**15.7**
16Cr-12Ni-2Mo-Ti	Tube	A213	TP316Ti	S31635	...	...	8	(30)	-325	75	30	20.0	*20.0*	*20.0*	**19.3**
18Cr-10Ni-Ti	Smls. pipe	A312	TP321	S32100	...	$>\frac{3}{8}$ thk.	8	(28) (30)	-425	70	25	16.7	*16.7*	*16.7*	*16.7*
18Cr-10Ni-Ti	Smls. pipe	A376	TP321	S32100	...	$>\frac{3}{8}$ thk.	8	(28) (30) (36)	-425	70	25	16.7	*16.7*	*16.7*	*16.7*
18Cr-10Ni-Ti	Smls. pipe	A312	TP321H	S32109	...	$>\frac{3}{8}$ thk.	8	(30)	-325	70	25	16.7	*16.7*	*16.7*	*16.7*
18Cr-10Ni-Ti	Smls. pipe	A376	TP321H	S32109	...	$>\frac{3}{8}$ thk.	8	(30) (36)	-325	70	25	16.7	*16.7*	*16.7*	*16.7*
25Cr-12Ni	...	A451	CPH8	J93400	...	...	8	(26) (28) (35)	-325	65	28	18.7	*18.7*	*18.5*	*18.0*
25Cr-20Ni	...	A451	CPK20	J94202	...	...	8	(12) (28) (35) (39)	-325	65	28	18.7	*18.7*	*18.5*	*18.0*
11Cr-Ti	Tube	A268	TP409	S40900	...	...	7	(35)	-20	60	30	20.0	...	...	...
18Cr-Ti	Tube	A268	TP430Ti	S43036	...	...	7	(35) (49)	-20	60	40	20.0	...	...	...
16Cr-14Ni-2Mo	...	A451	CPF10MC	J92971	...	...	8	(28)	-325	70	30	20.0	...	...	...
12Cr-Al	Tube	A268	TP405	S40500	...	...	7	(35)	-20	60	30	20.0	20.0	19.6	19.3
13Cr	Tube	A268	TP410	S41000	...	...	6	(35)	-20	60	30	20.0	20.0	19.6	19.3
17Cr	Tube	A268	TP430	S43000	...	...	7	(35) (49)	-20	60	35	20.0	20.0	19.6	19.3
18Cr-13Ni-3Mo	Pipe	A312	TP317L	S31703	...	...	8	...	-325	75	30	20.0	*20.0*	*20.0*	**18.9**
25Cr-20Ni	Pipe	A312	TP310S	S31008	...	...	8	(28) (35)	-325	75	30	20.0	*20.0*	*20.0*	*20.0*
25Cr-20Ni	...	A358	310S	S31008	...	...	8	(28) (35) (36)	-325	75	30	20.0	*20.0*	*20.0*	*20.0*
25Cr-20Ni	Pipe	A409	TP310S	S31008	...	...	8	(28) (31) (35) (36)	-325	75	30	20.0	*20.0*	*20.0*	*20.0*
18Cr-10Ni-Ti	Smls. pipe	A312	TP321	S32100	...	$\leq\frac{3}{8}$ thk.	8	(28)	-425	75	30	20.0	*20.0*	*20.0*	*20.0*
18Cr-10Ni-Ti	Wld. pipe	A312	TP321	S32100	...	...	8	(28)	-425	75	30	20.0	*20.0*	*20.0*	*20.0*
18Cr-10Ni-Ti	Wld. pipe	A358	321	S32100	...	...	8	(28) (36)	-425	75	30	20.0	*20.0*	*20.0*	*20.0*
18Cr-10Ni-Ti	Smls. pipe	A376	TP321	S32100	...	$\leq\frac{3}{8}$ thk.	8	(28) (36)	-425	75	30	20.0	*20.0*	*20.0*	*20.0*
18Cr-10Ni-Ti	Wld. pipe	A409	TP321	S32100	...	...	8	(28) (36)	-425	75	30	20.0	*20.0*	*20.0*	*20.0*
23Cr-12Ni	Pipe	A312	TP309	...	...	...	8	(28) (35) (39)	-325	75	30	20.0	*20.0*	*20.0*	*20.0*

ITALLICS - EXAMPLE OF ALLOWABLE STRESSES EXCEEDING 2/3 YIELD

BOLD - EXAMPLE OF ALLOWABLE STRESSES EQUAL TO 90% YIELD

TABLE 11.2

BASIC ALLOWABLE STRESSES IN TENSION METALS (Cont'd)

Numbers in Parentheses Refer to Notes for Appendix A Tables; Specifications Are ASTM Unless Otherwise Indicated

Basic Allowable Stress, S, ksi, at Metal Temperature, °F [Notes (1), (4a)]																					
500	600	650	700	750	800	850	900	950	1,000	1,050	1,100	1,150	1,200	1,250	1,300	1,350	1,400	1,450	1,500	Type/ Grade	Spec. No.
																Stainless Steel — Pipes and Tubes (3)(4a)					
16.1	**15.2**	**14.9**	**14.6**	**14.3**	**14.1**	**13.9**	**13.8**	**13.6**	**13.5**	9.6	6.9	5.0	3.6	2.6	1.7	1.1	0.8	0.5	0.3	TP321	A312
16.1	**15.2**	**14.9**	**14.6**	**14.3**	**14.1**	**13.9**	**13.8**	**13.6**	**13.5**	9.6	6.9	5.0	3.6	2.6	1.7	1.1	0.8	0.5	0.3	TP321	A376
14.7	**14.0**	**13.7**	**13.5**	**13.3**	**13.0**	**12.8**	*12.6*	*12.3*	12.0	6.3	5.1	4.0	3.2	2.6	2.1	1.7	1.1	1.0	0.9	TP304L	A213
14.7	**14.0**	**13.7**	**13.5**	**13.3**	**13.0**	**12.8**	*12.6*	*12.3*	12.0	6.3	5.1	4.0	3.2	2.6	2.1	1.7	1.1	1.0	0.9	TP304L	A269
14.7	**14.0**	**13.7**	**13.5**	**13.3**	**13.0**	**12.8**	*12.6*	*12.3*	12.0	6.3	5.1	4.0	3.2	2.6	2.1	1.7	1.1	1.0	0.9	TP304L	A270
14.7	**14.0**	**13.7**	**13.5**	**13.3**	**13.0**	**12.8**	*12.6*	*12.3*	12.0	6.3	5.1	4.0	3.2	2.6	2.1	1.7	1.1	1.0	0.9	TP304L	A312
14.7	**14.0**	**13.7**	**13.5**	**13.3**	**13.0**	**12.8**	*12.6*	*12.3*	12.0	6.3	5.1	4.0	3.2	2.6	2.1	1.7	1.1	1.0	0.9	304L	A358
14.8	**14.0**	**13.7**	**13.5**	**13.2**	**12.9**	**12.7**	**12.4**	**12.1**	**11.8**	**11.6**	*11.4*	*8.8*	6.4	4.7	3.5	2.5	1.8	1.3	1.0	TP316L	A213
14.8	**14.0**	**13.7**	**13.5**	**13.2**	**12.9**	**12.7**	**12.4**	**12.1**	**11.8**	**11.6**	*11.4*	*8.8*	6.4	4.7	3.5	2.5	1.8	1.3	1.0	TP316L	A269
14.8	**14.0**	**13.7**	**13.5**	**13.2**	**12.9**	**12.7**	**12.4**	**12.1**	**11.8**	**11.6**	*11.4*	*8.8*	6.4	4.7	3.5	2.5	1.8	1.3	1.0	TP316L	A270
14.8	**14.0**	**13.7**	**13.5**	**13.2**	**12.9**	**12.7**	**12.4**	**12.1**	**11.8**	**11.6**	*11.4*	*8.8*	6.4	4.7	3.5	2.5	1.8	1.3	1.0	TP316L	A312
14.8	**14.0**	**13.7**	**13.5**	**13.2**	**12.9**	**12.7**	**12.4**	**12.1**	**11.8**	**11.6**	*11.4*	*8.8*	6.4	4.7	3.5	2.5	1.8	1.3	1.0	316L	A358
17.8	**16.8**	**16.5**	**16.2**	**16.1**	**15.9**	**15.8**	**15.7**	**15.5**	**15.3**	**15.1**	12.3	9.8	7.4	5.5	4.1	3.1	2.3	1.7	1.3	TP316Ti	A213
16.1	**15.2**	**14.9**	**14.6**	**14.3**	**14.1**	**13.9**	**13.8**	**13.6**	**13.5**	12.3	9.1	6.9	5.4	4.1	3.2	2.5	1.9	1.5	1.1	TP321	A312
16.1	**15.2**	**14.9**	**14.6**	**14.3**	**14.1**	**13.9**	**13.8**	**13.6**	**13.5**	12.3	9.1	6.9	5.4	4.1	3.2	2.5	1.9	1.5	1.1	TP321	A376
16.1	**15.2**	**14.9**	**14.6**	**14.3**	**14.1**	**13.9**	**13.8**	**13.6**	**13.5**	12.3	9.1	6.9	5.4	4.1	3.2	2.5	1.9	1.5	1.1	TP321H	A312
16.1	**15.2**	**14.9**	**14.6**	**14.3**	**14.1**	**13.9**	**13.8**	**13.6**	**13.5**	12.3	9.1	6.9	5.4	4.1	3.2	2.5	1.9	1.5	1.1	TP321H	A376
17.7	**17.1**	**16.7**	**16.3**	**15.9**	**15.4**	*14.9*	*14.4*	*13.9*	11.1	8.5	6.5	5.0	3.8	2.9	2.3	1.8	1.3	0.9	0.8	CPH8	A451
17.7	**17.1**	**16.7**	**16.3**	**15.9**	**15.4**	*14.9*	*14.4*	*13.9*	11.3	9.8	8.5	7.3	6.0	4.8	3.5	2.4	1.6	1.1	0.8	CPK20	A451
...	...	...	...	...	...	...	...	...	...	...	...	...	...	...	...	...	...	...	...	TP409	A268
...	...	...	...	...	...	...	...	...	...	...	...	...	...	...	...	...	...	...	...	TP430Ti	A268
...	...	...	...	...	...	...	...	...	...	...	...	...	...	...	...	...	...	...	...	CPF10MC	A451
19.0	18.5	18.1	17.7	17.1	16.4	15.6	14.3	8.4	4.0	...	...	...	...	...	...	...	...	...	...	TP405	A268
19.0	18.5	18.1	17.7	17.1	16.4	15.6	12.3	8.8	6.4	4.4	2.9	1.8	1.0	...	...	...	...	...	...	TP410	A268
19.0	18.5	18.1	17.7	17.1	16.4	15.6	12.0	9.2	6.5	4.5	3.2	2.4	1.8	...	...	...	...	...	...	TP430	A268
17.7	**16.9**	**16.5**	**16.2**	**15.8**	**15.5**	**15.2**	...	...	...	...	...	...	...	...	...	...	...	...	...	TP317L	A312
19.3	**18.5**	**18.2**	**17.9**	**17.7**	**17.4**	*17.2*	*16.9*	*15.9*	9.9	7.1	5.0	3.6	2.5	1.5	0.8	0.5	0.4	0.3	0.2	TP310S	A312
19.3	**18.5**	**18.2**	**17.9**	**17.7**	**17.4**	*17.2*	*16.9*	*15.9*	9.9	7.1	5.0	3.6	2.5	1.5	0.8	0.5	0.4	0.3	0.2	310S	A358
19.3	**18.5**	**18.2**	**17.9**	**17.7**	**17.4**	*17.2*	*16.9*	*15.9*	9.9	7.1	5.0	3.6	2.5	1.5	0.8	0.5	0.4	0.3	0.2	TP310S	A409
19.3	**18.3**	**17.8**	**17.5**	**17.2**	**16.9**	**16.7**	**16.5**	**16.4**	**16.2**	9.6	6.9	5.0	3.6	2.6	1.7	1.1	0.8	0.5	0.3	TP321	A312
19.3	**18.3**	**17.8**	**17.5**	**17.2**	**16.9**	**16.7**	**16.5**	**16.4**	**16.2**	9.6	6.9	5.0	3.6	2.6	1.7	1.1	0.8	0.5	0.3	TP321	A312
19.3	**18.3**	**17.8**	**17.5**	**17.2**	**16.9**	**16.7**	**16.5**	**16.4**	**16.2**	9.6	6.9	5.0	3.6	2.6	1.7	1.1	0.8	0.5	0.3	321	A358
19.3	**18.3**	**17.8**	**17.5**	**17.2**	**16.9**	**16.7**	**16.5**	**16.4**	**16.2**	9.6	6.9	5.0	3.6	2.6	1.7	1.1	0.8	0.5	0.3	TP321	A376
19.3	**18.3**	**17.8**	**17.5**	**17.2**	**16.9**	**16.7**	**16.5**	**16.4**	**16.2**	9.6	6.9	5.0	3.6	2.6	1.7	1.1	0.8	0.5	0.3	TP321	A409
19.4	**18.8**	**18.5**	**18.2**	**18.0**	**17.7**	*17.5*	*17.2*	*16.9*	13.8	10.3	7.6	5.5	4.0	3.0	2.2	1.7	1.3	1.0	0.8	TP309	A312

BOLD - EXAMPLE OF ALLOWABLE STRESSES EQUAL TO 90% YIELD

ITALLICS - EXAMPLE OF ALLOWABLE STRESSES EXCEEDING 2/3 YIELD

The P-Numbers group allows for weld procedure qualification purposes, based on composition, weldability, and mechanical properties. These are assigned by ASME BPV Committee on Welding, Brazing and Fusing (IX).

Notes are typically provided. Prior to using a material, the notes should be reviewed. For example, Note (28) for Type 304 stainless steel indicates that the allowable stresses that are listed for temperatures above 538°C (1000°F) are only valid if the material has a carbon content of 0.04% or higher. There have been instances of systems designed to these allowable stresses without the material procured to the minimum carbon content requirements.

Some specific things about these tables described in the first few notes are:

1. The stress values given in ksi as shown in Tables A-1 and A-2,and given in MPa as shown in Tables A-1M and A-2M, may be used. The values stated in ksi are not exact equivalents to the values stated in MPa. Therefore, for any given material, the user of the Code should use only the ksi or the MPa values.
2. In Table A-1, stress values printed in *italics* exceed two thirds of the expected yield strength at temperature. Stress values in **boldface** are equal to 90% of expected yield strength at temperature.
3. In Table A-1M, stress values printed in *italics* are tensile controlled values. Yield-controlled stress values are in normal font and time-dependent controlled stress values are in **boldface.**

The Designer is permitted to interpolate linearly between temperatures for which allowable stresses are listed. As previously discussed, use of materials above the maximum temperature for which allowable stresses are provided is permitted as long as it is not specifically prohibited as described in the notes. An example of a prohibition is Note 8(a), which says "See para. 305.2.1; temperature limits are −29°C to 186°C (−20°F to 366°F)." The temperature limits are imposed because the material is permitted to be used only in Category D fluid service.

CHAPTER

12

FABRICATION, ASSEMBLY, AND ERECTION

12.1 OVERVIEW OF CHAPTER V OF ASME B31.3

Chapter V of ASME B31.3 covers the base Code rules for fabrication, assembly, and erection. It includes requirements for welding, details for specific types of welded joints, and conditions for preheating, heat treatment, bending and forming, brazing and soldering, assembly, and erection.

12.2 GENERAL WELDING REQUIREMENTS

A variety of welding processes are used with piping. These include shielded metal arc weld (SMAW), gas tungsten arc weld (GTAW or TIG), gas metal arc weld (GMAW or MIG), submerged arc weld (SAW), and flux cored arc weld (FCAW) processes. These are described below in Insert 12.1.

Welding involves a Welding Procedure Specification (WPS), which is qualified by a Procedure Qualification Test, which is documented in a Procedure Qualification Record (PQR). Welders are required to pass a Performance Qualification Test in order to be qualified to perform Code welding. This is documented in a Welder Qualification Record (WQR).

Materials are categorized by P-Numbers which are groupings of alloys for weld procedure qualification purposes; they group materials based on composition, weldability, and mechanical properties. P-Numbers are assigned in Section IX of the Boiler and Pressure Vessel Code. Qualification of a weld procedure for a particular base material qualifies that procedure for any other base material with the same P-Number. On the other hand, if a material does not have a P-Number, it requires its own procedure specification. This is one of the problems users have with materials that are in accordance with foreign specifications.

There are materials listed in ASME B31.3 that are not listed in Section II. Since Section IX does not provide P-Numbers for such materials, the B31.3 Committee had previously assigned P-Numbers for the convenience of the users.

Insert 12.1 Arc Welding Processes

Shielded Metal Arc Welding (SMAW)

Principles of Operation

Shielded metal arc welding (SMAW) is the most widely used of all arc welding processes (Fig. 12.1). It is a manual process, which employs the heat of an arc to melt the base metal and the tip of a consumable covered electrode. The bare end of the electrode is clamped in an electrode holder, which is connected to one terminal of a welding power source by a welding lead (cable). The base metal (work) is connected by a cable to the other terminal of the power source. The arc is initiated by touching the electrode tip against the work and then withdrawing it slightly. The heat of the arc melts the base metal in the immediate area, the electrode metal core, and the electrode covering. As the electrode is consumed, shielding of the arc and the weld metal may be provided by gases formed or by slag produced during decomposition of the covering, depending on the electrode type. Also, as the electrode melts, tiny droplets of molten metal are transferred to the molten weld pool by the arc stream. The welder moves the electrode progressively along the joint, and the deposited filler metal and the molten base metal coalesce to form the weld.

Process Capabilities and Limitations

The SMAW process is suitable for most of the commonly used metals and alloys. However, other processes, such as gas tungsten arc welding (GTAW) and gas metal arc (GMAW) welding, are much preferred for some metals, such as aluminum and aluminum alloys and copper and some copper alloys, and for certain applications, such as root pass welding of groove joints in pipe. The equipment required for SMAW is relatively portable and the process can often be used in areas of limited accessibility.

Because SMAW electrodes are produced in individual straight lengths, they can be consumed only to some minimum length. When this occurs, welding must stop while the electrode is replaced. This results in a loss of arc time; hence, the overall weld metal deposition rate tends to be lower than with a continuous electrode process. However, the deposition rate with SMAW is higher than with GTAW. With SMAW the slag which forms on the weld bead surface usually must be removed from the end of the weld bead before continuing with a new electrode. For many applications, the completed weld surface must be cleaned of slag or oxides, which could contaminate the process stream or cause corrosion. In SMAW the weld quality is highly dependent on the skill of the welder.

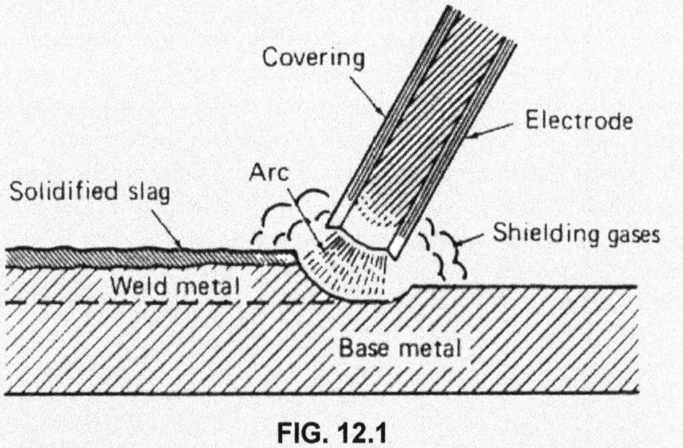

FIG. 12.1
SHIELDED METAL ARC WELDING (COURTESY OF THE JAMES F. LINCOLN FOUNDATION)

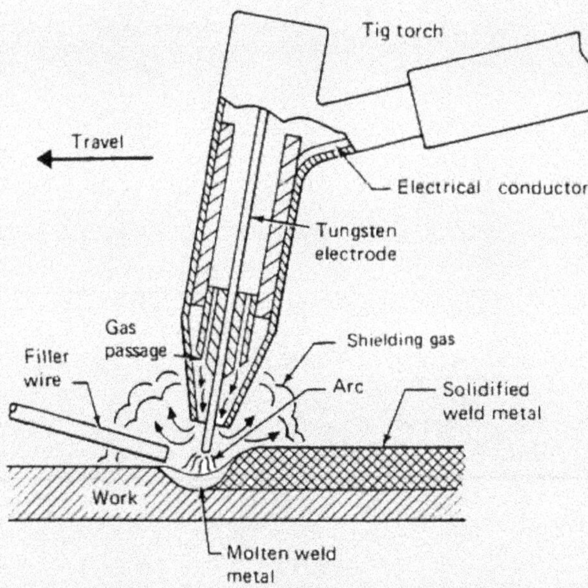

FIG. 12.2
GAS TUNGSTEN ARC WELDING (COURTESY OF THE JAMES F. LINCOLN FOUNDATION)

Gas Tungsten Arc Welding (GTAW)

Principles of Operation

Gas tungsten arc welding (GTAW) uses a special torch, which contains a nonconsumable tungsten or tungsten alloy electrode and provides a concentric flow of inert gas to shield the electrode, arc, and molten metal from the atmosphere (Fig. 12.2). The process is sometimes called TIG, for "tungsten inert gas." The heat for welding is produced by an electric arc between the nonconsumable electrode and the part to be welded. Filler metal is added by feeding a bare rod or wire into the zone of the arc and the molten weld metal, which is protected by the inert atmosphere. The shielding atmosphere consists of argon, helium, or mixtures thereof. Alternating current is preferred for manual welding of aluminum and magnesium, whereas direct current, electrode negative, is preferred for welding most other materials. A high-frequency pilot arc is usually used for arc starting in order to keep from damaging the electrode or contaminating the weld pool or the workpiece. A weld is produced by applying the arc so that the abutting workpieces and the added filler metal are melted and joined together as the weld metal solidifies.

GTAW equipment is portable, and the process is applicable to most metals in a wide range of thickness and in all welding positions. The process can be used to weld all types of joint geometries and overlays in plate, sheet, pipe, tubing, and structural shapes. It is especially suitable for welding sections of pipe less than 10 mm (3/8 in.) in thickness and also 25.4 mm to 152.4 mm (1 in. to 6 in.) in diameter. Thicker sections can be welded, but it is usually less expensive to use a consumable electrode process, which affords a higher deposition rate. For some applications the process can be used for welding by fusion alone without the addition of filler metal. For welding pipe the combination of GTAW for the root pass followed by either SMAW or GMAW is particularly advantageous. The GTAW process provides a root pass surface which is smooth and uniform on the inside of the pipe, whereas the fill and cap passes are made with a more economical process. High-quality welds can be produced with the GTAW process when adequately trained welders and proper procedures are used.

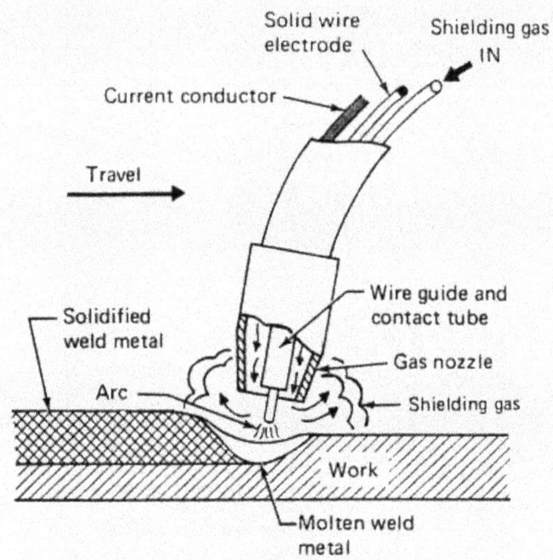

FIG. 12.3
GAS METAL ARC WELDING (COURTESY OF THE JAMES F. LINCOLN FOUNDATION)

For GTA welding it is essential to carefully clean the joint surfaces and use clean filler metal in order to avoid weld defects. The GTAW process is slower than consumable electrode processes. Argon and helium shielding gases are relatively expensive. Improper technique which causes transfer of tungsten into the weld or exposure of the hot filler rod to air can result in unsatisfactory welds.

Gas Metal Arc Welding (Gmaw)

Principles Of Operation

Gas metal arc welding, popularly known as MIG welding, uses a continuous solid-wire consumable electrode for filler metal and an externally supplied gas or gas mixture for shielding (Fig. 12.3). The shielding gas—helium, argon, carbon dioxide, or mixtures thereof—protects the molten metal from reacting with constituents of the atmosphere.

Process Capabilities and Limitations

Gas metal arc welding is operated in semiautomatic, machine, and automatic modes. It is utilized particularly in high-production welding operations. All commercially important metals, such as carbon steel, stainless steel, aluminum, and copper, can be welded with this process in all positions by choosing the appropriate shielding gas, electrode, and welding conditions.

Flux-Cored Arc Welding (FCAW)

Principles of Operation

Flux-cored arc welding uses a continuous tubular wire consumable filler metal. The filler metals are designed principally for joining carbon and low-alloy steels and stainless steels. However, some nickel-base filler metals of this type are also available. The flux core may contain minerals, ferroalloys, and ingredients that provide shielding gases, deoxidizers, and slag-forming materials. There are two basic types of cored electrodes, those designed to be used with an additional external shielding gas (Fig. 12.4), and those of self-shielding type, designed for use without external shielding gas. For carbon and low-alloy steels the external shielding gas is usually carbon dioxide or a mixture of argon and carbon dioxide. For stainless steels, argon and argon-oxygen or argon-carbon dioxide mixtures are often used. Self-shielded FCAW electrodes are designed to generate protective shielding gases from core ingredients, similar to the generation of gases by SMAW electrodes. Self-shielded electrodes do not require external shielding.

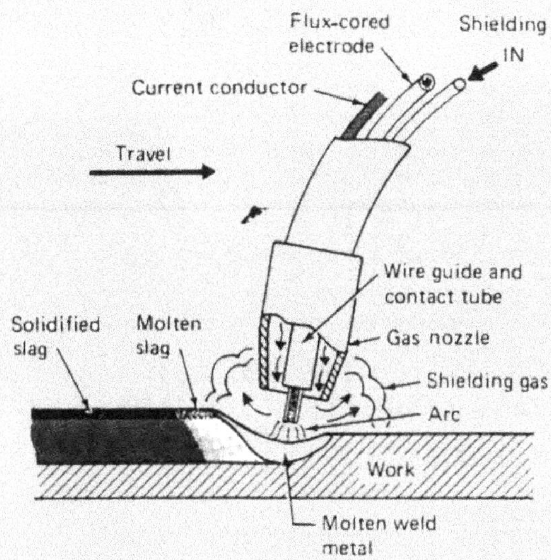

FIG. 12.4
GAS-SHIELDED FLUXED CORED ARC WELDING (COURTESY OF THE JAMES F. LINCOLN FOUNDATION)

Process Capabilities and Limitations

Flux-cored arc welding can be operated in either automatic or semi-automatic modes. Weld metal can be deposited at higher rates, and the welds can be larger and better contoured than those made with solid electrodes (GMAW), regardless of the shielding gas. FCAW lends itself to high-production welding applications, as in a manufacturing plant or fabrication shop.

Some classifications of FCAW electrodes of both the externally shielded and self-shielded types are designed for single-pass welding only. Such electrodes should not be used for multipass welding, because there is a risk of weld cracking. For low-temperature service applications, care should be exercised in selecting a classification of FCAW electrode which is designed and tested to ensure good low-temperature notch ductility.

Submerged Arc Welding (SAW)

Principles of Operation

Submerged arc welding uses an arc (or arcs) maintained between a bare electrode and the work (Fig. 12.5). The feature that distinguishes submerged arc welding from other arc welding processes is the blanket of granular, fusible material that covers the welding area. By common usage the material is termed a flux, although it performs several important functions in addition to those strictly associated with a fluxing agent. Flux plays a central role in achieving the high deposition rates and weld quality that characterize the SAW process in joining and surfacing applications.

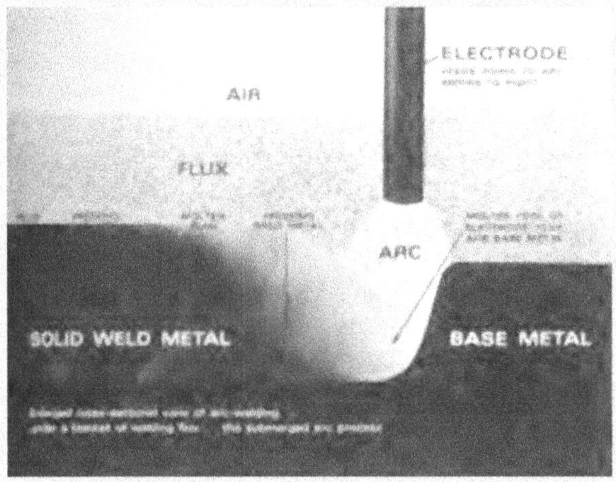

FIG. 12.5
SUBMERGED ARC WELDING (COURTESY OF THE JAMES F. LINCOLN FOUNDATION)

Process Capabilities and Limitations

With proper selection of equipment, SAW can be applied to a wide range of industrial applications. The high quality of welds, the high deposition rates, the deep penetration, and the adaptability to automatic operation make the process particularly suitable for fabrication of large weldments. It is used extensively in ship building, railroad car fabrication, pressure vessel fabrication, pipe manufacturing, and the fabrication of structural members where long welds are required.

SAW is not suitable for all metals and alloys. It is widely used for welding carbon steels, low-alloy structural steels, and stainless steels.

Submerged arc welding can be used for welding butt joints in the flat position, making fillet welds in the flat and horizontal positions, and surfacing in the flat position. With special tooling and fixturing, lap and butt joints can be welded in the horizontal position. The use of this process for welding piping systems is limited to fabrication shops where the pipe can be rolled.

12.3 WELDING PROCEDURE SPECIFICATION (WPS)

Welds are created in accordance with Welding Procedure Specifications (WPSs). A WPS provides direction for making production welds to ensure compliance with Code or other specification requirements. A WPS references one or more supporting Procedure Qualification Record(s) (PQR). ASME BPVC, Section IX, QW-482 gives a suggested format for WPSs.

A WPS consists of a number of items, called variables, that define the application of a welding process or combination of processes. Examples of variables are nature of the welding process, base metal type and thickness, filler metal type, welding position, preheat, postheat, electrical characteristics, and type of shielding gas. Variables are divided into three types; essential, nonessential, and supplementary essential.

An essential variable is one that, if changed to a value outside the limits permitted by the original Procedure Qualification Test, requires a new Procedure Qualification Test.

Nonessential variables include items in a WPS that may have to be changed in order to satisfy a particular welding application, but do not affect the properties of the weld. An example is a change in the groove design. A change in a nonessential variable requires revision or amendment of a WPS or a new WPS; however, requalification testing is not required.

A supplementary essential variable is one that comes into consideration when a particular fabrication Code, such as ASME BPVC, Section VIII or ASME B31.3, requires supplementary testing in addition to the customary tensile and bend tests required for procedure qualification (e.g., impact testing for low-temperature service).

ASME BPVC, Section IX, Article II lists the essential, nonessential, and supplementary essential variables for Welding Procedure Specifications and procedure qualification testing for each welding process used for joining and for special applications such as hard-facing overlay and corrosion-resistant overlay.

12.4 WELDING PROCEDURE QUALIFICATION RECORD (PQR)

The purpose of procedure qualification testing is to determine or demonstrate that the weldment proposed for construction is capable of providing the required properties for its intended application. It is presupposed that the welder or welding operator performing the Welding Procedure Qualification Test is a skilled worker. That is, the Welding Procedure Qualification Test establishes the properties of the weldment, not the skill of the welder or welding operator.

A PQR includes a record of the welding data used to weld a prescribed test coupon in accordance with a WPS and the results of specified mechanical and other tests. The completed PQR must document all essential and, when required, supplementary essential variables of ASME BPVC, Section IX. ASME BPVC, Section IX, QW-483 gives a suggested format for a PQR.

If any changes are made in a WPS that involve essential variables or, when required, supplementary essential variables, a new Procedure Qualification Test and a new PQR are required in order to document the changes and support a revision of the WPS or issuance of a new WPS.

The details of preparing a weld test assembly or assemblies, testing of the weld, and test results required for procedure qualification are given in ASME BPVC, Section IX. Customary tests for qualification of joint welding procedures are tensile tests and bend tests.

12.5 WELDER PERFORMANCE QUALIFICATION

The Manufacturer or Contractor is responsible for conducting tests to qualify welders and welding operators in accordance with qualified welding procedure specifications that the organization employs in the construction of weldments built in accordance with the Code. The purpose of welder and welding operator qualification tests is to ensure that the manufacturer or Contractor has determined that the welder(s) and welding operator(s) following the procedures are capable of developing the minimum requirements specified for an acceptable weldment. Performance qualification tests are intended to determine the ability of welders and welding operators to make sound welds.

ASME BPVC, Section IX lists and defines essential variables for each welding process for performance qualification of welders and welding operators. It should be noted that the essential variables for performance qualification are in many cases quite different from the variables for procedure qualification discussed in Section 12.3. For the SMAW process, for example, the essential variables for performance qualification are elimination of backing, change in pipe diameter, change in base material P-Number, change in filler metal F-Number, change in thickness of the weld deposit, addition of welding positions beyond that originally qualified, and change in direction of weld progression for vertical welding. A change beyond prescribed limits in an essential variable for performance qualification requires requalification testing and issuance of a new qualification record.

The details of preparing the weld test assembly or assemblies, testing of the weld, and the test results required for welder and welding operator performance qualification are given in ASME BPVC, Section IX. Customary tests for performance qualification are visual and bend tests. However, for some base materials and welding processes there are also alternate provisions for examination of the weld(s) by radiography in lieu of bend tests.

ASME BPVC Section IX, QW-484 gives a suggested format for recording Welder/Welding Operator Performance Qualification (WPQ). A WPQ for a specific process expires when the welder has not welded with that process during a period of 6 months or longer.

Within a 6-month period prior to expiration of a performance qualification a welder who welds using a manual or semiautomatic process maintains all of his or her qualifications for manual or semiautomatic welding with that process. For example, a welder qualified for manual SMAW welding of both carbon steel and stainless steel, but who welds only carbon steel with SMAW during a 6-month period, maintains his or her qualification for SMAW of stainless steel as well as for SMAW of carbon steel. This highlights the importance of keeping timely records of the welding process(es) used during a welder's production welding assignments.

When there is a specific reason to question the ability of a welder or welding operator to make welds that meet the specification, the performance qualification that supports that welding can be revoked. Other qualifications that are not in question remain in effect.

Renewal of performance qualification that has expired may be made for any process by welding a single test coupon of either plate or pipe, of any material, thickness, or diameter, in any position, and by testing that coupon by bending or by radiography (if the latter is permitted by ASME BPVC, Section IX for the process and material involved; see QW-304). Alternatively, where radiography is a permissible method of examination (QW-304), the renewal of qualification may be done on production work.

12.6 PREHEATING

Preheating requirements are provided in para. 330 of ASME B31.3. Preheating is used, along with heat treatment, to minimize the detrimental effects of high temperature and severe thermal gradients in welding, prevent hydrogen cracking, and improve the metalurgical properties. The effect of reducing hydrogen cracking is accomplished by a variety a factors, including driving off moisture, reducing the cooling rate, and increasing the rate of hydrogen diffusion in the material. The reduced cooling rate can also improve material properties; in some materials, too rapid a cooling rate results in brittle microstructural constituents.

The preheat requirements, which are applicable to all types of welding, including tack welds, repair welds, and seal welds of threaded joints, are provided in Table 330.1.1, reproduced here as Table 12.1. The requirements for preheat are a function of the P-Number of the base material, the nominal wall thickness, and the specified minimum tensile strength of the base metal. When welding dissimilar metals, the higher preheat temperature should be used. Preheat for unlisted materials must be as specified in the Welding Procedure Specification.

The preheat zone is required to extend at least 25 mm (1 in.) beyond each edge of the weld. The temperature must be obtained prior to, and maintained during, the welding. The temperature is required to be verified by checking with temperature-indicating crayons, thermocouple pyrometers, or other suitable means.

12.7 HEAT TREATMENT

A few efforts, the latest being in the 1990's, to make the heat treatment requirements the same for the various ASME pressure vessel and piping codes ended in failure because of the absence of enough data to show which set of requirements, if any, was superior. Recently completed studies compare as-welded properties of thick section welds to welds on the same material that were heat treated. Based on this research, the 2014 editions of both ASME B31.1 and ASME B31.3 included a major technical revision of the heat treatment requirements. Some of the changes were dramatic. For example, prior to the 2014 edition, ASME B31.3 required postweld heat treatment for carbon steels thicker than 19 mm (3/4 in.). The 2014 and later editions do not require heat treatment for any thickness of carbon steel with the application of specified preheat and the use of multi-layer welding.

These new rules have been refined and published in ASME B31P, *Standard Heat Treatments for Fabrication Processes*, first published in 2017. The Code now permits following the preheat and heat treatment rules in ASME B31P as an alternate to the rules described in paras. 330 and 331.

Post-weld heat treatment is performed to temper the weldment, relax residual stresses, and remove hydrogen. The consequential benefits are avoidance of hydrogen-induced cracking and improved ductility, toughness, corrosion resistance, and dimensional stability.

Heat treatment requirements are provided in para. 331 of ASME B31.3. The Code requires heat treatment after certain welding, bending, and forming operations. Specific requirements for postweld heat treatment are provided in Table 331.1.1 (Table 12.2). This table specifies the heat treatment time and temperature based on the P-Number, the material chemistry, the wall thickness, and the specified minimum tensile strength of the base material.

Table 331.1.3 (Table 12.3) provides exemptions to mandatory postweld heat treatment described in Table 331.1.1 (Table 12.2). For example, postweld heat treatment is not required for carbon steel of any thickness if

- A preheat of 95°C (200°F) is applied prior to welding on any nominal material thickness >25 mm (1 in.), and
- Multiple layer welds are used when the nominal material thickness >5 mm (3⁄16 in.)

Requirements for heat treatment after bending or forming are stated in para. 332.4. These depend on the material P-Number, whether the material is hot- or cold-bent or formed, and, for cold-bent or formed material, the fiber elongation.

Heat treatment in an enclosed furnace, local flame heating, electric resistance heating, electric induction heating, or exothermic chemical reactions are all permitted methods of supplying heat. However, the method is required to provide the required metal temperature, metal temperature uniformity, and temperature control. Similar to preheat, the metal temperature must be verified. The Code requires that at least a distance of three times the wall thickness on either side of the weld or bent or formed section be brought up to temperature. AWS D10.10, *Recommended Practices for Local Heating of Welds in Piping and Tubing,* provides useful guidance in the performance of local heat treating.

ASME B31.3, para. 331.2, permits exceptions to the heat treatment requirements of Table 331.1.1 (Table 12.2 here) where warranted based on knowledge or experience of the service conditions:

- Normalizing, or normalizing and tempering, or annealing may be used in lieu of the heat treatment of Table 331.1.1, provided the material properties of the weld and base material meet the specification requirements after heat treatment. This requires approval of the Designer.
- More or less stringent heat treatment may be specified by the Designer. If less stringent heat treatment is performed, the Designer must demonstrate to the Owner that the heat treatment is adequate, and the welding procedure qualification tests must be conducted with the heat treatment that will be used.

One of the reasons for selecting a lower heat treatment temperature is mentioned in para. F331.1 of Appendix F. It states the following.

"Heat treatment temperatures listed in Table 331.1.1 for some materials may be higher than the minimum tempering temperatures specified in the ASTM specifications for the base material. For higher-strength normalized and tempered materials, there is consequently a possibility of reducing tensile properties of the base material, particularly if long holding times at the higher temperatures are used."

TABLE 12.1
PREHEAT TEMPERATURES (ASME B31.3-2020 Table 330.1.1)

Base Metal P-No. [Note (1)]	Base Metal Group	Greater Material Thickness		Additional Limits [Note (2)]	Required Minimum Temperature	
		mm	in.		°C	°F
1	Carbon steel	≤25	≤1	None	10	50
		>25	>1	%C ≤ 0.30 [Note (3)]	10	50
		>25	>1	%C > 0.30 [Note (3)]	95	200
3	Alloy steel, Cr ≤ ½%	≤13	≤½	SMTS ≤ 450 MPa (65 ksi)	10	50
		>13	>½	SMTS ≤ 450 MPa (65 ksi)	95	200
		All	All	SMTS > 450 MPa (65 ksi)	95	200
4	Alloy steel, ½% < Cr ≤ 2%	All	All	None	120	250
5A	Alloy steel	All	All	SMTS ≤ 414 MPa (60 ksi)	150	300
		All	All	SMTS > 414 MPa (60 ksi)	200	400
5B	Alloy steel	All	All	SMTS ≤ 414 MPa (60 ksi)	150	300
		All	All	SMTS > 414 MPa (60 ksi)	200	400
		>13	>½	%Cr > 6.0 [Note (3)]	200	400
6	Martensitic stainless steel	All	All	None	200 [Note (4)]	400 [Note (4)]
9A	Nickel alloy steel	All	All	None	120	250
9B	Nickel alloy steel	All	All	None	150	300
10I	27Cr steel	All	All	None	150 [Note (5)]	300 [Note (5)]
15E	9Cr–1Mo–V CSEF steel	All	All	None	200	400
...	All other materials	...	...	None	10	50

NOTES:
(1) P-Nos. and Group Nos. from ASME BPVC, Section IX, QW/QB-422.
(2) SMTS = Specified Minimum Tensile Strength.
(3) Composition may be based on ladle or product analysis or in accordance with specification limits.
(4) Maximum interpass temperature 315°C (600°F).
(5) Maintain interpass temperature between 150°C and 230°C (300°F and 450°F).

TABLE 12.2
POSTWELD HEAT TREATMENT (ASME B31.3-2020 TABLE 331.1.1)

P-No. and Group No. (ASME BPVC, Section IX, QW/QB-420)	Holding Temperature Range, °C (°F) [Note (1)]	Minimum Holding Time at Temperature for Control Thickness [Note (2)]	
		Up to 50 mm (2 in.)	Over 50 mm (2 in.)
P-No. 1, Group Nos. 1 3	595 to 650 (1,100 to 1,200)	1 h/25 mm (1 hr/in.); 15 min min.	2 hr plus 15 min for each additional 25 mm (in.) over 50 mm (2 in.)
P-No. 3, Group Nos. 1 and 2	595 to 650 (1,100 to 1,200)		
P-No. 4, Group Nos. 1 and 2	650 to 705 (1,200 to 1,300)		
P-No. 5A, Group No. 1	675 to 760 (1,250 to 1,400)		
P-No. 5B, Group No. 1	675 to 760 (1,250 to 1,400)		
P-No. 6, Group Nos. 1–3	760 to 800 (1,400 to 1,475)		
P-No. 7, Group Nos. 1 and 2 [Note (3)]	730 to 775 (1,350 to 1,425)		
P-No. 8, Group Nos. 1–4	PWHT not required unless required by WPS		
P-No. 9A, Group No. 1	595 to 650 (1,100 to 1,200)		
P-No. 9B, Group No. 1	595 to 650 (1,100 to 1,200)		
P-No. 10H, Group No. 1	PWHT not required unless required by WPS. If done, see Note (4).		
P-No. 10I, Group No. 1 [Note (3)]	730 to 815 (1,350 to 1,500)		
P-No. 11A	550 to 585 (1,025 to 1,085) [Note (5)]		
P-No. 15E, Group No. 1	705 to 775 (1,300 to 1,425) [Notes (6) and (7)]	1 h/25 mm (1 hr/in.); 30 min min.	1 h/25 mm (1 hr/in.) up to 125 mm (5 in.) plus 15 min for each additional 25 mm (in.) over 125 mm (5 in.)
P-No. 62	540 to 595 (1,000 to 1,100)	...	See Note (8)
All other materials	PWHT as required by WPS	In accordance with WPS	In accordance with WPS

GENERAL NOTE: The exemptions for mandatory PWHT are defined in Table 331.1.3.

NOTES:
(1) The holding temperature range is further defined in para. 331.1.6(c) and Table 331.1.2.
(2) The control thickness is defined in para. 331.1.3.
(3) Cooling rate shall not be greater than 55°C (100°F) per hour in the range above 650°C (1,200°F), after which the cooling rate shall be sufficiently rapid to prevent embrittlement.
(4) If PWHT is performed after welding, it shall be within the following temperature ranges for the specific alloy, followed by rapid cooling:
Alloys S31803 and S32205 — 1 020°C to 1 100°C (1,870°F to 2,010°F)
Alloy S32550 — 1 040°C to 1 120°C (1,900°F to 2,050°F)
Alloy S32750 — 1 025°C to 1 125°C (1,880°F to 2,060°F)
All others — 980°C to 1 040°C (1,800°F to 1,900°F)
(5) Cooling rate shall be >165°C (300°F)/h to 315°C (600°F)/h.
(6) The minimum PWHT holding temperature may be 675°C (1,250°F) for nominal material thicknesses [see para. 331.1.3(c)] ≤13 mm ($\frac{1}{2}$ in.).
(7) The Ni + Mn content of the filler metal shall not exceed 1.2% unless specified by the designer, in which case the maximum temperature to be reached during PWHT shall be the lower transformation temperature of the filler metal, as determined by analysis and calculation or by test, but not exceeding 800°C (1,470°F). If the 800°C (1,470°F) limit was not exceeded but the lower transformation temperature of the filler metal was exceeded or if the composition of the filler metal is unknown, the weld must be removed and replaced. It shall then be rewelded with compliant filler metal and subjected to a compliant PWHT. If the 800°C (1,470°F) limit was exceeded, the weld and the entire area affected by the PWHT will be removed and, if reused, shall be renormalized and tempered prior to reinstallation. The lower transformation temperature is the steady-state temperature at which the austenite phase transformation occurs.
(8) Heat treat within 14 days after welding. Hold time shall be increased by 1.2 h for each 25 mm (1 in.) over 25 mm (1 in.) thickness. Cool to 425°C (800°F) at a rate ≤280°C (500°F)/h.

TABLE 12.3

EXEMPTIONS TO MANDATORY POSTWELD HEAT TREATMENT (ASME B31.3-2020 TABLE 331.1.3)

P-No. and Group No. (ASME BPVC, Section IX, QW/QB-420) [Note (1)]	Control Thickness, mm (in.) [Note (2)]	Type of Weld	Additional Limitations Required for Exemption From PWHT [Notes (3)–(5)]
P-No. 1, all Group Nos.	All	All	A preheat of 95°C (200°F) is applied prior to welding on any nominal material thickness >25 mm (1 in.) Multiple layer welds are used when the nominal material thickness >5 mm ($^3/_{16}$ in.) [Note (6)]
P-No. 3, Group Nos. 1 and 2	≤16 mm ($^5/_8$ in.)	All	A preheat of 95°C (200°F) is applied prior to welding on any nominal material thickness >13 mm ($^1/_2$ in.) A specified carbon content of the base materials ≤0.25% Multiple layer welds are used when the nominal material thickness >5 mm ($^3/_{16}$ in.) [Note (6)]
P-No. 4, Group No. 1	≤16 mm ($^5/_8$ in.)	Groove	Mandatory preheat has been applied Specified carbon content of the base materials ≤0.15% Multiple layer welds are used when the nominal material thickness >5 mm ($^3/_{16}$ in.) [Note (6)]
	≤16 mm ($^5/_8$ in.) except the thickness of a socket weld fitting or flange need not be considered	Socket and fillet welds	Mandatory preheat has been applied Throat thickness of the fillet weld or the socket weld ≤13 mm ($^1/_2$ in.) Specified carbon content of the pipe material ≤0.15% Nominal material thickness of the pipe ≤16 mm ($^5/_8$ in.) Multiple layer welds are used when the nominal material thickness >5 mm ($^3/_{16}$ in.) [Note (6)]
	≤16 mm ($^5/_8$ in.)	Seal welds and non-load-carrying attachments [Note (7)]	Mandatory preheat has been applied Multiple layer welds are used when the nominal material thickness >5 mm ($^3/_{16}$ in.) [Note (6)]
P-No. 5A, Group No. 1	≤16 mm ($^5/_8$ in.)	Groove	Mandatory preheat has been applied Specified carbon content of the base materials ≤0.15% Multiple layer welds are used when the nominal material thickness >5 mm ($^3/_{16}$ in.) [Note (6)]
	≤16 mm ($^5/_8$ in.) except the thickness of a socket weld fitting or flange need not be considered	Socket and fillet welds	Mandatory preheat has been applied Throat thickness of the fillet weld or the socket weld ≤13 mm ($^1/_2$ in.) Specified carbon content of the pipe material ≤0.15% Nominal thickness of the pipe ≤5 mm ($^3/_{16}$ in.) Multiple layer welds are used when the nominal material thickness >5 mm ($^3/_{16}$ in.) [Note (6)]
	≤16 mm ($^5/_8$ in.)	Seal welds and non-load-carrying attachments [Note (7)]	Mandatory preheat has been applied Multiple layer welds are used when the nominal material thickness >5 mm ($^3/_{16}$ in.) [Note (6)]
P-No. 5B, Group No. 1	...	...	No exemptions from PWHT
P-No. 6, Group Nos. 1–3	All	All	Specified carbon content of the base materials ≤0.08% Nominal material thickness ≤10 mm ($^3/_8$ in.) Weld filler metal is A-No. 8, A-No. 9, or F-No. 43 composition [Note (8)]

TABLE 12.3

(CONTINUED)

P-No. and Group No. (ASME BPVC, Section IX, QW/QB-420) [Note (1)]	Control Thickness, mm (in.) [Note (2)]	Type of Weld	Additional Limitations Required for Exemption From PWHT [Notes (3)–(5)]
P-No. 7, Group No. 1	All	All	Specified carbon content of the base materials ≤0.08% Nominal material thickness ≤10 mm ($^3/_8$ in.) Weld filler metal is A-No. 8, A-No. 9, or F-No. 43 composition [Note (8)]
P-No. 7, Group No. 2	...	...	No exemptions from PWHT
P-No. 8, all Group Nos.	All	All	PWHT neither required nor prohibited
P-No. 9A, Group No. 1	All	All	Specified carbon content of the pipe material ≤0.15% Nominal material thickness ≤13 mm ($^1/_2$ in.) Mandatory preheat has been applied
P-No. 9B, Group No. 1	All	All	Nominal material thickness ≤16 mm ($^5/_8$ in.) and the WPS has been qualified using a material of equal or greater thickness than used in the production weld
P-No. 10H, Group No. 1	All	All	PWHT neither required nor prohibited
P-No. 10I, Group No. 1	All	All	PWHT neither required nor prohibited for nominal material thickness ≤13 mm ($^1/_2$ in.)
P-No. 11A	≤50 mm (2 in.)	All	...
P-No. 15E	...	...	No exemptions from PWHT
P-No. 62	...	...	No exemptions from PWHT

NOTES:
(1) If differences with the P-No. listed in Appendix A are found, the P-No. listed in ASME BPVC, Section IX, Table QW/QB-422 applies.
(2) The control thickness is defined in para. 331.1.3.
(3) The nominal material thickness is defined in para. 331.1.3(c).
(4) No exemptions are permitted for PWHTs required by the designer or the WPS.
(5) Additional exemptions for welds made in accordance with para. 328.7 may be taken for the materials addressed.
(6) Single-layer or single-pass welds may be exempted from PWHT, provided the WPS has been qualified using single-pass welds with ±10% heat input and that all other conditions for exemption are met.
(7) Non-load-carrying attachments are defined as items where no pressure loads or significant mechanical loads are transmitted through the attachment to the pipe or pressure-containing material.
(8) The A-Nos. and the F-Nos. are found in ASME BPVC, Section IX, Tables QW-442 and QW-432, respectively.

12.8 GOVERNING THICKNESS FOR HEAT TREATMENT

When using Table 331.1.1 (Table 12.2 here), the *control thickness* to be used for groove welds is the thicker of the two components, measured at the joint, that are being joined by welding. For example, if a pipe is welded to a heavier wall valve, but the valve thickness is tapered to the pipe thickness at the welded joint, the governing thickness will be the nominal pipe wall thickness.

For other types of welds, the *control thickness* is

a. When the weld is attaching a nonpressure-containing material to a pressure-containing material, the lesser of the thickness of the weld and the thickness of the pressure-containing material
b. For fillet welds — the throat thickness of the weld
c. For material repair welds — the depth of the cavity to be repaired
d. For branch welds — the dimension existing in the plane intersecting the longitudinal axes, calculated as indicated for each detail using the thickness through the weld for the details shown in Figure 328.5.4D and Figure 328.5.4F.

When using Table 331.1.3 (Table 12.3 here) the term *nominal material thickness* is used. For the exemptions, the thickness to be used is the thicker of the materials being joined at the weld. The thickness of the weld is not considered for the exemptions.

12.9 PIPE BENDS

Pipe may be hot- or cold-bent. For cold-bending of ferritic materials, the temperature must be below the transformation range. For hot-bending, the temperature must be above the transformation range.

The thickness after bending must comply with the design requirements. As discussed in Section 4.8, the required thickness must be calculated.

When pipe is bent, it tends to ovalize (also termed flattening). Paragraph 332.2 requires that the flattening (the difference between the maximum and minimum diameters at any cross section) not exceed 8%, except when the pipe may be subject to external pressure, in which case the limit is 3%. Although this has often been treated as an absolute limit, it is possible to perform a detailed calculation in the engineering design per para. 300(c)(3) to establish a different, perhaps higher, limit. For example, a heavy-wall, small-diameter pipe may have a very substantial margin against buckling even if the 3% flattening is significantly exceeded.

In the 2004 edition, para, 306.2.1(b) was added to specifically permit qualification of bends with flattening exceeding these limits by para. 304.7.2. See Section 12.7 for heat treatment after bending.

12.10 BRAZING AND SOLDERING

Brazing procedures, brazers, and brazing operators are required to be qualified in accordance with ASME BPVC, Section IX, Part QB. The brazing process is described below, in Insert 12.2. An exception is for piping in Category D Fluid Service with a design temperature not exceeding 93°F, (200°F), for which the Owner can waive the requirements for such qualifications,

Solderers are required to be qualified in accordance with the requirements of ASTM B828, *Standard Practice for Making Capillary Joints by Soldering of Copper and Copper Alloy Tube and Fittings*.

Materials, preparation and cleaning requirements for brazing and soldering are specified in para. 333.

Insert 12.2 Brazing

Description of Process. Brazing includes a group of metal-joining processes which produce coalescence of materials by heating them to a suitable temperature and by using a filler metal which melts at a temperature above 427°C (800° F) and below the temperature at which the base metal starts to melt. The brazing filler metal is distributed between closely fitted surfaces of the joint by capillary attraction.

Various types of brazing are defined according to the method of heating employed, for example, dip brazing in a molten metal or salt bath, furnace brazing, induction brazing, infrared brazing, resistance brazing, and torch brazing. Torch brazing is the most commonly used brazing process for process plant construction and maintenance operations.

Torch Brazing: Principles of Operation. Torch brazing is accomplished by heating with one or more torches using a gas fuel (acetylene, propane, city gas, etc.) burned with air, compressed air, or oxygen. Brazing filler metal may be preplaced at the joint in the form of rings, washers, strips, powder, etc., or it may be fed in wire or rod form by hand. In any case, proper precleaning and fluxing are essential.

Capabilities and Limitations

Brazing can be employed for joining a wide variety of metals and alloys, and appropriate brazing filler metals are available for various applications. To achieve a good joint, the parts must be properly cleaned and must be protected during the heating process, usually by use of a flux in torch brazing, in order to prevent excessive oxidation. The parts must be designed and properly aligned to afford capillary flow of the filler metal. Heat must be applied so as to provide the required temperature range and uniform temperature distribution in the parts.

12.11 BOLTED JOINTS

Proper assembly of bolted joints is essential to avoid leakage during service. This includes not only visible leaks, but also fugitive emissions, Information on flange bolting is provided in ASME BPVC. Section VIII, Division 1, Appendix S. An ASME guideline on flange boltup procedures, PCC-1, *Guidelines for Pressure Boundary Bolted Flange Joint Assembly,* was first published in 2001.

ASME B31.3 provides some good practice guidelines with respect to flange boltup in para. 335. This includes requiring repair or replacement of flanges with damaged gasket seating surfaces, uniformly compressing the gasket during flange boltup, and using only one gasket between seating surfaces.

Paragraph 335.2.3 requires that the bolts extend completely through their threads. Perhaps one of the most frequently violated provisions of the Code is the flange alignment tolerance in para. 335.1(c). Regarding the out of parallel tolerance, the Code requires that one of two conditions be met.

1. Before bolting, mating gasket contact surfaces shall be aligned to each other within 1 mm in 200 mm (1⁄16 in./ft), measured across any diameter. This requirement is intended to be applied to larger diameter, more rigid piping.
2. The flanged joint shall be capable of being bolted such that the gasket contact surfaces bear uniformly on the gasket. This requirement would apply to smaller diameter, more flexible piping.

In both cases, the bolt holes must be aligned within 3 mm (1/8 in.) of maximum offset.

ASME PCC-1 provides a more comprehensive and somewhat more stringent flange alignment requirements.

Neither the Code nor the ASME PCC-1 alignment provisions are generally tight enough for larger piping connected to load-sensitive equipment. API 686, *Recommended Practice for Machinery Installation and Installation Design,* provides guidance on alignment and bolting practices to such equipment.

12.12 WELDED JOINT DETAILS

Welded joint details, including socket weld joints, socket weld and slip-on flanges, and branch connections are provided in ASME B31.3, Chapter V.

Standard details for slip-on and socket welding flange attachment welds are provided in Fig. 328.5.2B, reproduced here as Fig. 12.6. Points worth noting are the fillet weld size, which is 1.4 times the nominal pipe wall thickness, or the thickness of the hub, whichever is less, and the small gap shown between the flange face and the toe of the inside fillet for slip-on flanges. The small gap is intended to avoid damage to the flange face due to welding. It indicates a gap, but there is no specific limit. This differs from ASME BPVC, Section VIII, Division 1, which specifies the gap to be 6 mm (1/4 in.), and some Pipeline Codes that have limits.

The question arose as to whether a specific limit to the gap between the fillet weld and flange face was appropriate. Studies, including finite-element analysis (Becht et al, 1992) and earlier Markl fatigue testing, indicated that it essentially did not matter how much the pipe was inserted into the flange. Insertion by an amount equal to the hub height was optimal for fatigue life, but there was not a significant difference.

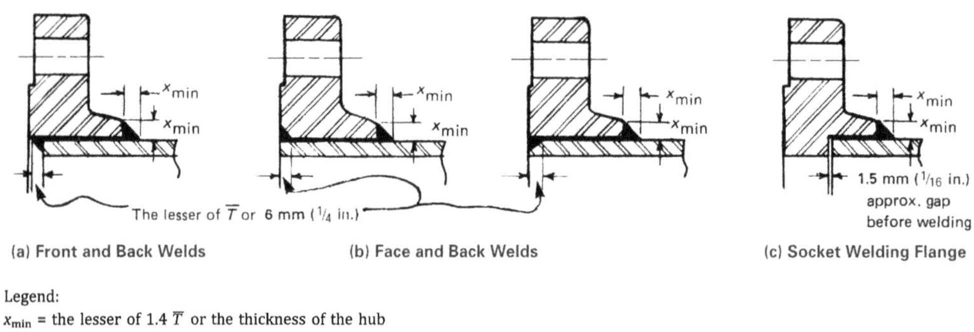

FIG. 12.6
TYPICAL DETAILS FOR DOUBLE-WELDED SLIP-ON AND SOCKET WELDING FLANGE ATTACHMENT WELDS (ASME B31.3-2020, FIG. 328.5.2B)

The required fillet weld size for socket welds other than socket weld flanges is specified in Fig. 328.5.2C, reproduced here as Fig. 12.7.

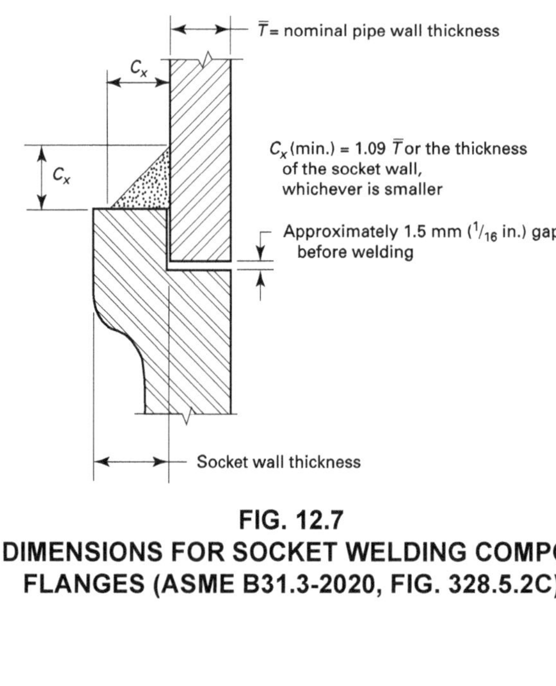

FIG. 12.7
MINIMUM WELDING DIMENSIONS FOR SOCKET WELDING COMPONENTS OTHER THAN FLANGES (ASME B31.3-2020, FIG. 328.5.2C)

An issue that has caused considerable controversy is the 1.5-mm (1/16-in.) approximate gap before welding indicated on the figure. This is a requirement for a gap before welding, so that weld shrinkage will be less likely to cause small cracks in the root of the fillet weld. Whether or not such cracks cause problems is questionable, and fatigue testing has shown that socket welds that are welded after jamming the pipe into the socket have longer fatigue lives than ones welded with a gap. There is no requirement for a gap after welding, and weld shrinkage can close a gap that was present prior to welding.

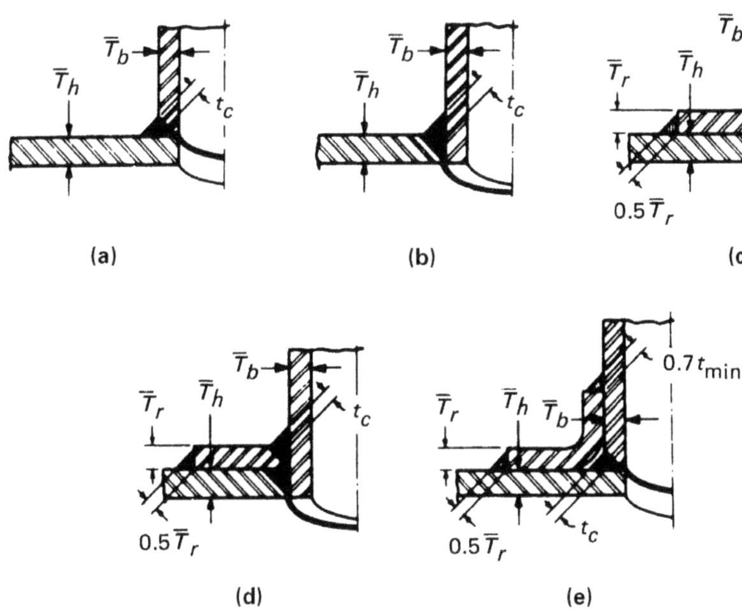

GENERAL NOTE: These illustrations show minimum acceptable welds. Welds may be larger than those shown here.

FIG. 12.8
ACCEPTABLE DETAILS FOR BRANCH ATTACHMENT WELDS
(ASME B31.3-2020, FIG. 328.5.4)

Some owners require random radiographic examination to ensure proper socket welding practice. One of the items checked for is the presence of a gap. An argument for doing this is that it is not possible to determine if there was a gap prior to welding unless there is a gap shown by radiography. If desired, the requirement that there be a gap after welding should be specified as part of the engineering design. It is not a Code requirement.

Acceptable details for fabricated branch connections are provided in Fig. 328.5.4D, reproduced here as Fig. 12.8. ASME B31.3 does not include calculations for required weld sizes for these connections; instead, the minimum weld sizes are specified in this figure.

The weld requirements are the branch to run weld be a full penetration weld, with a cover fillet with the dimension t_c which is specified in the Code. The requirement for the fillet weld attaching a reinforcing plate to the run pipe is that the weld have a throat dimension not less than half of the nominal thickness of the reinforcing plate.

Similarly, acceptable details for attaching branch connection fittings to headers are shown in Fig. 328.5.4F, reproduced here as Fig 12.9.

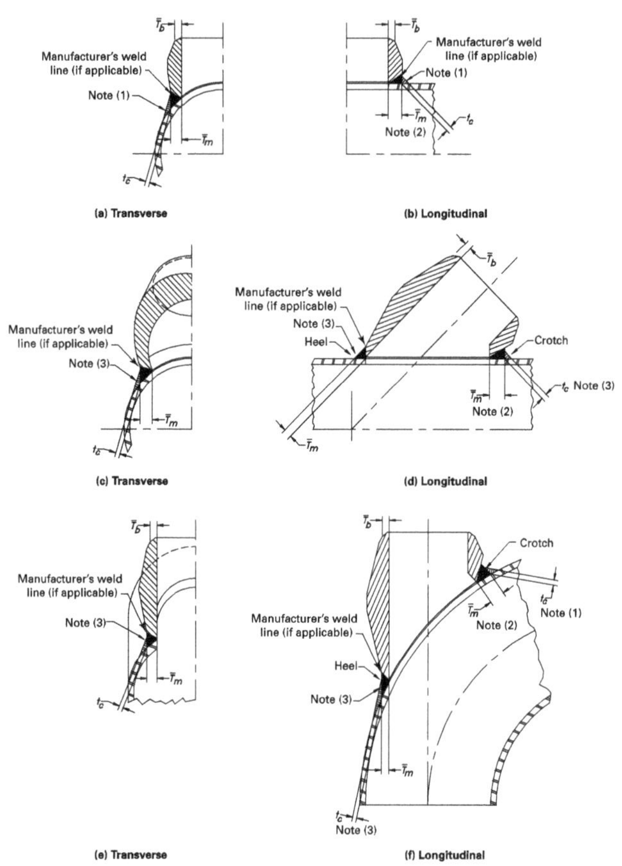

GENERAL NOTE: Welds shall be in accordance with para. 328.5.4(i).

NOTES:

(1) Cover fillet weld shall provide a smooth transition to the run pipe with an equal leg fillet at the longitudinal section to an equal leg fillet, unequal (oblique) leg fillet, or groove butt joint at the transverse section (depending on branch connection size).

(2) Heat treatment requirements shall be in accordance with para. 331.1.3(a).

(3) Cover fillet weld shall provide a smooth transition to the run pipe with an equal leg fillet at the crotch in the longitudinal section to an equal leg fillet, unequal (oblique) leg fillet, or groove butt joint at the transverse section (depending on branch connection size) to nothing at the heel of the branch connection fitting in the longitudinal section.

FIG. 12.9

ACCEPTABLE DETAILS FOR INTEGRALLY REINFORCED BRANCH CONNECTIONS (ASME B31.3-2020, FIGURE 328.5.4)

12.13 MISCELLANEOUS ASSEMBLY REQUIREMENTS

Threaded joints should be assembled with a thread compound or lubricant that is suitable for the service conditions and does not react unfavorably with either the service fluid or the piping material. However, if the joint is intended to be seal-welded, the use of thread compound or tape is prohibited. The material can result in a poor-quality seal weld. There is one exception to this. If a joint is made up with thread tape or compound and it leaks during the leak test, it can be seal-welded without disassembling it, provided that all the material is removed from the exposed threads. As discussed in the Design Section, seal welds are not considered to contribute to the joint strength. Per para. 328.5.3, seal welds are required to cover all exposed threads.

Various good-practice requirements are provided in para. 335 for assembly of straight-threaded joints, tubing joints, caulked joints, expanded joints, and packed joints.

CHAPTER

13

EXAMINATION

13.1 OVERVIEW OF EXAMINATION REQUIREMENTS

ASME B31.3 requires that examination of the piping be performed by the piping Manufacturer (for components only), Fabricator, and/or Erector as a quality control function. These examinations include radiography, for which ultrasonic examination or in-progress examination can be used as substitutes, visual examination, and, under certain conditions, magnetic particle or liquid penetrant examination.

Examination personnel are required to be qualified and certified for the method to be utilized following a procedure as described in ASME BPVC, Section V, Article 1, T-120(e) or (f). Section V requires that qualification of NDE personnel be in accordance with their employer's written practice, and that the written practice be in accordance with either

1. SNT-TC-1A, *Personnel Qualification and Certification in Nondestructive Testing*, as amended by Section V; or
2. ANSI/ASNT CP-189, *ASNT Standard for Qualification and Certification of Nondestructive Testing Personnel*, as amended by Section V.

National or international central certification programs, such as that described in ISO 9712 may be used to fulfill the training, experience, and examination requirements of these two documents.The owner's Inspector oversees the work performed by the Examiner. It is the Inspector's responsibility to verify that all the required examinations have been completed and to inspect the piping to the extent necessary to be satisfied that it complies with all of the applicable examination requirements of the Code and of the engineering design. Note that the process of inspection does not relieve the Manufacturer, Fabricator, or Erector of their responsibilities for complying with the Code.

The owner's Inspector is also required to be qualified to perform the work. The inspector may be qualified by

1. having not less than 10 years of experience in the design, fabrication, or inspection of industrial pressure piping, with each 20% of completed work toward an engineering degree recognized by the Accreditation Board for Engineering and Technology counting as 1 year of experience, up to a maximum of 5 years.
2. having a professional engineering registration or nationally recognized equivalent with at least 5 years of experience in the design, fabrication, or examination of industrial pressure piping.
3. being a certified welding inspector or a senior certified welding inspector as defined in AWS QC1, *Specification for AWS Certification of Welding Inspectors*, or nationally recognized equivalent with at least 5 years of experience in the design, fabrication, or examination of industrial pressure piping.
4. being an authorized piping inspector as defined in API 570, *Piping Inspection Code: In-service Inspection, Rating, Repair, and Alteration of Piping Systems*, with at least 5 years of experience in the design, fabrication, or examination of industrial pressure piping.

The owner's Inspector may be an employee of the owner, of an engineering or scientific organization, or of a recognized insurance or inspection company acting as the owner's agent. Unless the owner is the Manufacturer, Fabricator, or Erector, the Inspector cannot be an employee of those organizations. To be so would of course create a conflict of interest.

Requirements for the examination processes are described in ASME BPVC, Section V, with limited exceptions and additions. The required degree of examination and the acceptance criteria for the examinations are provided in Chapter VI of ASME B31.3. Examiners are required to have a written procedure for their examination work (para. 343).

It is not the intent of ASME B31.3 that the examination ensure that the constructed piping system will be free of defects, even ones that are rejectable if found. Nor is it required that welds that are not examined be free of rejectable defects. Instead, the examination work is part of an overall quality assurance procedure. The leak test is used to test the overall system. Interpretation 5-18, issued on May 5, 1987, addresses this point:

"Question (2): Assuming the Engineering Design requires fabrication to ANSI/ASME B31.3, requiring the minimum 5% visual examination and random radiography of girth butt welds, what is the minimum weld quality required by the Code for butt welds which have not been individually examined, but were represented by a successful random visual and radiographic examination?"

"Reply (2): The Code accepts such welds subject to leak testing. Note: The Code assumes that the 5% which were required to be examined will be representative of the entire lot, but a guarantee of such is a contractual matter to be specified by the engineering design."

If a piping system passes the examination requirements of the Code, it is accepted (complies with the Code) if it passes the leak test. If additional examinations are performed that reveal defects, it is beyond Code requirements; the system already complied with the Code, and the issues of whether to repair these defects and who is to pay for the repair is purely a contractual issue. The defect can be left in and the piping system will still comply with the Code. Of course, it would be judicious to at least perform a fitness-for-service evaluation of the defect if it is intended to be left unrepaired.

Records of the examination procedures and examination personnel qualifications are required to be retained for at least 5 years, unless otherwise specified in the engineering design.

13.2 PROGRESSIVE EXAMINATION

ASME B31.3 includes the concept of progressive examination. The concept applies to all random examinations, but will be discussed here in the context of random radiography. In such examinations, the items to be examined (e.g., welds to be radiographed) are separated into lots. This separation is normally done in terms of time periods, with some number of welds included in each lot as the fabrication/erection continues. Some percentage (e.g., 5% of girth welds for Normal Fluid Service) of the items are selected at random and examined. For each item found to be defective, two more are selected from the same lot and given the same type of examination. These two additional samples must be of the same kind (if welded or bonded joints, by the same welder, bonder, or operator). For these additional items, for each one that is found to be defective, two additional items of the same kind must be examined. At that point, if any defects are found, the entire lot that is being randomly examined must be 100% examined, repaired, redone, or replaced.

If, at any of the above steps, no further defective work was found, the entire lot would be accepted contingent on successfully passing the leak test. The defects that were found must be repaired and reexamined. Note that once a weld is found to be defective, it must be repaired and reexamined until it passes the examination. However, defects found in the repaired weld do not count as additional defects in the progressive examination procedure.

It would generally be a poor decision, except for rather small projects, to include all of the work in one lot or to wait until the end of the construction to perform the random radiography. One possible result Is that all of the welds would have to be examined and defective welds repaired or redone. In addition, performing the examinations as the work progresses can help eliminate poorly performing welders earlier in the project. Interpretation 2-25 addresses lot sizes.

> "Question (1): A certain piping system has 100 welds made by the same welder and welding procedure. Ten percent radiography was required. Ten welds were radiographed, of which one was defective. The subsequent second and third progression groups also had defective welds. Does ASME B31.3 require radiographic examination, repair, and radiographic reexamination of all the welds other than the 9 found acceptable or of the 10 welds represented by the one original defective weld?"
>
> "Reply (1): The first alternative is correct if the lot size has been established as 100 and the selection is truly random. The second is correct if the lot size has been established as 10 and each is represented by 1 of the 10 welds radiographed at random."

For welds, the work of each welder or welding operator must be included in the random examination. This may affect the selected lot size and/or the percentage of the welds that must be examined. For example, if six welders worked on a lot of 100 welds, with only one welder working on any given weld, a minimum of 6 welds would have to be examined for Normal Fluid Service, or 6%. If there were five or fewer welders, only 5 welds would have to be examined.

In order to avoid disputes during the work, the quantity or extent of a designated lot should be established by agreement between the contracting parties before the start of work. More than one kind of designated lot may be established for different kinds of piping work. PFI ES-48, *Random Examination*, suggests that these different types of lots be considered:

- By daily production
- By welder
- By truck load
- By all welds on a job
- By spool grouping
- By piping system
- By all welds in an easily defined space

Many owner's specifications require that the lots be selected by the welder, sometimes specifying lots be for each welder for each week of work. This approach simplifies selection of welds to be examined and avoids the potential for one welder's work resulting in extra examination of other welder's work in the same lot. Using other types of lots may be better depending on how the work is being done. For example, defining the lot by the truck load may be a good choice for shop fabricated spools being shipped to the jobsite by trucks.

Random examination will not detect every defect. It is a quality control procedure that is intended to produce an appropriate level of quality in welds. Part of the effectiveness of the procedure derives from the fact that the welders know that some of their welds will be radiographed, but do not know which ones will be examined.

Because there have been a number of inquiries, 341.3.4 (f) was added. This did not add a new requirement, but was an addition to make the requirements of the Code very clear. If a defective item is found in progressive examination, it is required to be replaced or repaired, and then re-examined. At the point of re-examination, it is no longer part of the progressive examination sampling process. The defective item is repaired or replaced and re-examined until it is found to be acceptable. Further defects found in the repaired or replaced item do not count as additional defects in the progressive examination process, and further examinations of items that are repaired or replaced do not count towards the random examination requirement.

13.3 TYPES OF EXAMINATION

Visual examination (VT) uses the unaided eye (except for corrective lenses) to examine the exterior and readily accessible internal surface areas of piping assemblies or components. It does not include nor require remote examination, such as by use of horoscopes. Visual examination is used to check materials and components for conformance to specifications and freedom from defects; fabrication including welds; assembly of threaded, bolted, and other joints; piping during erection; and piping after erection. Furthermore, visual examination can be substituted for radiography, as described later, which is called in-process examination. Requirements for visual examination are provided in ASME BPVC, Section V, Article 9. Records of visual examinations are not required other than those of in-process examination.

Radiographic examination (RT) uses X-ray or gamma-ray radiation to produce a picture of the subject part, including subsurface features, on radiographic film for subsequent interpretation. It is a volumetric examination procedure that provides a means of detecting defects that are not observable on the surface of the material. Requirements for radiographic examination of welds are provided in ASME BPVC, Section V, Article 2.

Ultrasonic examination (UT) detects defects using high-frequency sound impulses. The defects are detected by the reflection of sound waves. Ultrasonic examination is also a volumetric examination method, which can be used to detect subsurface defects. It can be used as an alternative to radiography for weld examination. The requirements for ultrasonic examination of welds are provided in ASME BPVC, Section V, Article 4, with an alternative for basic calibration blocks provided in para. 344.6 of ASME B31.3.

The acceptance criteria for ultrasonic examination in para. 344.6.2 are based on the amplitude of the reflected sound wave. This type of criteria is suitable for pulse-echo type ultrasonic examination. While more advanced techniques, such as time of flight diffraction are permitted by the Code, via its reference to Section V for examination methods, appropriate acceptance criteria were not provided. Ultrasonic methods that actually size the indication require an acceptance criteria based on the dimensions of the indication.

Appendix R was added to provide specific rules to facilitate the use of these more advanced ultrasonic techniques. Appendix R also requires that the examination be automated and demonstrated on a qualification block(s). Automated ultrasonic examination has a much greater probability of detection than manual UT.

In-process examination involves the visual examination of the entire joining process, as described in para. 344.7 of ASME B31.3. It is applicable to welding and brazing for metals and bonding for nonmetals. Since radiographic examination is not considered to provide useful results in brazing and bonding, in-process examination is used for these instead of radiography. For welding, it is permitted as a substitute for radiographic examination if specified in the engineering design or specifically authorized by the Inspector. In-process examination, however, is not as effective a quality control procedure as random radiography, and it should only be used for welds when special circumstances warrant.

Liquid-penetrant examination (PT) detects surface defects by spreading a liquid dye penetrant on the surface, removing the dye after sufficient time has passed for the dye to penetrate into any surface defect, and applying a thin coat of developer to the surface, which draws the dye from defects. The defects are observable by the contrast between the color of the dye penetrant and the color of the developer. Liquid-penetrant examination is used for the detection of surface defects, for the examination of socket welds and branch connections in severe cyclic service that cannot be radiographed, and for the examination of all welds, including structural attachment welds, that are not radiographed when the alternative leak test (ASME B31.3, para. 345.9) is used. In addition, liquid-penetrant examination of metallic bellows is required by Appendix X, para. X302.2.2. The requirements for liquid-penetrant examination of welds and components other than castings are provided in ASME BPVC, Section V, Article 6.

Magnetic-particle examination (MT) employs either electric coils wound around the part or prods to create a magnetic field. A magnetic powder is applied to the surface, and defects are revealed by patterns the powder forms in response to the magnetic field disturbances caused by defects. This technique reveals surface and shallow subsurface defects. As such, it can provide more information than liquid-penetrant examination. However, its use is limited to magnetic materials. Magnetic-particle examination is an alternative to liquid-penetrant examination wherever such an examination is required in ASME B31.3 (except in the case of metallic bellows). The requirements for magnetic-particle examination of welds and components other than castings are provided in ASME BPVC, Section V, Article 7.

Hardness testing is only required by the Code for certain materials via Note 60 in Appendix A. It may be required as a supplementary requirement for certain fluid services especially following heat treatment.

In addition to the above, ASME B31.3 permits increasing the quality factor for castings with supplementary examination of the casting. Liquid-penetrant, magnetic-particle, ultrasonic, and radiographic examinations may be used in these supplementary examinations. In this case, the requirements for the examination and acceptance criteria are stated and/or referenced in para. 302.3.3 of ASME B31.3.

13.4 REQUIRED EXAMINATION

The required examination depends on the fluid service. Different degrees of examination are required for Category D, Normal, Elevated Temperature, Category M, High Pressure, and High Purity Fluid Services. More examination is required for more hazardous services. In addition, piping under severe cyclic conditions and piping that will be subjected to an alternative leak test require even more examination. Table 13.1 summarizes the required examination for metallic piping except that examination requirements for High Pressure Fluid Service are described in Chapter 17 and examination requirements for High Purity Fluid Service are described in Chapter 18.

13.5 ACCEPTANCE CRITERIA

The acceptance criteria for the various examination techniques are provided as follows:

- For RT and VT, in Table 341.3.2
- For MT, in para. 344.3.2
- For PT, in para. 344.4.2
- For UT, in para. 344.6.2

Table 341.3.2 differentiates between Normal Fluid Service, severe cyclic conditions, and Category D Fluid Service. The acceptance criteria for Category M Fluid Service is the same as for Normal Fluid Service. There is no such differentiation in the acceptance criteria for MT, PT and UT examination.

The difference between the acceptance criteria for severe cyclic conditions and Normal Fluid Service relates to the potential for fatigue failure. For example, although some undercutting is permitted for Normal Fluid Service, none is permitted for severe cyclic conditions.

13.6 EXAMINATION OF NONMETALLICS

Nonmetallic piping systems have the same visual examination requirements as the base Code, as follows:

- VT requirements for normal fluid service follow the base Code.
- VT requirements for Category D Fluid Service follow the base Code.
- VT requirements for Category M Fluid Service follow the Chapter VII requirements for Normal Fluid Service plus an additional requirement for 100% visual examination of all fabrication as well as all bolted and mechanical joints.

Because radiography is not a generally accepted practice for nonmetallics, 5% in-process examination is substituted for 5% random radiography for Normal Fluid Service.

TABLE 13.1

REQUIRED EXAMINATION FOR MOST METALLIC PIPING

	Normal	Severe Cyclic Conditions [Note (1)]	Elevated Temperature	Category D	Category M
Materials and components [Note (2)]	Random VT	Random VT	Random VT	VT to the extent necessary [Note (11)]	Random VT
Fabrication [Note (3)]	Min. 5% VT	100% VT	100% VT		100% VT
Welds made by the fabricator	100% VT	100% VT	100% VT		100% VT
Bolted, threaded and other mechanical joints [Note (5)]	Random VT [Note (13)]	100% VT	100% VT		100% VT
Longitudinal welds [Note (4)]	100% VT	100% VT	100% VT [Note (8)]		100% VT
During erection [Note (6)]	Random VT	Random VT	Random VT		Random VT
Erected piping [Note (7)]	VT	100% VT	100% VT		VT
Circumferential butt and miter groove welds	Min. 5% random RT or UT[Note (14)]	100% RT [Note (12)]	Min. 5% random RT or UT [Note (14)]		100% RT or UT [Note (12)]
Socket welds	NA	100% MT or PT	100% MT or PT for P-No. 4 & P-No. 5 materials	NA	NA
Non-radiographed branch connections [Note (9)]	NA	100% MT or PT	100% MT or PT for P-No. 4 & P-No. 5 materials	NA	NA
Structural attachments	Note (10)	Note (10)	Note (10)	NA	Note (10)
Brazed joints	5% IP	NP	NP	VT to the extent necessary [Note (11)]	NP
Solder joints	NP	NP	NP		NP

LEGEND:

IP = in-process examination
MT =magnetic-particle examination
NA = no additional requirements (beyond otherwise specified VT)
NP = not permitted
PT = liquid-penetrant examination
RT = radiographic examination (UT is always an acceptable alternative and IP is an acceptable alternative if specified in the engineering design or approved by the inspector)
UT = ultrasonic examination
VT = visual examination

NOTES:

1. "Severe cyclic conditions" requirements apply only for piping that is in Normal or Elevated Temperature Fluid Service. There are no provisions in ASME B31.3 for severe cyclic conditions in the other fluid services.
2. Enough materials and components must be examined to satisfy the examiner that they conform to specifications and are free from defects.
3. Fabrication includes cutting, threading, grooving, forming, bending, and joining of components into subassemblies.
4. Welds made by the fabricator. Welds in components made in accordance with a listed specification are not required to be examined.
5. Assembly of threaded, bolted, and other mechanical joints such as grooved joints.
6. Examination during erection of piping, including checking alignment, supports, and cold spring.
7. Examination of erected piping for evidence of defects that would require repair or replacement, and for other evident deviations from the intent of the design.
8. Welds in P-No. 4 and P-No. 5 materials require 100% RT or UT.
9. Branch connection welds that are not radiographed.
10. 100% PT or MT is required if the alternative leak test is to be performed.
11. Visual examination to the extent necessary to satisfy the examiner that components, materials, and workmanship conform to the requirements of ASME B31.3 and the engineering design.
12. In-progress examination, if used, must be supplemented by other nondestructive examination.
13. 100% VT is required if pneumatic testing is to be performed.
14. 100% RT or UT is required if the alternative leak test is to be performed.

CHAPTER

14

LEAK TESTING

14.1 OVERVIEW OF LEAK TEST REQUIREMENTS

While the exercise of pressurizing a piping system and checking for leaks is sometimes called pressure testing, the Code refers to it as leak testing. The main purpose of the test is to demonstrate that the piping can confine fluid without leaking. When the piping is leak tested at pressures above the design pressure, the test also demonstrates that the piping is strong enough to withstand the pressure. For large bore piping where the pipe wall thickness is close to the minimum required by the Code, being strong enough to withstand the pressure is an important test. For small bore piping that typically has a significant amount of extra pipe wall thickness, being strong enough is not in question. Making sure that the piping is leak free is important for all piping systems.

ASME B31.3 requires leak testing of all piping systems. For piping in Category D Fluid Service, the piping may (at the Owner's option) be put in service and examined for leakage during the initial operation of the system. This is an initial service leak test. For all other piping, the following options are available:

- Hydrostatic test
- Pneumatic test
- Hydropneumatic test
- Alternative leak test

An alternative leak test is only permitted by ASME B31.3 when:

- Exposure of the piping to water via a hydrostatic test would damage the linings or internal insulation, or contaminate a process that would be hazardous, corrosive, or inoperative in the presence of moisture; or
- A hydrostatic test would require significant support modifications for the hydrostatic test load; or
- A pneumatic test is considered by the owner to entail an unacceptable risk due to the potential release of stored energy in the system (the danger of a pneumatic test increases with the pressure and contained volume); or
- A hydrostatic test or pneumatic test would present the danger of brittle fracture due to low metal temperature during the test

The leak test is required to be conducted after any heat treatment has been completed. If repairs or additions are made following the leak test, the affected piping must be retested unless the repairs or additions are minor and the owner waives retest.

All joints and connections, except those previously tested, must be left uninsulated and exposed for the leak test. ASME B31.3 permits painting of the joints prior to the test except when the joint is to be sensitive leak-tested; however, paint can effectively seal small leaks to extremely high pressures. These small leaks can then, in fact, start leaking in service if the paint film fails, such as due to the presence of solvents in the line.

The pressure is required to be held for at least 10 minutes and as long as is required to completely examine the piping system. All joints and connections must be visually examined for leakage. There is no provision in ASME B31.3 for substituting a monitoring of pressure decay for the 100% visual examination of the pipe joints during the hydrotest. Intent interpretation 19-24 states that structural attachment welds are included in the joints requiring visual examination during leak testing.

Piping subassemblies may be tested either separately or as assembled piping. This permits, for example, testing subassemblies and insulating them prior to leak testing the entire system. Interpretation 20-11 states that a piping component may be considered to be a piping subassembly.

Prior to the 2008 edition, para. 345.2.3(b) provided that flanged joints at which a blank is inserted to isolate other equipment need not be leak tested. However, in practice, there are other circumstances when the flanged joint cannot be included in the leak test. A very common occurrence is when temporary spools are included instead of control valves until after the hydrostatic testing and flushing is complete. In the 2008 edition, this paragraph was revised to state that flanged joints at which a blank or blind is used to isolate equipment or other piping during a test need not be leak tested.

Test records are required, per para. 345.2.7 of ASME B31.3, but need not be retained after completion of the test if a certification by the Inspector that the piping has satisfactorily passed leak testing as required by ASME B31.3 is retained.

In attempts to make the leak testing rules more flexible without compromising safety, the ASME B31.3 Section Committee added some provisions since the 2008 edition:

1. Lines open to the atmosphere, such as vents or drains downstream of the last shutoff valve, need not be leak tested. Relief valve tail pipes are another example of piping that need not be leak tested.
2. Threaded joints, tubing joints, or a combination of these joints used to connect instruments to previously leak tested piping need not be leak tested. This provision allows removal of pressure gages for the leak test and replacing them without leak testing the newly made joint.
3. Piping systems designed for vacuum service only may be subjected to a vacuum leak test method, technique, and acceptance criteria specified by the owner.
4. As an exception to the requirement that all joints be exposed for the leak test, the Code permits the owner to allow, joints in Category D Fluid Service that are subject to a hydrostatic leak test or an initial service leak test to be insulated and have protective weather sheathing installed prior to leak testing. The Code goes on to say that consideration shall be given to increasing the test period to allow time for possible leakage to pass through the insulation and weather sheathing.

14.2 HYDROSTATIC TEST

A hydrostatic test is generally the preferred alternative, because it is conducted at a higher pressure, which has beneficial effects, such as crack blunting and warm pre-stressing, and entails substantially less risk of injury or facility damage than a pneumatic test. The hydrostatic test reduces the risk of crack growth and brittle fracture when the pipe is placed in service. The test is generally conducted at a pressure, P_T, of 1.5 times the design pressure, P, times a temperature correction factor. The temperature correction factor compensates for the fact that the test may be conducted at a lower temperature, where the material has a higher strength than at the design condition. The equation is[19]

$$P_T = 1.5\ PS_T / S \tag{14.1}$$

where

P = internal design gage pressure

P_T = minimum test gage pressure

[19] ASME B31.3, Eq. (24).

S = allowable stress at component design temperature for the prevalent pipe material
S_T = allowable stress at test temperature for the prevalent pipe material

If the test pressure determined in the above equation results in a nominal pressure stress or sustained stress in excess of the yield strength of the material at the test temperature or a pressure more than 1.5 times the component rating at the test temperature, the test pressure may be reduced to a value that does not exceed those limits.

If the piping leak test is combined with a pressure vessel, it is also permitted to reduce the test pressure if it is necessary to do so to avoid overpressuring the vessel when it is not practicable to isolate the piping from the vessel for the purpose of testing them separately and if the owner approves. In that case, the test pressure may be reduced to the vessel test pressure, but must be at least 77% of the hydrotest pressure required by Eq. (14.1).

Any spring supports should generally be left with their travel stops in place through the hydrotest. Also, if the line normally contains vapor or a fluid with a lower density than water, the need for supplemental temporary supports must be considered due to the higher fluid weight than in normal operation.

14.3 PNEUMATIC TEST

A pneumatic test is considered to potentially entail a significant hazard due to the amount of stored energy in the compressed gas. A rupture could result in an explosive release of this energy. For example, an explosion of 60 m (200 ft) of NPS 36 line containing air at 3500 kPa (500 psi) can create a blast wave roughly equivalent to 35 kg (80 lb) of TNT. The hazard is proportional to both the volume and the pressure.

Because of this concern, the pneumatic test pressure is specified to be in the range of 1.1 to 1.33 times the design pressure and not corrected for temperature. Further, specific precautions are required. These include provision of an adequate pressure relief device to prevent pressurizing the line to too high a pressure. An intermediate hold at the lesser of one-half of the test pressure or 170 kPa (25 psi) with a visual examinination of all joints is required prior to bringing the pressure gradually up to the full test pressure.

Specific guidance regarding precautions to take for pneumatic testing can be found in ASME PCC-2, *Repair of Pressure Equipment and Piping*, Article 5.1.

14.4 HYDROPNEUMATIC TEST

A hydropneumatic test entails having part of the system filled with water and part of the system filled with pressurized air. Basically, the portions of the system filled with water can be tested at a higher pressure than would be permitted for a pneumatic test. The portions of the system filled with compressed gas are subjected to the limitations and requirements for a pneumatic test.

14.5 ALTERNATIVE LEAK TEST

An alternative leak test is permitted, with the owner's approval, when neither a hydrostatic nor a pneumatic leak test would be possible or safe. An alternative leak test contains the following required elements:

- A detailed examinination of all weld joints, including 100% radiographic or ultrasonic examination of all circumferential and longitudinal seam welds made by the fabricator and liquid-penetrant or magnetic-particle examination of all welds, including structural attachment welds, that cannot be radiographed
- A formal flexibility analysis
- A sensitive leak test

14.6 SENSITIVE LEAK TEST

A sensitive leak test is a required part of an alternative leak test and is also required for Category M Fluid Service piping. This is a test performed at low pressures to test for leaks. Methods for performing this type of test are described in ASME BPVC, Section V, Article 10. One approach is to pressurize the piping system with air and check for leaks using a bubble test solution.

The test pressure for the sensitive leak test is required to be at least 105 kPa (15 psi) or 25% of the design pressure, whichever is less. Similar to a pneumatic test, the pressure is to be increased gradually up to an intermediate hold at the lesser of one-half of the test pressure or 170 kPa (25 psi), at which point a preliminary check for leaks must be made. The pressure is then gradually increased to the test pressure, at which point the sensitive leak test is performed.

In addition to the above Code-required circumstances, a sensitive leak test is used by some owners as a supplemental tightness check of other piping systems prior to placing them in operation and/or to check for leaks in piping systems assembled from shop-fabricated and tested subassemblies.

14.7 JACKETED AND VACUUM PIPING

Piping that is designed for external pressure conditions is generally tested with internal pressure. The test pressure is required to be the greater of 1.5 times the design differential pressure or 105 kPa (15 psi). As an alternative, piping systems designed for vacuum service only may be leak tested using a vacuum leak test for which the technique and acceptance criteria are specified by the owner.

Jacketed or other double-wall piping requires leak testing of both the inner pipe and the jacket. The test of the inner pipe is based on the more severe of the internal pressure or the jacket pressure condition. The inner pipe welds must be visible for this leak test, which creates a very significant consideration in the design of these systems. The jacket is tested as normal piping based on its design pressure, unless otherwise specified in the engineering design. The inner pipe is normally designed to carry the jacket pressure with no internal pipe pressure, without buckling.

The requirement that all joints be visually observed during a leak test can be problematic for jacketed or double containment piping. It generally requires the jacket piping to be only partly fabricated when the inner pipe is being leak tested. Code Case 180 provides some relief in the form of alternative rules. For welded joints and connections on the inner pipe, it is not required to visually observe the joints during the leak test if 1) the leak test is otherwise conducted in accordance with code rules and 2) an additional sensitive leak test is conducted to demonstrate the leak tightness of the joints that were not visually examined during the leak test of 1).

14.8 INITIAL SERVICE LEAK TEST

For piping in Category D Fluid Service, ASME B31.3 permits an initial service leak test in lieu of other leak tests, such as hydrostatic or pneumatic tests. In this test, the system is pressurized to operating pressure with the process fluid and the joints are examinined for leaks.

If the process fluid is a vapor, provisions for gradual increase in pressure, with an examinination at an intermediate pressure, apply, similar to pneumatic test requirements. For piping filled with liquid, ASME B31.3 requires that the pressure be gradually increased. All joints and connections must be observed for leakage.

14.9 CLOSURE WELDS

In the 1996 edition, addenda c (1998), closure welds were added [para. 345.2.3(c)] as an acceptable exemption from leak testing. A closure weld is a final weld connection piping systems or components which have been successfully leak tested. The closure weld does not require leak testing if it passes 100% radiographic or ultrasonic examination and is in-process examined.

Closure welds are not used in the Code in the context of a connection to an existing pipe, since that weld is considered to be outside of the scope of ASME B31.3 (although you could look to the closure weld requirements for guidance). Rather, it is a connection between new components. For example, consider a large diameter vapor line, for which providing additional support for the fluid weight in a hydrotest is impractical, and for which a pneumatic test would entail undue hazard. The line could be leak tested, as a subassembly, at grade, erected, and connected to equipment with a closure weld.

14.10 REQUIREMENTS FOR NONMETALLIC PIPING

The leak test rules in the base Code, described in the prior paragraphs, are generally applicable to non-metallic piping, with a few exceptions. The hydrotest pressure for nonmetallics other than thermoplastics (e.g., RTR, fiberglass pipe) and metallic piping lined with nonmetals is 1.5 times the design pressure, but not more than 1.5 times the maximum rated pressure of the lowest rated component in the system. There is no temperature correction factor. It is particularly important not to overpressure fiberglass piping systems. Excessive hydrotest pressures in fiberglass systems have caused subsequent failures in service. The overload condition can damage the material without evidence of a leak during the test itself. This damage has led to subsequent failures in service.

For thermoplastic piping, a temperature correction factor is used, the same as for metallic piping, except that the allowable stresses for thermoplastics in Appendix B of ASME B31.3 are used.

For metallic piping lined with nonmetals, the Code test pressures for the metallic portion of the pipe apply.

Some thermoplastics can behave in a brittle manner when they fail under compressed gas service. The Code specifically prohibits polyvinyl chloride (PVC) and chlorinated polyvinyl chloride (CPVC) from pneumatic testing, because of this hazard. The owner's approval is required for pneumatic testing of any other non-metallic piping system.

The alternative leak test is not permitted for nonmetallic piping and piping lined with nonmetals.

CHAPTER

15

NONMETALLIC PIPING

15.1 ORGANIZATION

The rules for nonmetallic piping and piping lined with nonmetals are given in Chapter VII of ASME B31.3. The paragraphs in that chapter follow the same numbering as the base Code, Chapters I through VII, but start with the letter A. If requirements located elsewhere in the Code apply to these piping systems, they are referenced by a paragraph in Chapter VII.

The behavior of nonmetallic piping is more complex than that of metallic piping, and the design criteria are significantly less well developed. As a result, designers are left to use their best judgment in many circumstances. For example, although a formal flexibility analysis is required, no methods are provided for performing one.

For metallic piping lined with nonmetals, the base Code requirements for metallic piping are applied to the metallic portion. Supplemental rules are provided for the nonmetallic lining. Background on the Chapter VII rules is provided by Short (1989, 1992).

15.2 DESIGN CONDITIONS

The base Code requirements with respect to design pressure and temperature are generally applicable. However, no credit is permitted for ambient cooling for uninsulated components. In other words, the design temperature is not permitted to be less than the fluid temperature.

15.3 ALLOWABLE STRESS

Various nonmetals have different established methods of determining allowable stresses. Some limited allowable stress values are provided in ASME B31.3, Appendix B for thermoplastic and reinforced thermosetting resin pipe. For the most part, allowable stresses or pressure ratings must be determined from tests performed by the manufacturer. Allowable pressures are provided for reinforced concrete pipe in Table B-4 and borosilicate glass pipe in Table B-5.

The methods of determining the allowable stresses in thermoplastics and reinforced thermosetting resins are provided in ASTM specifications, as follows.

a. *Thermoplastic:* Hydrostatic design stress (HDS) is determined in accordance with ASTM D2837. Note that the strength is determined based on time-dependent properties (long-term tests extrapolated to longer design times 100,000 hours and 50 years). This is because creep is significant for these materials even at ambient temperature. In addition, the strength drops significantly with even a modest increase in temperature.

b. *Reinforced Thermosetting Resin (Laminated):* Design stress (DS) is taken as one-tenth of the minimum tensile strengths specified in Table 1 of ASTM C582. This is also called hand layup or contact molding. The strength of RTR is not particularly temperature-sensitive in the range of application, so this allowable stress is considered to be valid from –29°C (–20°F) through 82°C (180°F).
c. *Reinforced Thermosetting Resin and Reinforced Plastic Mortar (Filament-Wound and Centrifugally Cast):* The hydrostatic design stress (HDS) used in design is the hydrostatic design basis stress (HDBS) times a service factor, F (often taken as 0.5), which is selected in accordance with ASTM D2992. The HDBS is determined in accordance with ASTM D2992. The Code states that the HDBS is considered to be valid only at 23°C (73°F). However, the strength of reinforced thermosetting resin piping is relatively temperature-insensitive within the normal range of use temperatures. The HDBS is determined from long-term testing.

15.4 PRESSURE DESIGN

The philosophy of the base Code with respect to metallic piping applies to nonmetallic piping. The primary differences are that the table of listed components for nonmetallic piping is Table A326.1 rather than Table 326.1, and there are substantially fewer pressure design equations provided in the Code.

Listed components with established ratings are accepted at those ratings. Listed components without established ratings, but with allowable stresses listed, can be rated using the pressure design rules of para. A304; however, these are very limited. In the cases of listed components without allowable stresses or unlisted components, components must be rated per para. A304.7.2.

The variations permitted in the base Code (para. 302.2.4) are not permitted for nonmetallic piping. They are not permitted to exceed the maximum permissible pressure, even during relief, considering accumulation. They are permitted for metallic piping with nonmetallic lining, provided the lining material is established to be suitable for the increased conditions through prior successful service experience or tests under comparable conditions.

The equations that are available for sizing nonmetallic components are very limited in ASME B31.3. The equations are for straight pipe, flanges, and blind flanges. The use of the referenced flange design method (per ASME BPVC, Section VIII, Division 1, Appendix 2) is questionable for many non-metallics. As a result, for pressure design, most nonmetallic piping components must be either per a listed standard (i.e., listed in Table A326.1) or qualified per A304.7.2.

Paragraph A304.7.2, on pressure design of unlisted components and joints, differs from the base Code in that neither experimental stress analysis nor numerical analysis (e.g., finite-element analysis) are listed as an acceptable alternative for qualifying components. The two methods that are considered acceptable for substantiating the pressure design are (1) extensive successful service experience under comparable design conditions with similarly proportioned components made of the same or like material and (2) a performance test. The performance test must include the effects of time, because failure of nonmetallic components can be time-dependent.

For straight pipe, Eq. (26a) for thermoplastic, Eq. (26b) for RTR laminated pipe, and Eq. (26c) for RTR filament-wound and RPM centrifugally cast pipe are provided in ASME B31.3. They are of the form

$$t = \frac{PD}{2SF + P} \tag{15.1}$$

where

D = outside diameter of pipe
F = service factor, which is only used for filament-wound and centrifugally cast pipe
P = internal design gage pressure
S = design stress from applicable table in Appendix B of ASME B31.3
t = pressure design thickness

15.5 LIMITATIONS ON COMPONENTS AND JOINTS

Fluid service requirements for nonmetallic piping components are covered in ASME B31.3, Chapter VII, Part 3. Fluid service requirements for nonmetallic piping joints are covered in Chapter VII, Part 4. For the most part, the requirements are similar to the base Code requirements, with relevant paragraphs on non-metallic components and joints substituted for paragraphs on their metallic counterparts.

15.6 FLEXIBILITY AND SUPPORT

Rules regarding flexibility and support for nonmetallic piping are provided in ASME B31.3, Chapter VII, Part 5. ASME B31.3 does not provide detailed rules for evaluation of nonmetallic piping systems for thermal expansion. However, it requires a formal flexibility analysis when the following exemptions from formal flexibility analysis are not met:

- The system duplicates, or replaces without significant change, a system operating with a successful service record.
- The system can readily be judged adequate by comparison with previously analyzed systems.
- The system is laid out with a conservative margin of inherent flexibility, or employs joining methods or expansion joint devices, or a combination of these methods, in accordance with Manufacturer's instructions.

As in metallic piping, a formal analysis is not necessarily a computer analysis. It can be any appropriate method, including charts and simplified calculations. The objectives in the design of a piping system for thermal expansion are the same as for metallic piping systems. Specifically, they are to prevent:

- Failure of piping or supports from overstrain or fatigue
- Leakage at joints
- Detrimental stresses or distortion in piping or in connected equipment (e.g., pumps) resulting from excessive thrusts and moments in the piping (ASME B31.3, para. A319.1.1).

One of the significant differences from metallic systems is that fully restrained designs are commonly used, that is, systems where the thermal expansion is offset by elastic compression/extension of the piping between axial restraints. This is possible due to the relatively low elastic modulus of plastic piping. The resulting loads are generally reasonable for the design of structural anchors. Note, however, that in a computer flexibility analysis of such systems, the axial load component of thermal expansion stress must be included. See Section 8.1 for a discussion of stresses due to axial loads in flexibility analysis.

Other significant differences from metallic piping include the following:

a. Most RTR and RPM systems are nonisotropic. That is, the material properties are different in different directions, depending on the orientation of the reinforcing fibers.
b. Axial extension of the pipe due to internal pressure can be significant and should be considered. Note that filament-wound RTR pipe can either extend or contract due to internal pressure, depending on the orientation of the reinforcing fibers.
c. Plastic materials creep at ambient temperature. For example, a plastic pipe that is fully restrained and is compressed as it heats up can experience compressive creep strain during operation. When it cools back to ambient temperature, this can result in tension in the pipe and a load reversal on the restraints.
d. In plastic piping, particularly RTR systems, the limiting component is often a fitting or a joint. For such systems, the result of the flexibility analysis can be an evaluation of the loads versus the allowable loads on components, rather than a comparison of stress with allowable stress.
e. Material properties, even for nominally the same material, are often Manufacturer-specific. Thus, design of plastic systems generally requires interaction and consultation with the Manufacturer of the pipe and

information on the resin. This is particularly so for RTR and RPM piping, which also include the consideration of the fiber reinforcing.

f. Stress intensification factors have not been developed for nonmetallic piping. For many non-metallic components (RTR in particular), the design is manufacturer-specific, so the development of industry standard stress intensification factors is problematic.

In general, design of RTR and RPM piping considers the material to be brittle. Thus, there is essentially no difference between stresses due to thermal expansion and those due to weight or pressure. Except for filament wound and centrifugally cast RTR piping, an allowable stress that is comparable to that permitted for pressure is commonly used as an allowable value for the total weight plus pressure plus thermal expansion stress. For laminated RTR, the allowable pressure stress is one-tenth of the tensile strength, and a commonly used allowable value for longitudinal stress due to combined loads is one-fifth of the tensile strength. For filament wound and centrifugally cast RTR, the allowable stress in the longitudinal direction has to be determined based on the difference in physical properties from the circumferential direction.

Although the behavior of thermoplastics is generally not brittle, so that the shakedown concepts of metallic piping may be applicable, this technology is not developed. Thus, the allowable value for longitudinal stresses due to combined loads is often taken conservatively as the allowable stress for internal pressure.

Some methods have been developed for evaluation of plastic piping. An industry survey is provided in WRC 415 (Short et al, 1996).

With respect to support, ASME B31.3 highlights some specific concerns for nonmetallic piping in para. A321.5. In nonmetallic piping that has limited ductility (e.g., generally with RTR and always with borosili-cate glass), avoidance of point loads can be critical to system performance. Although local loads may be accommodated in ductile systems by local plastic deformation, such loads can result in brittle fracture in materials that are brittle. Another consideration is that deformation can accumulate over time due to creep. Support spacing must be sufficiently close to avoid excessive long-term sagging due to creep.

15.7 MATERIALS

Thermoplastic materials may be used for flammable fluid service when they are underground. Thermoplastics are permitted, as of the 2004 edition, in above ground flammable fluid service, provided all of the following are satisfied.

1. The size of the piping does not exceed DN 25 (NPS 1).
2. Owner's approval is obtained.
3. Safeguarding per Appendix G is provided.
4. The precautions of Appendix F, para. F 323.1(a) through F 323.1(c) are considered.

In any use other than Category D Fluid Service, thermoplastic piping is required to be safeguarded (see Section 2.5 for information on safeguarding). Poly(vinyl chloride) and chlorinated poly(vinyl chloride) are prohibited from compressed air or other compressed gas service due to the potential for brittle failure.

Reinforced plastic mortar (RPM) piping is required to be safeguarded when used in other than Category D Fluid Service. Reinforced thermosetting resin (RTR) piping may be used in toxic or flammable fluid service, but requires safeguarding. It is generally acceptable for other services, subject to suitability of the material.

Borosilicate glass and porcelain are brittle materials and therefore are required to be safeguarded against large, rapid temperature changes (i.e., thermal shock). Further, these materials are required to be safeguarded when used in toxic or flammable fluid services.

Note that the above limitations only apply when the material is used as the outer pressure-containing element. They do not apply to materials used as linings within metallic piping.

Recommended maximum and minimum temperatures are generally provided. However, similar to the requirements of the base Code, the maximum temperature may be exceeded (unless there is a specific prohibition) if the Designer verifies the serviceability of the material at the temperature. If a material is to be used at a temperature below the minimum temperature listed in ASME B31.3, Appendix B, the Designer must have some test results at or below the lowest use temperature that ensure that the materials and bonds will have adequate toughness and are suitable at the design minimum temperature. Unlike metallic materials, for nonmetallics, specific tests such as Charpy are not specified.

15.8 BONDING OF PLASTICS

One of the key elements to successful construction of a plastic piping system is the joints. ASME B31.3 requires a formal process of developing, documenting, and qualifying bonding procedures and personnel performing the bonding. The joints in RTR, RPM and thermoplastic piping are called bonds. The requirements are similar to the requirements for qualification of welds and welders.

The first step is to have a documented bonding procedure specification (BPS). The specification must document the procedures for making the joint, as set forth in para. A328.2.1. This procedure must be qualified by a bonding procedure qualification test. Once it is so qualified, it may be used by personnel to bond nonmetallic ASME B31.3 piping systems. Those bonders, however, must also be qualified to perform the work.

It is the responsibility of the employer of the bonder who will be joining plastic piping to have a BPS for the joining process that has been qualified by that employer by testing. Per ASME B31.3, para. A328.2.1(b), the BPS is required to specify at least the following:

- All materials and supplies (including storage requirements)
- Tools and fixtures (including proper care and handling)
- Environmental requirements (e.g., temperature, humidity, and methods of measurement)
- Joint preparation
- Dimensional requirements and tolerances
- Cure time
- Protection of work
- Tests and examinations other than those required by ASME B31.3, para. A328.2.5
- Acceptance criteria for the completed test assembly

The bonding procedure must be qualified by test, as described below, to be used by the organization for which the bonder works. The BPS and records of the BPS qualifications must be maintained by the employer of the bonders and be available for review by the Owner or Owner's agent and the Inspector.

The qualification test is the same for the bonding procedure and the bonder. The bonder must fabricate an assembly and pressure-test it. There are two options for the pressure test. The first is a burst test in accordance with ASTM D1599. This requires increasing the pressure uniformly at a rate that will result in failure in 60 seconds to 70 seconds. The acceptance criteria when using this burst test procedure to qualify the bond is that the failure must initiate outside of the joint (the joint must not be the weak link). The second option is a hydrostatic test with a duration of 1 hour with no leakage or separation of joints.

The hydrostatic test pressure depends on the material of construction. For thermoplastics, it is calculated based on the short-term (about 1 minute) and long-term (50 years) burst strengths. These are interpolated by ASME B31.3, Eq. (27) in a manner appropriate to evaluation of thermoplastic creep data to arrive at a 1-hour strength. This is used to test the pipe. Prior rules that used an arbitrary factor of three times the manufacturer's rated pressure did not work, because quite often the pipe would not satisfy this test, so the bond could not be tested. The factor for RTR and RPM remains at three times the manufacturer's allowable pressure for the components being joined. ASME B31.3, Eq. (27) is

$$P_T = 0.80\bar{T}\left(\frac{S_S + S_H}{D - \bar{T}}\right) \tag{15.2}$$

where

D = outside diameter of pipe

P_T = hydrostatic test pressure

S_H = mean long-term hydrostatic design strength (LTHS) in accordance with ASTM D2837; at the test temperature.

S_s = mean short-term burst stress in accordance with ASTM D1599; at the test temperature.

T = nominal thickness of pipe

To qualify the BPS, at least one of each joint type covered by the BPS must be included in the test(s). With respect to size, if the largest joint is DN 100 (NPS 4) or smaller, the test assembly is required to be the largest size to be joined. If the largest pipe to be joined is greater than DN 100 (NPS 4), the size needs to be DN 100 (NPS 4) or 25% of the largest pipe to be joined, whichever is greater.

The same as for welding, the employer of the bonder is responsible for performing the bonding procedure qualification test, qualifying bonders, and maintaining records of the specifications and test. Under certain circumstances as described in ASME B31.3, including approval of the (owner's) Inspector, use of a BPS qualified by others and bonders qualified by others is permitted. If a bonder or bonding operator has not used a specific bonding process for a period of 6 months, requalification is required.

Again, similar to welding, the bonds that are made are required to be identified with a symbol that indicates which joints are made by which bonder. As an alternative, appropriate records that provide this information may be used instead of a physical mark on each joint.

General requirements are provided in ASME B31.3, para. A328.5 for hot-gas-welded joints in thermoplastic piping, solvent-cemented joints in thermoplastic piping, heat-fusion joints in thermoplastic piping, electrofusion joints in thermoplastic piping, adhesive joints in RTR and RPM piping, and butt-and-wrapped joints in RTR and RPM piping. Descriptions of some bonding processes are provided below in Insert 15.1.

Insert 15.1 Bonding Processes

Thermoplastic Piping

A thermosetting plastic material is one that is polymerized (cured) by application of heat or chemical means and is thus changed into an infusible or insoluble final product. Fiberglass pipe is an example of a reinforced thermosetting resin (RTR) pipe. Unlike thermoplastic materials, they cannot be readily reshaped by heating them to their melting range once they are cured. The thermosetting plastic products will usually degrade once they are heated to such temperatures. The two types of thermosetting resin for piping are (1) RTR, which is composed of a fibrous reinforcement material such as glass or carbon fiber in a thermosetting plastic resin such as epoxy, polyester, furan. or phenolic, and (2) reinforced plastic mortar (RPM), which also includes a filler material such as sand. Bonding processes for thermoplastic pipe include adhesive joints and various types of butt-and-wrap joints.

Adhesive Joints

Principles of Operation

Adhesive joining for RTR piping uses adhesive to join the pipes (Fig. 15.1). The pipe joint usually has a bell-and-spigot or a tapered-joint configuration. Depending on the adhesive used and the piping materials, the joint may require curing by one of several methods. These include using chemical heat wraps, applying heat to the pipe joint by other means, or allowing the joint to cure naturally.

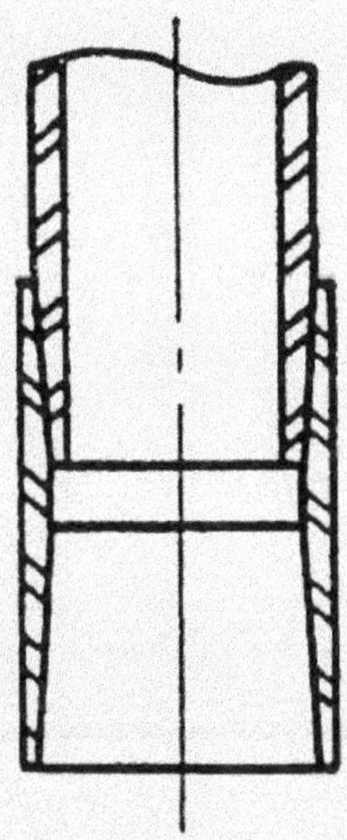

FIG. 15.1
FULLY TAPERED THERMOSETTING ADHESIVE JOINT (ASME B31.3, FIG. A328.5.6)

Process Limitations

The application of the adhesive to the surfaces to be joined and subsequent assembly of the surfaces should result in a continuous bond. These joints require the pipe to overlap in order to provide a sufficient area of glued surfaces. As a result, in assembling a piping system, the piping may need to have sufficient flexibility for the final pipe joint or joints to be made. Strength of the glued joint may be less than that of the pipe, although the strength of a properly made tapered joint can be as high as that of the pipe.

Butt-and-Wrap Joints

Principles of Operation

Butt-and-wrap pipe joints may be applied to a variety of RTR piping (Fig. 15.2). This joining technique can be applied to pipe joints by butting two pipes end to end or where the joints have a bell-and-spigot arrangement. In either case, layers of re sin-impregnated glass fiber cloth (or other reinforcing fiber consistent with the pipe) overwrap the pipe joint area. The cloths are applied in layers on the joint, building it up. Ideally, the finished joint should be capable of withstanding internal pressure, longitudinal force, and bending moments.

Cuts of the pipe should always be sealed so the fiber reinforcement can be protected from contacting the service fluid. Branch connections may be made using similar hand layup techniques.

Process Limitations

The strength of the pipe joint may be designed such that it equals or exceeds that of the pipe. However, the joints can also be a weak link in the system, with butt-and-wrap strength (using square-ended pipes butted together) is sometimes considered to be one-half the strength of the pipe.

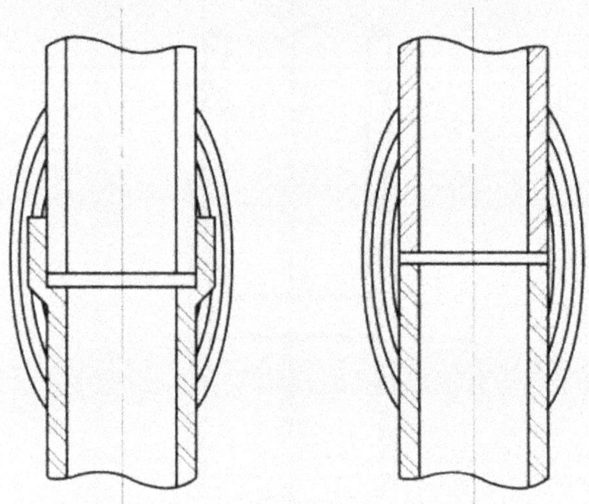

FIG. 15.2
THERMOSETTING WRAPPED JOINTS (ASME B31.3, FIG. A328.5.7)

Thermoplastic Piping

Thermoplastic piping is typically unreinforced and composed of thermoplastic material such as polyethylene (PE), high-density polyethylene (HDPE), polypropylene (PP), ethylene-chlorotrifluoroethylene (ECTFE), ethylene-tetrafluoroethylene (ETFE), poly(vinyl chloride) (PVC). chlorinated poly(vinyl chloride) (CPVC), acrylonitrile-butadiene-styrene (ABS), and poly (vinylidene fluoride) (PVDF). These materials soften when heated and can be shaped and fused in the heated state and cooled without degradation of the material. As a result, fusion bonding of thermoplastic is a possible and often preferred option for bonding.

Solvent-Cement Joining

Principles of Operation

Solvent-cement joints are used on some common types of thermoplastic piping materials, including PVC, CPVC, and ABS (Fig. 15.3). Joining methods are covered by ASTM standards. Usually, the pipe and socket fitting to be joined have a slight interference fit. Proper joining relies on softening and fusing of the pipe and fitting material and should result in a bond that is stronger than the pipe. Both parts of the joint to be made are first cleaned by wiping with an appropriate cleaning agent. The joint to be made is prepared by applying a suitable primer before the solvent cement is applied. The primers are described in an ASTM specification that applies to the material being solvent-cemented. Primer and cement are applied to both surfaces to be joined according to recommended procedures. A continuous bond must be produced, with a small bead of excess cement appearing on the outer limit of the joint, when the fully wetted male end is inserted into the fully wetted female socket.

Process Limitations

Some materials, such as PP, cannot be suitably joined using solvent-cement techniques. Further, a good solvent-cement joint must have a slight interference fit between the outside diameter of the pipe and the inside diameter of the mating socket. The diametrical clearance between the pipe and the entrance of the fitting socket should never exceed 0.04 in. With a large gap, heavier cement, which has a greater film thickness to bridge the gap, is used. However, such joints can end up relying on the shear strength of the glue rather than a bonding of the mating plastic parts, which is significantly weaker than proper solvent-cement joints with interference fits.

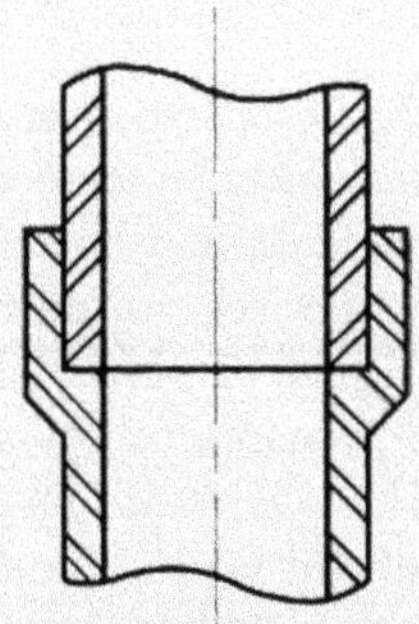

Socket Joint

FIG. 15.3
THERMOPLASTIC SOLVENT-CEMENTED JOINT (ASME B31.3, FIG. A328.5.3)

Hot Gas Welding

Principles of Operation

Hot gas welding is used exclusively for thermoplastic materials; it is not suitable for ther-mosetting resin materials (Fig. 15.4). It employs hot air or inert gas to melt the base material and tip of the welding rod. The melted welding rod fuses with the base material to form a weld joint. Although it is common practice to use partial-penetration welds, it is essential to use full-penetration welds to properly develop the joint strength, which will be weaker than the base material in any case.

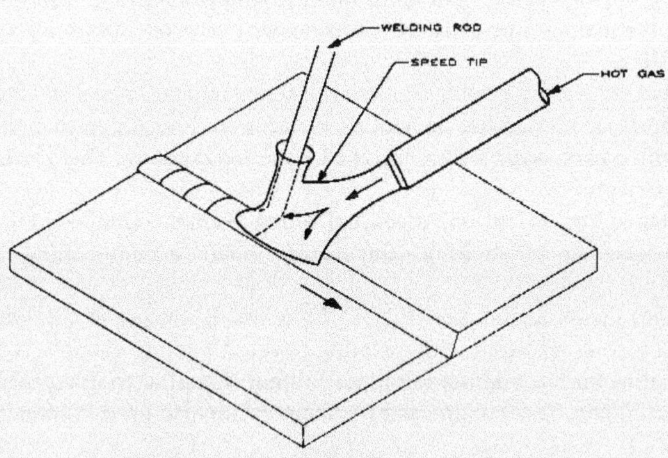

FIG. 15.4
HOT GAS WELDING

Extrusion welding is similar to hot gas welding except that the weldment is extruded from a screw-driven nozzle. The equipment required for extrusion welding is much larger than that for hot gas welding. The nozzle must be large enough to hold an extrusion screw and a motor must be attached to drive the screw. As a result, extrusion welding requires a great deal of space to conduct the welding; approximately 3 ft of clearance in all directions is the usual requirement.

The advantage of extrusion welding over hot gas welding is the application of a thick weld bead in a single pass (usually the equivalent to 5 to 12 passes of hot gas welding). The result is a substantially tower degree of residual stress and lesser effects of oxidation.

Process Limitations

Proper joining by this technique is more difficult for some materials, such as PVC, and should not be used for those materials except in special circumstances. In those circumstance, special precautions, such as use of an inert gas such as nitrogen for the hot gas, may be necessary.

Due to a number of factors, including imperfections, residual stress, and oxidation, the strength of hot-gas-welded joints and extrusion-welded joints is lower than that of the base material. According to German Standard DVS 2203.the long-term strength of a hot-gas-welded joint is 40% of that of the base material for HDPE, PP. PVC, and PVDF. The long-term strength of extrusion welds is, according to the same standard, 60% of that of the base material for HDPE and PP. The relatively common practice of making partial-penetration welds using hot gas welding techniques results in substantial further reduction of joint strength.

Heat Fusion

Principles of Operation

Heat fusion relies on using a suitable heating element to bring the mating surfaces up to the melting temperature and then forcing the two surfaces together to cause them to flow and fuse. It is the most common techniques for bonding all polyolefins, such as HDPE and PE, and fluo-ropolymers, s uch as PVDF, ECTFE. and ETFE. Available techniques include heat-element butt fusion, heat-element socket fusion, electrical resistance fusion, and branch fabrication techniques such as side-wall fusion.

In heat-element butt fusion, the mating pipe ends are forced against a heating element with a specific pressure to bring the mating surfaces up to melting temperature (Fig. 15.5). The heating element is removed and then the mating pipe ends are forced against each other with a specified pressure for a specified duration of time.

In heat-element socket fusion, a heating element is inserted into the socket and the pipe end is inserted into a heating element (Fig, 15.6). After the mating surfaces are brought up to melting temperature, the pipe is inserted into the socket, again with a specified force and duration. There is an interference fit and the pipe fuses to the socket.

In electric resistance socket fusion, electric heating element embedded in the socket fitting wall brings the surfaces up to melting temperature, again accomplishing the same objective (Fig. 15.7),

In sidewall (saddle) fusion techniques (Fig. 15.7), a hole is cut in the sidewall of the pipe with a bevel and the branch pipe is prepared with a mating bevel. A heating element is used to heat the end of the branch and mating surfaces in the run pipe. Joining is similar to other heat fusion techniques. Other heat fusion techniques based on the same principles are also used in branch and reinforcement bonding.

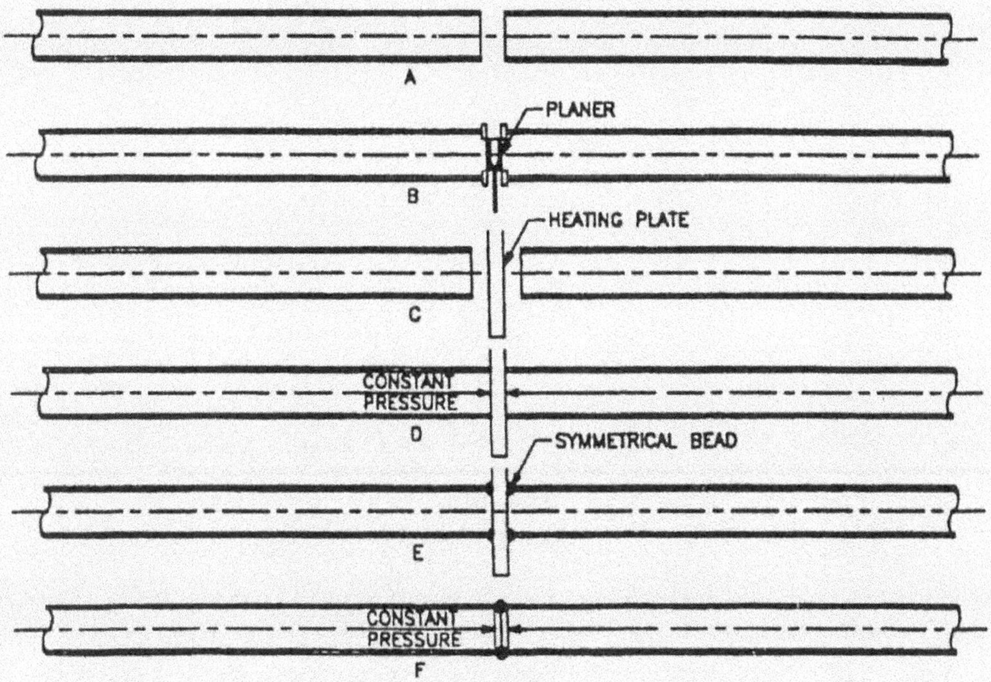

FIG. 15.5
STEPS FOR HEAT-ELEMENT BUTT FUSION (COURTESY OF CHRIS ZIU)

Process Limitations

Heat fusion provides superior joints to hot gas and extrusion welding. However, special equipment is required to heat the surfaces and draw the mating components together with the appropriate force (pressure on the surfaces being bonded). The equipment can be large, so sufficient working space is required. Also, the mating parts must be physically separated to provide space for the heating element and then pressed together. As such, sufficient pipe flexibility or other means must be provided to make the final joint in piping systems between anchors or other fixed points.

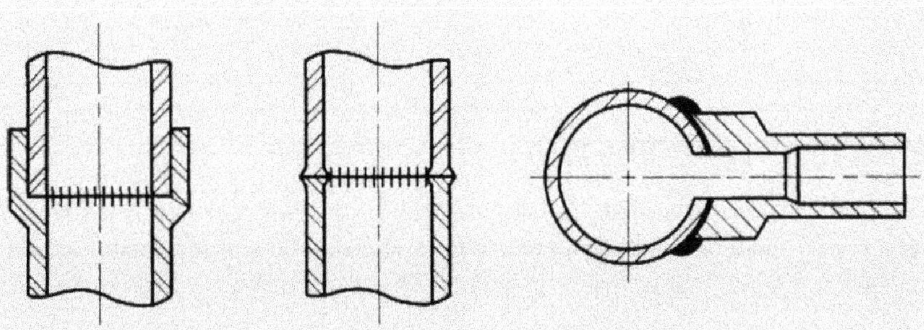

FIG. 15.6
THERMOPLASTIC HEAT FUSION JOINTS (ASME B31.3, FIG. A328.5.4)

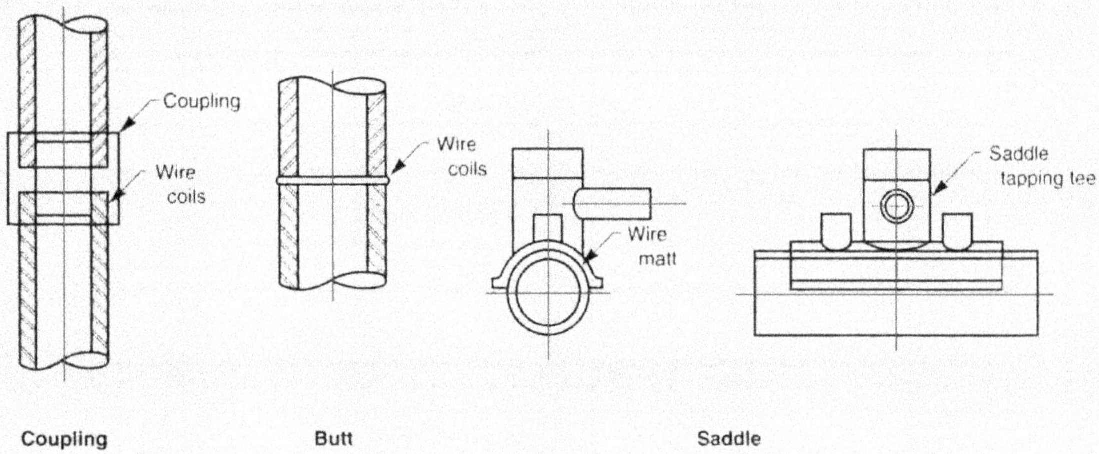

FIG. 15.7
THERMOPLASTIC ELECTROFUSION JOINTS (ASME B31.3, FIG. A328.5.5)

According to German Standard DVS 2203, the long-term strength of a properly made heat fusion butt joint is 80% of that of the base material for HDPE and PP and 60% of that for PVC and PVDF.

15.9 OTHER FABRICATION, ASSEMBLY, AND ERECTION REQUIREMENTS

Requirements for fabrication of piping lined with nonmetals (para. A329), joining of nonplastic piping (e.g., borosilicate glass) (para. A334), and assembly and erection of nonmetallic piping (para. A335) are provided in ASME B31.3. These are not comprehensive requirements, but instead they address some specific considerations that apply to nonmetallic piping in addition to the base Code requirements.

Welding of metallic piping lined with nonmetals generally follows the base Code requirements for welding of metallic piping. Precautions are provided in ASME B31.3, para. A329. For example, precautions are required to avoid damage to the nonmetallic lining. If such damage occurs, it must be repaired. Further, qualification of a welder or welding operator for a WPS for lined pipe is specific to the lining; a different qualification test is required for each lining material. This, of course, only applies to pipe that has already been lined, not welding of piping prior to lining it.

15.10 EXAMINATION AND TESTING

The nondestructive examination techniques for nonmetallic piping are not nearly as well developed as for metallic piping. As a result, the techniques that are used are visual and in-process examination. Specific examination requirements for nonmetallic piping are included with the discussion of examination for metallic piping in Section 13.6.

Leak testing requirements for nonmetallic piping are covered in Section 14.10.

CHAPTER 16

Category M Piping

16.1 WHEN TO USE THE RULES FOR CATEGORY M FLUID SERVICE

The rules in Chapter VIII of ASME B31.3 are used when the owner designates a piping system to be in Category M Fluid Service. The owner is guided in the classification for the piping system by the definition of Category M Fluid Service in Chapter I of ASME B31.3. This definition is the Code rule relative to classification. A guide to the application of these rules is provided in ASME B31.3, Appendix M, which contains a flow chart to assist the owner in classifying fluid services.

The definition of Category M Fluid Service is as follows.

Category M Fluid Service: a fluid service in which both of the following apply:

1. the fluid is so highly toxic that a single exposure to a very small quantity of the fluid, caused by leakage, can produce serious irreversible harm to persons on breathing or bodily contact, even when prompt restorative measures are taken
2. after consideration of piping design, experience, service conditions, and location, the owner determines that the requirements for Normal Fluid Service do not sufficiently provide the leak tightness required to protect personnel from exposure

The first part of the definition involves how toxic the fluid is and the second part involves whether people will be adequately protected from harm with a piping installation following Normal Fluid Service requirements. The owner has the right and responsibility to designate a fluid service as Category M.

Note that both parts are subjective. The Code does not define "very small quantity" or "serious irreversible harm". But the Code does say that "restorative measures" may include flushing with water, administration of antidotes, or medication. And the owner must somehow "determine" whether the requirements for Normal Fluid Service will adequately protect people. Because the definition is subjective, different owners reach different conclusions regarding Normal Fluid Service versus Category M Fluid Service for similar installations. Thus, there cannot be a list of chemical compounds that states the fluid service irrespective of the conditions of installation.

Category M Fluid Service was incorporated into ASME B31.3 in the 1976 edition when the draft Chemical Plant Piping Code was incorporated into ASME B31.3. Because the draft was similar to the Refinery Piping Code, the ASME B31 Committee decided to the merge the two codes rather than publish a separate document. While Category M Fluid Service is found in some chemical plants, fuels refinery piping is rarely, if ever, designated as Category M. A common toxic substance in refineries is H_2S which causes asphyxiation. But, in very small quantities it is not lethal. However, an owner could decide that piping containing H_2S that is in an occupied building warrants Category M classification.

Following the requirements in the Chapter VIII of the Code for Category M Fluid Service generally result in an installed piping system that is less likely to leak. When Category M Fluid Service is invoked by the owner, all the requirements of Chapter VIII are applied to the designated fluid service piping. Rather than so designating fluid services as Category M, some owners prefer keeping the designation as Normal and specifying additional design, construction, examination, and testing requirements that are described in Chapter VIII.

Frequently there is a lot of attention on the first part of the definition and little attention to the second part. The question of whether people will be protected with the Normal Fluid Service designation needs to be addressed. If, for example, the piping is installed in an area where people are prohibited while the piping is pressurized, a small leak will not harm any person. Another example is a double containment piping with leak detection. In these cases, for chemicals such as chlorine, phosgene and methyl isocyanate, an owner might decide against designating the fluid service as Category M even though the first part of the definition was satisfied.

16.2 ORGANIZATION OF CHAPTER VIII OF ASME B31.3

Chapter VIII follows the same paragraph numbering as the base Code, Chapters I through VI, except that in Chapter VIII the paragraphs start with M for metallic piping and MA for nonmetallic piping.

To determine the rules for piping systems in Category M Fluid Service, one simply refers to Chapter VIII. If rules elsewhere in the Code apply, they are referred to in Chapter VIII. In general, Chapter VIII refers to the base Code for metallic piping and to Chapter VII for nonmetallic piping.

Chapter VIII makes no provision for severe cyclic conditions, as stated in para. M300(e). Severe cyclic conditions should be avoided by design in these systems. If, for some reason, this is not feasible, ASME B31.3 requires that the engineering design specify any necessary provisions.

16.3 OVERVIEW OF METALLIC RULES

The rules for metallics prohibit the use of certain components considered to have lower integrity, and they require additional design considerations, additional examination, and additional testing. The measures are intended to result in a piping system that is less likely to leak.

The following are highlights of some of these requirements. It is not an all-inclusive list; refer to ASME B31.3 for the complete requirements.

- The presumptive degree of ambient cooling (e.g., 5% for uninsulated pipe) provided in the base Code is not permitted. Instead, the design metal temperature, if less than the fluid temperature, must be substantiated by heat transfer calculations confirmed by tests or by experimental measures.
- Increased pressure and/or temperature for short-term variations (allowances for variations) are not permitted.
- Lower integrity piping and components are prohibited.
- Special consideration is required for prevention of valve stem leakage to the environment. Specific requirements are provided for valve bonnet or cover plate closures (para. M307.2).
- Single-welded slip-on, expanded-joint, and threaded (with certain exceptions) joint flanges are prohibited (para. M308.2).
- Expanded joints are prohibited (para. M313).
- Additional limitations are provided for threaded joints (para. M314).
- Joints such as caulked (para. M316), soldered and brazed (para. M317), and bell-type joints (para. M318) are prohibited.
- Pipe supports are required to be constructed of listed materials (para. M321).
- Specific provisions are given for instrument piping [e.g., limiting tubing to 15-mm (5/8-in.)-diameter maximum, accessible block valves required to be available to isolate the instrument piping from pipeline] (para. M322.3).

- The design pressure is not permitted to be exceeded by more than 10% during pressure relief (para. M322.6.3).
- The low-stress exemption from impact testing is not permitted (para. M323.2).
- Gray iron and ductile iron are not permitted for pressure-containing parts, and lead and tin may only be used as linings (para. M322.4.2).
- Less stringent heat treatments than required in the base Code are not permitted (para. M331).
- Additional examination is required. Although the acceptance criteria of the base Code are applicable, the amount of radiography is increased from 5% to 100% and random visual examination is generally increased to 100% visual examination. See Section 13.4 for specific requirements.
- In addition to the testing required for Normal Fluid Service, an additional sensitive leak test is required to ensure that the piping is free of small leaks. See Section 14.6 for specific requirements.

16.4 OVERVIEW OF NONMETALLIC RULES

The rules for nonmetallics prohibit the use of certain components considered to have lower integrity, and they require additional design considerations, additional examination, and additional testing. The measures are intended to result in a piping system that is less likely to leak. These rules generally refer to Chapter VII.

The following are highlights of some of these requirements. It is not an all-inclusive list; refer to ASME B31.3 for the complete requirements.

- The piping is not permitted to exceed the design pressure under any circumstance, including pressure relief conditions; this is the same as the rule for nonmetallic piping (para. MA302.2.4).
- Nonmetallic fabricated branch connections are prohibited (para. MA306.5).
- Nonmetallic valves and specialty components are prohibited (para. MA307).
- Hot-gas-welded, heat-fusion, solvent-cemented, and adhesive-bonded joints are not permitted, except in linings (para. MA311.2).
- Expanded, nonmetallic threaded, and caulked joints are prohibited (paras. MA313, MA314, MA316).
- Thermoplastics and reinforced plastic mortar are permitted only as linings and, for thermoplastics, gaskets (para. MA323.4.2).
- The examination and testing rules of Chapter VII apply with an additional requirement of 100% visual examination of all Fabrication, including all bolted and mechanical joints.
- In addition to the testing required for Normal Fluid Service, an additional sensitive leak test is required to ensure that the piping is free of small leaks. See Section 14.6 for specific requirements.

16.5 GENERAL COMMENTS

The rules of ASME B31.3, Chapter VIII are intended to provide greater assurance of leak-tightness. An example is the requirement that a sensitive leak test be performed in addition to the standard leak test (e.g., hydrotest). The ASME B31.3 Section Committee does not generally spend a great deal of time on this section. In this author's opinion, the rules are not sufficient for the extremely dangerous chemicals that fit the definition of Category M. Additional precautions should be considered, such as double-containment (which could be considered to remove the piping system from Category M).

CHAPTER 17

HIGH-PRESSURE PIPING

17.1 SCOPE OF CHAPTER IX OF ASME B31.3

Chapter IX of ASME B31.3 only applies when the owner specifies its use. It applies to piping in High-Pressure Fluid Service. Note that the definition of High-Pressure Fluid Service simply requires that the owner specify use of Chapter IX. Some guidance is provided in FK300, which states "Although High Pressure Fluid Service is often considered to be service exceeding that allowed by the ASME B16.5, Class 2500 pressure-temperature rating for a particular material group, there are no pressure limitations for the application of these rules." This is not a requirement, and the base Code may be satisfactorily used at pressures higher than ASME B16.5, Class 2500. However, the base Code rules become increasingly conservative and, in fact, impossible to use as the pressure approaches the allowable stress (including quality factors). See Section 17.3 for a discussion of pressure design of straight pipe. By the same token, Chapter IX may be used at lower pressures; it has been used with high-strength steels with pressures as low as 28,000 kPa (5,000 psi).

The rules provide a combination of considerations. Although reduced wall thicknesses and provisions that are specific to the needs of high pressure (e.g., not including thread depth as an allowance under specific conditions) are provided, additional material toughness, analysis during design, examination, and testing are required. For background on these rules see Sims (1986).

Chapter IX makes no provision for Category D, Category M, Elevated Temperature and High Purity Fluid Services. Considering the pressure, Category D is not applicable. Elevated-temperature creep effects are not included; thus, the Chapter IX excludes Elevated Temperature Fluid Service. The concept of severe cyclic conditions is not used in Chapter IX because the rules are more stringent than the base Code rules for severe cyclic conditions.

This section provides an overview of Chapter IX. See the ASME B31.3 Code for specific requirements.

ASME BPVC, Section VIII, Division 3, *Pressure Vessels, Alternative Rules for Construction of High Pressure Vessels,* was completed after Chapter IX. As a result, there are various references to the requirements of ASME, BPVC, Section VIII, Division 2 that have been changed to either include Division 3 as an acceptable alternative or to simply require the Division 3 rules rather than the Division 2 rules. The Division 3 rules are generally more applicable, because they were developed for high-pressure equipment.

17.2 ORGANIZATION OF CHAPTER IX OF ASME B31.3

Chapter IX follows the same paragraph numbering as the base Code, Chapters I through VI, except that the paragraphs in Chapter IX start with K. To determine the rules for high-pressure piping systems, simply refer to Chapter IX. If rules elsewhere in the Code apply, they are referenced in Chapter IX.

17.3 PRESSURE DESIGN OF HIGH-PRESSURE PIPING

Beyond a certain pressure, it is not possible to design piping in accordance with the basic wall thickness equation in the base Code, Eq. (3a). When Y is assumed equal to zero (note that Y approaches zero as the inside diameter approaches zero; see definition of Y in para. 304.1.1), the required wall thickness is equal to the outside radius of the pipe when the pressure is equal to the allowable stress times the quality factor *(SE;* see Section 4.2). However, heavy wall pipe has substantial pressure capacity beyond the point where the circumferential stress at the bore reaches yield. Further, for high internal pressure, the radial stresses due to the surface traction of internal pressure significantly affect yielding of the material on the inside of the pipe, in accord with the Von Mises or the Tresca yield theory.

Equations (34a) and (34b) in Chapter IX provides the required thickness for high-pressure straight pipe. Rather than being based on maximum circumferential stress, as in the base Code, they are based on limit load pressure. The equations for calculation of required wall thickness are [Eqs. (35a) and (35b) in the Code provide the allowable pressure based on available thickness][20]

$$t_m = t + c \tag{17.1}$$

$$t = \frac{D - 2c_o}{2}\left[1 - e^{(-P/S)}\right] \tag{17.2}$$

$$t = \frac{d + 2c_i}{2}\left[e^{(-P/S)} - 1\right] \tag{17.3}$$

where

D = outside diameter of the pipe (maximum permitted by specifications)
S = allowable stress from Table K-l of ASME B31.3
$c = c_i + c_o$
c_i = sum of internal mechanical plus corrosion and erosion allowances
c_o = sum of external mechanical plus corrosion and erosion allowances including, except under certain conditions, thread depth
d = inside diameter of the pipe (maximum permitted by specifications)
t = pressure design wall thickness
t_m = minimum required wall thickness including mechanical, corrosion, and erosion allowances

These equations provide a margin against burst at design pressure of 1.732 (which is $\sqrt{3}$) and a lower limit margin against through thickness yielding during leak testing of 1.15. The 1.732 margin is based on experimental burst tests of closed end cylinders as described by Terada in three referenced papers.

No quality factor is included, because the minimum permitted quality factor in Chapter IX is 1.0. When the mechanical allowances are not specified to be internal or external, they are to be assumed to be internal. The external thread on pipe can be neglected in the mechanical allowances when, for straight-threaded connections, the following are satisfied:

- The thread depth does not exceed 20% of the wall thickness
- The ratio of outside to inside diameter, *D/d,* is greater than 1.1
- The internally threaded attachment provides adequate reinforcement
- The thread plus the undercut area, if any, does not extend beyond the reinforcement for a distance more than the nominal wall thickness of the pipe

[20] Equations (17.1) through (17.3) correspond to ASME B31.3 Eqs. (33), (34a), and (34b), respectively.

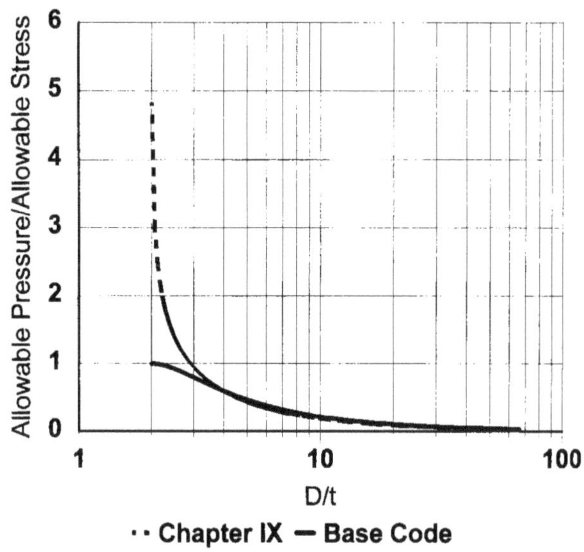

FIG. 17.1
COMPARISON OF WALL THICKNESS REQUIRED BY BASE CODE PRESSURE DESIGN RULES [ASME B31.3, EQ. (3A)] VERSUS THOSE OF ASME B31.3, CHAPTER IX [ASME B31.3, EQS. (34A), (34B)]

This recognizes that the threaded fitting, typically a threaded flange, to which the pipe is attached can effectively reinforce the pipe for internal pressure under certain conditions.

Figure 17.1 provides a comparison of the base Code pressure design rules with those of Chapter IX. It provides the ratio of allowable pressure to allowable stress as a function of the *D/t* ratio. It can be observed that the base Code provides a higher allowable pressure for *D/t* ratios greater than about four. However, because the allowable stress in Chapter IX is based on yield strength and not tensile strength, there will be a greater advantage with respect to wall thickness if Chapter IX is used for steels with high yield-to-tensile strength ratios. This is because the base Code allowable stress would be controlled by one-third the tensile strength.

Provisions similar to, but more limited than those of the base Code are provided for design of specific types of components, and listed components are accepted. The listed components are given in Table K326.1.

The comparable paragraph to 304.7.2 is K304.7.2, Unlisted Components and Elements. The options for proving the pressure design of unlisted components are more limited than in the base Code. Extensive service experience, performance testing or detailed stress analysis (with the results evaluated in accordance with BPVC Section VIII, Div 3, Article KD-2, with two exceptions for linear elastic analysis) are permitted for qualifying components. The performance testing must substantiate both the static pressure design and the fatigue life.

Short-term variations above the design pressure and temperature are not permitted, except for accumulation during pressure-relieving events.

17.4 EXTERNAL PRESSURE

Buckling due to external pressure is not generally a concern for high-pressure piping. However, the limit pressure determined using Eqs. (34a), (34b), (35a), and (35b) of ASME B31.3 is not always conservative for external pressure on straight pipe, considering buckling, and under rare circumstances when external pressure does not cause axial compression of the pipe. In the latter case, collapse can be predicted by triaxial stress states and yield theory.

When $D/t < 3.33$ and at least one end of the pipe is exposed to full external pressure, which produces compressive axial stress, Eqs. (34a), (34b), (35a), and (35b) of ASME B31.3 for internal pressure can be used. In all other circumstances, the base Code rules for external pressure design of straight pipe are used.

17.5 DESIGN FOR SUSTAINED AND OCCASIONAL LOADS

The criterion for sustained loads is the same as in the base Code. The stress due to pressure, weight, and other sustained loads must be less than S_h.

The criterion for occasional loads is more conservative than in the base Code. A factor of 1.2 times the allowable stress (the same as ASME BPVC, Section VIII, Divisionl) is used, rather than a factor of 1.33.

17.6 DESIGN FOR THERMAL EXPANSION AND FATIGUE

Flexibility analysis is conducted similar to the base Code. However, only the more conservative equation for S_A, which assumes that $S_L = S_h$, is used [Eq. (32) of ASME B31.3]. The allowable stresses from Appendix K are used, rather than the allowable stresses from Appendix A. These will be higher when tensile strength controls the allowable stresses in Appendix A.

Chapter IX (para. 304.8) requires a detailed fatigue analysis in addition to the flexibility analysis provided in the base Code. The fatigue analysis is based on ASME BPVC, Section VIII, Div 3, Article KD-3.

The fatigue analysis should include the typically high strains due to internal pressure, in particular those at the bore of the pipe, as well as the stress due to thermal expansion. Since the calculated stress is compared to the polished bar fatigue curve rather than a butt-welded pipe fatigue curve, the stresses calculated in accordance with the flexibility analysis rules of ASME B31.3 need to be multiplied by a factor of two. The reason for doing this is explained in Section 8.4.

The requirements for fatigue analysis pose some challenging problems to the Designer. Some of these are highlighted below.

a. The Designer must consider both pressure and displacement cycles. Thus, a load histogram and a procedure such as rainflow counting is appropriate for determining the variety of stress ranges and number of cycles at each stress range for which the piping must be designed. For example, there may be normal pressure cycles, which may be more numerous than thermal displacement cycles, and also pressure pulsations. There will be some number of cycles at a maximum stress range, which could be the normal pressure plus pulsation plus displacement stress, as well as many cycles with smaller stress ranges to consider.
b. Fatigue analysis per ASME BPVC, Section VIII, Division 2 deals with stresses at a point, whereas the flexibility analysis rules of ASME B31.3 do not inform the Designer as to the location and direction of stresses that are calculated. For example, the stresses in an elbow due to in-plane bending are through-wall bending and greatest in the circumferential direction and should be directly added to the pressure stress, whereas the stresses due to bending in straight pipe are gross bending on the pipe section and longitudinal and should be combined with the pressure stress to determine Tresca stress intensity.
c. The stresses due to internal pressure vary through the thickness and this should be considered in determining what pressure stress to combine with the deflection stresses (e.g., bore stress should be considered for an elbow and the outside stress should be considered for a threaded joint).
d. The radial stress through the wall of the pipe [compressive and equal to internal pressure on the inside surface and generally considered to be zero on the outside surface (for a pipe under internal pressure)] can be a significant component of the Tresca stress intensity.

All of these mean that a Designer tackling a fatigue analysis for Chapter IX piping should be an expert, intimately familiar with the stress distributions in thick-wall piping components.

ASME B31.3 provides the equation below for the stress intensity on the inside surface of straight pipe due to internal pressure. Note that this is not likely to be the controlling location in the system. As such, its usefulness is limited. Also note that this equation is only considered to be valid if the calculated stress per this equation does not exceed three times the allowable stress from Table K-l (this is two times the yield strength). If this stress is exceeded, an inelastic analysis is required to determine the strain range due to internal pressure. This provision assumes that the pipe will shake down to elastic action if the stress intensity range is less than two times the yield strength. The equation is[21]

$$S = \frac{PD^2}{2(T-c)\left[D-(T-c)\right]} \tag{17.4}$$

where

D = outside diameter
P = pressure [definition in nomenclature, Appendix J, is not correct, because a variety of pressure conditions may be considered]
S = stress intensity on the inside surface of the pipe
c = sum of mechanical plus corrosion and erosion allowances

17.7 MATERIALS

The allowable stress is provided in Appendix K, Table K-1. It is two-thirds of the material yield strength, to be consistent with the pressure design equation. For solution-heat-treated austenitic stainless steels and certain nickel alloys with similar stress-strain behavior, the minimum of two-thirds the specified minimum yield strength and 90% of the yield strength at temperature is used. Similar to the base Code, this is because the material has significant strength beyond the nominal 0.2% offset yield stress. As in the base Code, these higher stress values provided for stainless steel and similar materials are not recommended for flanges and similar components where slight deformation can cause leakage or other malfunction.

As in the base Code (as of the 2000 Addenda), materials may be used above the maximum temperature for which allowable stresses are provided. Previously, this had been prohibited because of concerns such as temper embrittlement. Per para. K323.2.1, there must be no prohibition in the Code, the Designer must verify the serviceability of the material, and the temperature must be below that at which creep properties would govern the allowable stress that would be determined using the base Code allowable stress criteria.

There is a continuing issue in keeping the editions of material standards in Appendix E current with the current editions of the listed standards. Per K323.1.1, if the material is from a different edition than the listed one, it may be used if aspects such as chemical composition, heat-treatment, strength, and testing is the same as the edition of the specification. Otherwise, the material may be considered as an unlisted material in accordance with K323.1.2. Unlisted materials may be used provided they conform to a published specification covering chemistry, physical and mechanical properties, method and process of manufacture, heat treatment, and quality control, and otherwise meet the requirements of Chapter IX.

The impact test requirements are a very important part of the material requirements in Chapter IX for high-pressure piping. Essentially all high-pressure piping materials and welds must be impact-tested to determine that they have sufficient notch toughness for any temperature condition at which stresses exceed 41 MPa (6 ksi). Impact test requirements are provided in para. K323.3. Transverse specimens are required, unless the component size or shape does not permit cutting transverse (to the longitudinal axis of the pipe, or the maximum direction of elongation during rolling, or the direction of major working during forging) specimens. In that case, longitudinal specimens may be used; however, the required impact energy absorption is higher.

[21] ASME B31.3, Eq. (37).

For materials, at least one set of impact tests per lot is required. For impact tests on welds, significantly more testing is required than in the base Code. Whereas the base Code only requires impact testing as part of the weld procedure qualification, Chapter IX requires impact testing for each welder and welding procedure, type of electrode, filler metal, and flux to be used. For tests on welds, separate tests are not required for each lot of material. Test specimens for the welds and heat-affected zones are required.

The minimum permissible temperature for a material is the minimum temperature at which an impact test that satisfies the Code requirements was performed. The only exception to this is the 41-MPa (6-ksi) exemption, but that exemption may only be used down to –46°C (–50°F). Impact testing, regardless of stress, is required for use at temperatures below that temperature.

Nonmetallic parts, other than gaskets, packing and valve seats are not permitted.

17.8 FABRICATION

Requirements for qualification of welding procedures and welders or welding operators follow ASME BPVC, Section IX. However, there are additional requirements and limitations, including the following:

- Qualification test weldments are required to be made using the same specification and type or grade of base metal(s) and the same specification and classification of filler metal(s) as will be used in production welding.
- Qualification by testing weldments made with materials with the same P-Number is not permitted.
- Test weldments are required to be subjected to essentially the same heat treatment, including cooling rate and cumulative time at temperature, as the production welds.
- Mechanical testing is required for all performance qualification tests; testing by radiography is not permitted.

Branch connection fittings are required to be designed to permit 100% radiography. Acceptable details are provided in Fig. K328.5.4 of ASME B31.3.

General fabrication requirements pertaining to end preparation, alignment, welding, preheat, and post-weld heat treatment are provided. They are largely similar to those in the base Code, with some variations. For example, the exemptions to mandatory postweld heat treatment in Table 331.1.3 for P-No. 1 materials are not permitted when the control thickness exceeds 19 mm (3/4 in.).

17.9 EXAMINATION

In general, Chapter IX requires 100% examination. This includes 100% visual examination of materials and components; fabrication; threaded, bolted, and other joints; piping erection; and pressure-containing threads.

All girth, longitudinal, and branch connection welds are required to be 100% examined by radiography. When specified in the engineering design and with the owner's approval, ultrasonic examination of welds may be substituted for radiographic examination when the wall thickness is equal to or greater than 13mm (1/2 in.). In-process examination is not permitted to be be substituted for radiographic or ultrasonic examination of welds. The acceptance criteria for radiography of welds are provided in Table K341.3.2 and for ultrasonic examination of welds, are described in para. K344.6.3.

17.10 TESTING

All elements, including all components and welds, of a high-pressure piping system, except for bolting and gaskets used during final system assembly and pressure relieving devices to be used during operation, are required to be subjected to a full-pressure test. This pressure test need not be performed on the installed piping system, but can be done on piping subassemblies prior to erection. The test requirement is the same as for the base Code hydrostatic test, except the test pressure is 1.25 times the design pressure corrected for temperature instead of 1.5 times. Furthermore, if this test is performed as a pneumatic test, the same full pressure as for a hydrotest is required. However, the test pressure can be reduced, if necessary, to limit the pressure to that at which through-thickness yielding (calculated based on specified minimum yield stress) occurs in any component. For closure welds in systems that are all welded, examination in accordance with K345.2.3(c) can be substituted for the leak test. This paragraph requires in process examination and 100% radiographic examination.

After the piping is assembled, an additional leak test of the installed piping system is required. This is conducted at 110% of the design pressure. However, if the full-pressure test described in the preceding paragraph was performed on the installed piping, the additional leak test is not required.

K345.1(d) was added with the 2008 edition. This permits reduction of the test pressure, with the owner's approval, when pressure relieving devices to be used during operation are included in the leak test of the installed piping. The leak test may be reduced to prevent the operation of, or damage to, the pressure relieving devices, but must be at least 90% of the lowest set pressure of the pressure relieving devices in the system.

Several additions and alternatives were added in the 2008 edition in para. K345.2.3. The provision for closure welds was added. A provision that piping components and subassemblies may be tested either separately or as assembled piping was included. Further, the provisions that were added to the base Code that provides that flanged joints used to connect piping components that have previously been leak tested, and flanged joints at which a blank or blind flange is used to isolate equipment or other piping during the leak test, need to be leak tested.

17.11 RECORDS

Chapter IX contains more substantive requirements for record transfer to the owner and the retention of records than the base Code. For example, the base Code does not contain requirements relative to documentation of the engineering design. However, Chapter IX requires records of the following to be provided to the owner or the (owner's) Inspector:

- The engineering design
- Material certifications
- Procedures used for fabrication, welding, heat treatment, examination, and testing
- Repair of materials including the procedure used for each and location of repairs
- Performance qualifications for welders and welding operators
- Qualification of examination personnel
- Records of examination of pipe and tubing for longitudinal defects as specified in paras. K344.6.4 and K344.8.3

These records are required to be retained by the owner for at least 5 years after they have been received.

17.12 REFERENCES

Terado, S; Development of Alternate Methods for Establishing Design Margins for ASME Section VIII Division 3; PVP 2008-61146; 2008 PVP Chicago, USA.

Terado, S; Development of Alternate Methods for Establishing Design Margins for ASME Section VIII Division 3 (Part 2); PVP 2009-77209; 2009 PVP Prague, Czech Republic.

Terado, S; Proposal of New Equations for Cylindrical and Spherical Shell of ASME Section VIII Division 3 for High Pressure Vessels; ICPVT 2009; 2009 Jeju Island, Korea.

CHAPTER

18

HIGH-PURITY PIPING

18.1 WHEN TO USE THE RULES FOR HIGH PURITY FLUID SERVICE

The rules in Chapter X of ASME B31.3 are used when the owner designates a piping system to be in High-Purity Fluid Service. The owner is guided in the classification for the piping system by the definition of High-Purity Fluid Service in Chapter I of ASME B31.3. This definition is the Code rule relative to classification. A guide to the application of these rules is provided in ASME B31.3, Appendix M, which contains a flow chart to assist the owner in classifying fluid services.

The definition of High Purity Fluid Service is:

> a fluid service that requires alternative methods of fabrication, inspection, examination, and testing not covered elsewhere in the Code, with the intent to produce a controlled level of cleanness. The term thus applies to piping systems defined for other purposes as high purity, ultra high purity, hygienic, or aseptic.

Examples of industries using this fluid service include food, drug and silicon wafer manufacturing.

Regarding Code coverage of piping in High-Purity Fluid Service:

- Chapter X provides rules for piping that is also designated Category M Fluid Service.
- Chapter X provides rules for piping that is also designated Elevated Temperature Fluid Service.
- Chapter X refers to Chapter VII for requirements for nonmetallic piping and piping lined with nonmetals.
- The Code does not provide rules for piping that is in both High-Purity and Category D Fluid Service.
- The Code does not provide rules for piping that is in both High-Purity and High-Pressure Fluid Service.

18.2 ORGANIZATION OF CHAPTER X OF ASME B31.3

Chapter X follows the same paragraph numbering as the base Code, Chapters I through VI, except that in Chapter X the paragraph numbers start with U. For piping designated as also being in Category M Fluid Service, the paragraph numbers start with UM. To determine the rules for piping systems in High Purity Fluid Service, one simply refers to Chapter X. If rules elsewhere in the Code apply, they are referred to in Chapter X.

18.3 GENERAL REQUIREMENTS

Chapter X provides some specific requirements along with some recommendations:

- Flanges should be avoided whenever possible.
- Expanded joints, flared tube fittings and caulked joints are prohibited.
- Compression-type tube fittings may be used provided the gripping action of the fitting provides vibration resistance.
- Face seal or hygienic clamp-type fittings in which the tightness of the joint is provided by a seating surface other than the threads may be used.
- The welding process shall be orbital GTAW, except tack welds made prior to orbital welding may be manual GTAW, and manual GTAW may be used on welds where the orbital welding equipment cannot be used because the geometry of the installation won't permit placement of the welding head.

18.4 EXAMINATION AND LEAK TESTING

Materials commonly used in high purity process piping systems include austenitic, ferritic, and duplex stainless steels, and nickel and nickel alloys. The wall thickness of the pipe and fittings is typically thinner than most process piping. For this piping, visual examination of the inside and outside surfaces of the weld is believed to provide the same assurance of structural integrity as radiographic or ultrasonic examination. Chapter X describes weld coupon examination that may be used in lieu of the 5% random radiography or ultrasonic examination required for Normal Fluid Service production welds that are made by orbital GTAW. Chapter X does not permit weld coupon examination to be substituted for High-Purity Fluid Service piping that is also designated as Category M Fluid Service.

Weld coupons are required to be made and examined when any of the following conditions exist:

a. beginning of shift
b. change of purge source
c. change of power supply
d. change of equipment, e.g., weld head, weld-head extensions, tungsten
e. any time there is a weld defect

With the owner's approval, the personnel performing the production work are permitted to perform the examination, provided the personnel meet the personnel qualification and certification requirements.

Wels coupons are examined for compliance with para. U341.3.2

1. alignment
2. weld penetration
3. weld bead width variation
4. weld bead meander
5. discoloration
6. weld defects, e.g., cracks, porosity, or sulfur stringers

To allow direct visual examination of the inside surfaces, the weld coupon may be cut or a suitable indirect visual examination such as using a borescope, may be performed.

The Leak testing requirements are the same as for Normal Fluid Service except that at the owner's option, a helium mass spectrometer leak test may be substituted for the hydrostatic leak test or a sensitive leak test or both.

CHAPTER

19

BELLOWS EXPANSION JOINTS

19.1 OVERVIEW OF BELLOWS REQUIREMENTS

Prior to 1988, coverage of metallic bellows expansion joints in ASME B31.3 was very limited. Bellows were required to be designed in accordance with the rules for pressure design of unlisted components (par. 304.7.2) and the Standards of the Expansion Joint Manufacturers Association, Inc. (EJMA Standards). Appendix X was added in 1988, and has undergone several significant revisions in subsequent years. It provides specific Code coverage for metallic bellows expansion joints by adopting, with exceptions and additions, a widely used industry standard for expansion joints, the EJMA Standards.

The overall philosophy of Appendix X was to adopt the EJMA Standards with modifications that the Code Committee felt were necessary to make the EJMA Standards into Code requirements suitable for ASME B31.3 piping. This differs somewhat from the approach taken in ASME BPVC, Section VIII, Division 1, Appendix 26 (formerly Appendix BB), Pressure Vessel and Heat Exchanger Expansion Joints, where the design equations from the EJMA Standards were direcdy incorporated into the Code. The background behind these bellows stress analysis equations and comparison with detailed analysis and test results are provided in Becht (1985), Becht and Skopp (1981a), Osweiller (1989), and Broyles (1994).

The most significant of these modifications to the EJMA Standards were in requirements for pressure design, fatigue design, and hydrotest. It should be noted that both the EJMA Standards and the ASME B31.3 Code are living documents. For example, EJMA revised the pressure design requirements in the Sixth Edition, issued in 1993. This Edition was adopted by ASME B31.3 in the 1994 Addenda. Some of the pressure design requirements in Appendix X could now be dropped, because they are now adequately addressed in the EJMA Standards. The 10th edition of the EJMA Standards changed the fatigue design rules, resulting in the ASME B31.3 Code to drop its fatigue design rules in Appendix X.

19.2 RULES FOR PRESSURE DESIGN

It was shown by Becht (1980) and Becht and Skopp (1981b) that the design rules in the Fifth Edition of the EJMA Standards for internal pressure were, in some cases, nonconservative. In fact, it was possible to design a bellows that would fail at a pressure lower than the design pressure. As a result, both ASME BPVC, Section VIII, Division 1 and ASME B31.3 contain an allowable stress basis that differs from the 5th Edition of the EJMA Standards.

The source of the nonconservatism was an inconsistency between the allowable stress basis and empirical test data. Based on test data, the EJMA Standards included a design approach that permitted a pressure-induced bending stress which was significantly greater than the specified minimum yield strength of the material. It permitted a stress as high as 2.86 times the ASME BPVC, Section VIII, Division 1 allowable stress in tension, which could be as high as 1.9 times the minimum yield strength (or 2.6 times for austenitic stainless steel).

The reason that bellows generally do not fail at this high level of stress is that the material, in the as-formed condition, is cold-worked and therefore has a yield strength two or more times higher than the specified minimum yield strength upon which the Code-allowable stress is based. The bellows that were tested in developing the pressure stress limits in the EJMA Standards were all in the as-formed condition, and therefore had the benefit of a higher than specified minimum yield strength. Thus, a high multiple of the Code-allowable stress, which was based on specified minimum yield strength values, was found to be acceptable.

If a bellows is annealed, as is sometimes required for corrosive or high-temperature service, the benefit of the cold work in terms of increased yield strength is no longer available, and the EJMA pressure design equations became nonconservative. Figure 19.1 illustrates how the root in a test bellows bulged outward to failure when the meridional stress due to internal pressure exceeded 1.5 times the actual yield strength in an annealed bellows (Becht and Skopp, 1981b).

The ASME B31.3 approach is to require the meridional membrane plus bending stresses for annealed bellows to be limited to 1.5 times the basic allowable stress in tension. This took care of the known non-conservatism for the case of annealed bellows.

The degree of conservatism for other materials in the as-formed condition was not known and depended on the degree of strengthening that occurs in the forming process. However, the Committee did not want to create excessively restrictive rules that would unduly impact bellows designs that have historically worked. Also, for bellows, a balance must be maintained between design for pressure and deflection. Therefore, no specific stress limits were provided for bellows in the as-formed condition (not heat-treated). The design defaulted to the rules in the EJMA Standards. However, the shop leak test was changed so that it was essentially a proof test of the bellows, so the pressure capacity of a bellows design is proven by the shop leak test.

Note that the philosophy in ASME BPVC, Section VIII, Division 1, Appendix 26 has the same intent. The meridional membrane plus bending stress due to internal pressure is limited to 1.5 times the Code-allowable stress. However, "alternate factors may be used if substantiated by test data (e.g., by hydrotest for design temperatures below the creep range)."

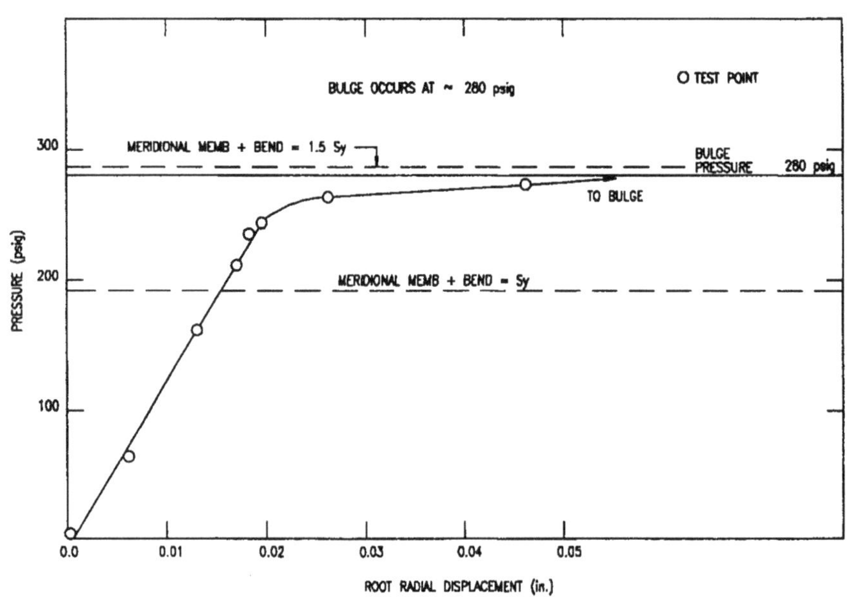

FIG. 19.1
BELLOWS PRESSURE TEST DATA

New design equations have been developed by Broyles (1989, 1994) that address this noncon-servatism in the pressure design equations; these equations are in the Sixth and later Editions of the EJMA Standards which have been adopted by ASME B31.3. The specific stress limit for annealed bellows could have been dropped from Appendix X, because it is now in the EJMA Standards. However, they were left in place for additional emphasis.

19.3 FATIGUE DESIGN

Prior to the 10th edition of the EJMA Standards, the fatigue curve provided was an average failure curve without a safety factor. However, the advantage of the fatigue curves of the type provided in the EJMA Standards is that they were based on component testing. There are a number of complications in bellows performance, such as strain concentration that occurs when bellows are deflected beyond the elastic limit. Bellows can be very sensitive to strain concentration with even slight geometric irregularities (Becht, 1985). These effects are inherently included in empirical fatigue curves and need not be determined analytically. This simplifies the consideration of a number of complex phenomena.

To develop design fatigue curves, a bellows fatigue database provided to the Committee by the EJMA was evaluated and fatigue curves were developed that included typical Code safety margins. Since the fatigue curves in the 10th edition provided a means to include a safety margin, the fatigue curves in Appendix X were deleted, and the user is instead referred to the EJMA Standards for fatigue design. Description of the Appendix X fatigue curves development is included in prior editions of this book but is removed from this edition as it is no longer relevant.

Prior to the 10th edition and in the prior Appendix X rules different fatigue curves are provided for reinforced and unreinforced bellows. In the 10th edition, only one fatigue curve is provided for both. This is because they behaved similarly. This author believes it is for the following reason. The prior EJMA database had a significant number of high profile bellows fatigue tests. High profile is generally defined as bellows pitch divided by bellows convolution height less than 0.9. As shown by Becht (1999, 2000a-c, 2001), high profile bellows have significant plastic strain concentration when deflected beyond the elastic limit. What this means is that high profile bellows have a lower fatigue life with the same calculated elastic displacement stress range. Low profile and reinforced bellows (which are generally not high profile) do not exhibit the same degree of plastic strain concentration.

The unreinforced fatigue curves that were previously used were driven down relative to the reinforced fatigue curve because of the inclusion of high profile bellows fatigue data. Becht (1999, 2000a-c, 2001) found that if the high profile bellows fatigue data were not included, reinforced and unreinforced bellows followed the same fatigue curve. The new dataset of bellows fatigue data used to develop the 10th edition fatigue curve did not have high profile bellows. As a result, the reinforced and unreinforced bellows ended with the same fatigue curve.

The fatigue curves in the 10th edition are average failure curves, but the rules provide for inclusion of a factor, f_c, which introduces a safety factor. ASME B31.3 requires this factor to be 75% unless use of an alternate factor is approved by the Owner. Figure 19.2 compares the fatigue curves of EJMA prior to the 10th edition, the prior Appendix X fatigue curves, and the 10th edition fatigue curve for austenitic stainless steel bellows. Note that while the absence of high profile fatigue curve data in the development of the 10th edition fatigue curve may lead to some concern with respect to possible conservatism of the curve for high profit bellows, comparison of the 10th edition curve (with the 75% safety factor required by ASME B31.3) with the unreinforced curve previously in Appendix X shows that the design fatigue curve is sufficiently conservative for high profile bellows.

19.4 ELEVATED-TEMPERATURE DESIGN

The ASME B31.3 Code and ASME BPVC, Section VIII, Division 1 do not address elevated-temperature design of bellows. Design of bellows in the creep regime can often be avoided in piping by designing the expansion joint such that the bellows element operates at a significantly lower temperature than the process temperature. Where this is not possible, additional design considerations beyond the rules provided in the Code are required.

In ASME B31.3, the Designer is referred to para. 304.7.2, Unlisted Components and Elements. However, even there, no specific guidance is provided for elevated-temperature design, and design for deflection stresses, a critical aspect of bellows design, is not covered.

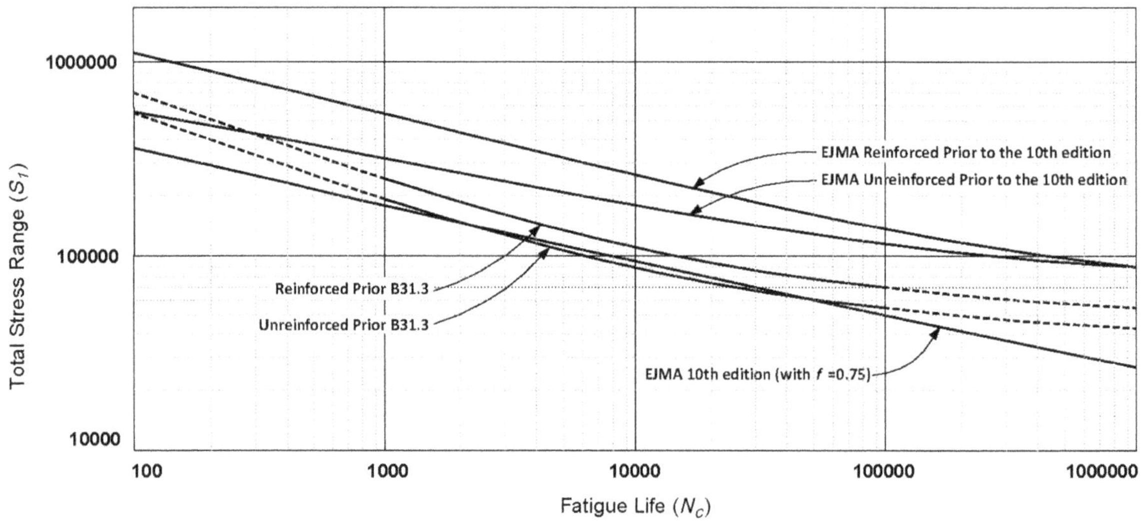

FIG. 19.2
FATIGUE CURVE COMPARISON

The EJMA Standards started incorporating specific rules in the Seventh Edition that address the design of bellows in the creep range of the material of construction. For bellows displacement conditions, elevated-temperature cyclic testing is required.

Hold times are critical in creep fatigue. For example, a 1-hour hold time each cycle for Type 304 stainless steel at 550°C (1022°F) was shown to reduce cyclic life by as much as a factor of four compared to zero hold time (Yamamoto et al, 1986; Kobatake et al, 1986). Significantly greater factors have been shown in polished bar tests at higher temperatures. Some guidance for elevated-temperature design of metallic bellows expansion joints is provided by Becht (1989a). Criteria developed in the Japanese Liquid Metal Fast Breeder Reactor Program are described by Tsukimori et al (1989). Broyles (1994) developed a simplified design method based on elevated-temperature bellows creep fatigue testing, which was incorporated in the Seventh Edition of the EJMA Standards.

19.5 SHOP TEST OF EXPANSION JOINTS

Because of concerns regarding the EJMA pressure design rules, the leak test for bellows was changed so that it is effectively a proof test. The purpose was to demonstrate a minimum design margin of 1.5 against gross deformations, with specific concerns with respect to in-plane squirm and high meridional bending stresses due to internal pressure. Therefore, even pneumatic tests are required to be performed at or above 1.5 times the design pressure. Specifically, including the temperature correction factor, the test pressure is given by the equation

$$P_T = \frac{1.5PS_T}{S} \tag{19.1}$$

where

P_T = minimum test gage pressure
S_T = allowable stress at test temperature
S = allowable stress at design temperature
P = design gage pressure

However, an exception to the above is required, because a problem can be caused by the temperature adjustment to the test pressure (Becht, 1989b). It can cause bellows that would perform satisfactorily in service to fail by buckling during the test. The test pressure in the Code is normally increased by the ratio of the allowable stress at the test temperature to the allowable stress at the design temperature. However, the strength properties on which the allowable stress is based vary to a much greater extent with temperature than the elastic modulus, which governs column squirm-type buckling of bellows. Therefore, the temperature adjustment for strength properties will overadjust the test pressure with respect to stability. Because of this, the Code provides that the test pressure need not exceed

$$\text{Minimum } P_T = \frac{1.5 P_S E_A}{E} \tag{19.2}$$

where

P_S = limiting design gauge pressure based on column instability
E_A = Young's modulus at test temperature
E = Young's modulus at design temperature

The test pressure is required to be a minimum of 1.5 times the design pressure.

Hardware on the expansion joint that is intended to resist pressure thrust forces must be included in the test and used to restrain the expansion joint without additional axial restraint during the pressure test. This essentially qualifies the hardware. It is to a certain extent more severe than actually necessary, because the temperature correction factor applied to the test pressure may not be applicable to the hardware, which may operate at temperatures closer to ambient. Therefore, the hardware must be designed with the test pressure condition in mind.

19.6 FIELD TEST OF PIPING SYSTEMS WITH EXPANSION JOINTS

One of the problems in the piping system caused by insertion of an expansion joint is the need to carry pressure thrust loads (end loads) in hardware on the expansion joint or restraints on the piping system. These same restraints must carry the pressure thrust of the expansion joint in the hydrotest condition, if the expansion joint is installed in the system during the test. The temperature correction factor for the system test pressure can result in a pressure thrust force that would fail system restraints. This can be readily observed from the following example.

Consider a system with an allowable stress of 10 at the design temperature and an allowable stress of 20 at the test temperature. The system test pressure could then be three times the design pressure of the system (1.5 x 20/10). During the test, the pipe may be at a stress as high as 1.5 times the allowable stress at the test temperature; the result will generally be a test at yield stress or less for the piping. Consider that a bellows is included in the system and structural members are used to restrain the pressure thrust forces in the piping. These members would generally be designed for ambient temperature operation in accordance with the American Institute of Steel Construction *Manual of Steel Construction.* Axial stress in the structural members caused by these pressure thrust forces may be 60% of yield stress at the design pressure. Stress at the system test pressure could then be 1.8 (3 times 0.6) times the material's yield strength. Thus, the system structural restraints, properly designed for the design condition, could fail at the test pressure condition due to the temperature correction factor, which is not really applicable to the ambient temperature structural members.

However, because the resistance of the structural members to the pressure thrust force is essential for system integrity, and because inadequate design for pressure thrust has caused significant system failures, the structural members should be tested as part of the system leak test. Therefore, the ASME B31.3 Code requires that expansion joints in systems in which anchors are used to restrain the pressure thrust force be included in the system pressure test, without temporary restraints, up to 150% of the design pressure. The structural members should be acceptable for this test load. For continuation of the test to above 150% of the design pressure, the expansion joint may be provided with temporary restraints or removed from the system. Expansion joints should never be subjected to a test pressure above the shop test pressure.

Insert 19.1 Pressure Thrust Considerations

One of the key characteristics of expansion joints in design is pressure thrust. The failure to properly consider pressure thrust has perhaps led to most of the system-design-error-related failures of piping systems with expansion joints.

Pressure thrust is simply the internal pressure in the pipe pushing on the ends, at changes of direction such as elbows and tees, and on partial surfaces such as reducers, orifice plates, and pressure letdown valves. It is the force that causes the $Pr/2t$ stress discussed in Chapter 4, Insert 4.2. In a straight pipe without expansion joints, it is simply the internal pressure times the internal area of the pipe. The inclusion of an expansion joint does not so much create a pressure thrust as cut the means that normally carries it, the pipe. Because the bellows or slip joint is not normally capable of carrying this pressure thrust load, some other means must be provided. This can be hardware, such as tie rods, hinges, or gimbals, provided with the expansion joint, or external anchors. Failure to provide such restraint typically results in a failed expansion joint.

The actual magnitude of the pressure thrust load in a metallic bellows expansion joint is actually higher than the pressure times the internal area of the pipe. This is because the included area is larger at the expansion joint. The pressure thrust force is the pressure times the bellows effective area, which is the area included in a circle defined by the mean diameter of the bellows (roughly the average of the inside and the outside diameters of the bellows). Note that the pressure times the internal area of the pipe acts at changes in cross section, such as at changes in direction. The difference between the bellows pressure thrust and this force acts at the ends of the bellows. When the bellows is under internal pressure, this force compresses the pipe on either end of the bellows if the load is taken by some external anchor remote from the expansion joint. As a result, guides may be required to prevent buckling of the pipe.

In some systems, it is important to understand the locations of the forces and how they are carried. For example, consider a pipe connected to a vessel nozzle. The pipe extends straight from the vessel to an elbow. Without an expansion joint in that straight run. there is a pressure thrust load of pressure times the pipe inside area, pulling (with internal pressure) on the nozzle. The net load on the vessel is zero, because this load is offset by the pressure acting on the wall of the vessel opposite the nozzle. If an expansion joint is placed in the straight run, with an external anchor at the change of direction, the situation is different. The load on the vessel nozzle will very likely be significantly reduced. It will be the pressure times the difference between the effective bellows area and the pipe inside area. Further, it will be compression rather than tension. The vessel, however, will see an unbalanced load equal to the pressure times the effective bellows area. This is the sum of the compressive load acting on the nozzle and the unbalance pressure acting on the vessel wall opposite the nozzle. Note that under conditions of vacuum, the pressure thrust acts in reverse, tending to compress rather than stretch a bellows.

Detailed guidance on design of piping systems with expansion joints can be found in the Standards of the Expansion Joint Manufacturers Association.

APPENDIX

I

Properties of Pipe and Pressure Ratings of Listed Piping Components

I-1 THREAD AND GROOVE DEPTHS

The following table provides the nominal thread depth, *h,* from Table 2 of ASME B1.20.1 *Pipe Threads, General Purpose (Inch)*. It is applicable for pressure design of pipe when ASME B1.20.1 threads are specified.. It is 0.8 times the thread pitch. The thread depth must be considered as a mechanical allowance in the pressure design of straight pipe in accordance with the base Code.

Nominal Pipe Size	Height of Thread, /7(in.)
$^1/_{16}$	0.02963
$^1/_8$	0.02963
$^1/_4$	0.04444
$^3/_8$	0.04444
$^1/_2$	0.05714
$^3/_4$	0.05714
1	0.06957
$1^1/_4$	0.06957
$1^1/_2$	0.06957
2	0.06957
$2^1/_2$	0.100
3	0.100
$3^1/_2$	0.100
4	0.100
5	0.100
6	0.100
8	0.100
10	0.100

(Continued)

12	0.100
14 O.D.	0.100
16 O.D.	0.100
18 O.D.	0.100
20 O.D.	0.100
24 O.D.	0.100

The following table provides the thread depth from Table 1 of ISO 7-1 *Pipe threads where pressure-tight joints are made on the threads - Part 1: Dimensions, tolerances and designation.* It is applicable for pressure design of pipe when ISO 7 threads are specified. The thread depth must be considered as a mechanical allowance in the pressure design of straight pipe in accordance with the base Code.

Nominal Pipe Size	Height of Threads, h (mm)
1/16	0.581
1/8	0.581
1/4	0.856
3/8	0.856
1/2	1.162
3/4	1.162
1	1.479
1 ¼	1.479
1 ½	1.479
2	1.479
2 ½	1.479
3	1.479
4	1.479
5	1.479
6	1.479

The following table provides standard cut groove depths for pipe from Victaulic®. Roll grooving does not remove metal from the pipe and therefore need not be considered in the allowances in pressure design. The cut groove depth must be considered in the allowances in pressure design.

Nominal Size (in.)	Groove Depth, *D* (in.)	Minimum Allowable Pipe Wall Thickness, *T* (in.)
3/4	0.056	0.113
1	0.063	0.133
1 1/4	0.063	0.140
1 1/2	0.063	0.145
2	0.063	0.154
2 1/2	0.078	0.188

3 O.D.	0.078	0.188
3	0.078	0.188
$3^1/_2$	0.083	0.188
4	0.083	0.203
$4\ ^1/_4$ O.D.	0.083	0.203
$4^1/_2$	0.083	0.203
$5\ ^1/_4$ O.D.	0.083	0.203
$5\ ^1/_2$ O.D.	0.083	0.203
5	0.084	0.203
6 O.D.	0.085	0.219
$6\ ^1/_4$ O.D	0.109	0.249
$6\ ^1/_2$ O.D.	0.085	0.219
6	0.085	0.219
8 O.D.	0.092	0.238
8	0.092	0.238
10 O.D.	0.094	0.250
10	0.094	0.250
12 O.D.	0.109	0.279
12	0.109	0.279
14 O.D.	0.109	0.281
15 O.D.	0.109	0.312
16 O.D.	0.109	0.312
18 O.D.	0.109	0.312
20 O.D.	0.109	0.312
22 O.D.	0.172	0.375
24 O.D.	0.172	0.375
30 O.D.	0.250	0.625

The groove depth, *D,* is the nominal groove depth which is required to be uniform. The thickness, *T,* is the minimum pipe wall thickness that may be cut grooved. Note that the pressure rating of the fitting, provided by the manufacturer, may govern the pressure design. Also note that the pressure rating of Victaulic® couplings on roll and cut grooved standard wall pipes are the same.

I-2 PROPERTIES OF PIPE

The following table provides properties for nominal pipe. The schedule numbers are provided as follows.

a. Pipe schedule numbers per ASME B36.10, Welded and Seamless Wrought Steel Pipe.
b. Pipe nominal wall thickness designation per ASME B36.10.
c. Pipe schedule numbers per ASME B36.19, Stainless Steel Pipe.

The weight per foot of pipe is based on carbon steel.

PROPERTIES OF PIPE
(Courtesy of Anvil International)

The following formulas are used in the computation of the values shown in the table:

†weight of pipe per foot (pounds)	$= 10.6802t\,(D - t)$
weight of water per foot (pounds)	$= 0.3405d^2$
square feet outside surface per foot	$= 0.2618D$
square feet inside surface per foot	$= 0.2618d$
inside area (square inches)	$= 0.785d^2$
area of metal (square inches)	$= 0.785\,(D^2 - d^2)$
moment of inertia (inches⁴)	$= 0.0491\,(D^4 - d^4)$
	$= A_m R_g^{\,2}$
section modulus (inches³)	$= \dfrac{0.0982\,(D^4 - d^4)}{D}$
radius of gyration (inches)	$= 0.25\,\sqrt{D^2 + d^2}$

A_m = area of metal (square inches)
d = inside diameter (inches)
D = outside diameter (inches)
R_g = radius of gyration (inches)
t = pipe wall thickness (inches)

† The ferritic steels may be about 5% less, and the austenitic stainless steels about 2% greater than the values shown in this table which are based on weights for carbon steel.

***schedule numbers**

Standard weight pipe and schedule 40 are the same in all sizes through 10-inch; from 12-inch through 24-inch, standard weight pipe has a wall thickness of ⅜-inch.

Extra strong weight pipe and schedule 80 are the same in all sizes through 8-inch; from 8-inch through 24-inch, extra strong weight pipe has a wall thickness of ½-inch.

Double extra strong weight pipe has no corresponding schedule number.

a: ANSI B36.10 steel pipe schedule numbers

b: ANSI B36.10 steel pipe nominal wall thickness designation

c: ANSI B36.19 stainless steel pipe schedule numbers

nominal pipe size *outside diameter*, in.	**schedule number***			**wall thickness, in.**	**inside diameter, in.**	**inside area, sq. in.**	**metal area, sq. in.**	**sq ft outside surface, per ft**	**sq ft inside surface, per ft**	**weight per ft, lb†**	**weight of water per ft, lb**	**moment of inertia, in.⁴**	**section modulus, in.³**	**radius gyration, in.**
	a	**b**	**c**											
⅛	—	—	10S	0.049	0.307	0.0740	0.0548	0.106	0.0804	0.186	0.0321	0.00088	0.00437	0.1271
0.405	40	Std	40S	0.068	0.269	0.0568	0.0720	0.106	0.0705	0.245	0.0246	0.00106	0.00525	0.1215
	80	XS	80S	0.095	0.215	0.0364	0.0925	0.106	0.0563	0.315	0.0157	0.00122	0.00600	0.1146
¼	—	—	10S	0.065	0.410	0.1320	0.0970	0.141	0.1073	0.330	0.0572	0.00279	0.01032	0.1694
0.540	40	Std	40S	0.088	0.364	0.1041	0.1250	0.141	0.0955	0.425	0.0451	0.00331	0.01230	0.1628
	80	XS	80S	0.119	0.302	0.0716	0.1574	0.141	0.0794	0.535	0.0310	0.00378	0.01395	0.1547
⅜	—	—	10S	0.065	0.545	0.2333	0.1246	0.177	0.1427	0.423	0.1011	0.00586	0.01737	0.2169
0.675	40	Std	40S	0.091	0.493	0.1910	0.1670	0.177	0.1295	0.568	0.0827	0.00730	0.02160	0.2090
	80	XS	80S	0.126	0.423	0.1405	0.2173	0.177	0.1106	0.739	0.0609	0.00862	0.02554	0.1991
	—	—	5S	0.065	0.710	0.3959	0.1583	0.220	0.1859	0.538	0.171	0.0120	0.0285	0.2750
	—	—	10S	0.083	0.674	0.357	0.1974	0.220	0.1765	0.671	0.1547	0.01431	0.0341	0.2692
½	40	Std	40S	0.109	0.622	0.304	0.2503	0.220	0.1628	0.851	0.1316	0.01710	0.0407	0.2613
0.840	80	XS	80S	0.147	0.546	0.2340	0.320	0.220	0.1433	1.088	0.1013	0.02010	0.0478	0.2505
	160	—	—	0.187	0.466	0.1706	0.383	0.220	0.1220	1.304	0.0740	0.02213	0.0527	0.2402
	—	XXS	—	0.294	0.252	0.0499	0.504	0.220	0.0660	1.714	0.0216	0.02425	0.0577	0.2192
	—	—	5S	0.065	0.920	0.665	0.2011	0.275	0.2409	0.684	0.2882	0.02451	0.0467	0.349
	—	—	10S	0.083	0.884	0.614	0.2521	0.275	0.2314	0.857	0.2661	0.02970	0.0566	0.343
¾	40	Std	40S	0.113	0.824	0.533	0.333	0.275	0.2157	1.131	0.2301	0.0370	0.0706	0.334
1.050	80	XS	80S	0.154	0.742	0.432	0.435	0.275	0.1943	1.474	0.1875	0.0448	0.0853	0.321
	160	—	—	0.218	0.614	0.2961	0.570	0.275	0.1607	1.937	0.1284	0.0527	0.1004	0.304
	—	XXS	—	0.308	0.434	0.1479	0.718	0.275	0.1137	2.441	0.0641	0.0579	0.1104	0.2840
	—	—	5S	0.065	1.185	1.103	0.2553	0.344	0.310	0.868	0.478	0.0500	0.0760	0.443
	—	—	10S	0.109	1.097	0.945	0.413	0.344	0.2872	1.404	0.409	0.0757	0.1151	0.428
1	40	Std	40S	0.133	1.049	0.864	0.494	0.344	0.2746	1.679	0.374	0.0874	0.1329	0.421
1.315	80	XS	80S	0.179	0.957	0.719	0.639	0.344	0.2520	2.172	0.311	0.1056	0.1606	0.407
	160	—	—	0.250	0.815	0.522	0.836	0.344	0.2134	2.844	0.2261	0.1252	0.1903	0.387
	—	XXS	—	0.358	0.599	0.2818	1.076	0.344	0.1570	3.659	0.1221	0.1405	0.2137	0.361
	—	—	5S	0.065	1.530	1.839	0.326	0.434	0.401	1.107	0.797	0.1038	0.1250	0.564
	—	—	10S	0.109	1.442	1.633	0.531	0.434	0.378	1.805	0.707	0.1605	0.1934	0.550
1¼	40	Std	40S	0.140	1.380	1.496	0.669	0.434	0.361	2.273	0.648	0.1948	0.2346	0.540
1.660	80	XS	80S	0.191	1.278	1.283	0.881	0.434	0.335	2.997	0.555	0.2418	0.2913	0.524
	160	—	—	0.250	1.160	1.057	1.107	0.434	0.304	3.765	0.4580	0.2839	0.342	0.506
	—	XXS	—	0.382	0.896	0.0631	1.534	0.434	0.2346	5.214	0.2732	0.341	0.411	0.472
1½	—	—	5S	0.065	1.770	2.461	0.375	0.497	0.463	1.274	1.067	0.1580	0.1663	0.649
1.900	—	—	10S	0.109	1.682	2.222	0.613	0.497	0.440	2.085	0.962	0.2469	0.2599	0.634

PROPERTIES OF PIPE (CONTINUED)
(Courtesy of Anvil International)

nominal pipe size *outside diameter,* in.	schedule number* a	b	c	wall thick-ness, in.	inside diam-eter, in.	inside area, sq. in.	metal area, sq. in.	sq ft outside surface, per ft	sq ft inside surface, per ft	weight per ft, lb†	weight of water per ft, lb	moment of inertia, in.⁴	section modul-lus, in.³	radius gyra-tion, in.
	40	Std	40S	0.145	1.610	2.036	0.799	0.497	0.421	2.718	0.882	0.310	0.326	0.623
	80	XS	80S	0.200	1.500	1.767	1.068	0.497	0.393	3.631	0.765	0.391	0.412	0.605
1½	160	—	—	0.281	1.338	1.406	1.429	0.497	0.350	4.859	0.608	0.483	0.508	0.581
1.900	—	XXS	—	0.400	1.100	0.950	1.885	0.497	0.288	6.408	0.412	0.568	0.598	0.549
	—	—	—	0.525	0.850	0.567	2.267	0.497	0.223	7.710	0.246	0.6140	0.6470	0.5200
	—	—	—	0.650	0.600	0.283	2.551	0.497	0.157	8.678	0.123	0.6340	0.6670	0.4980
	—	—	5S	0.065	2.245	3.96	0.472	0.622	0.588	1.604	1.716	0.315	0.2652	0.817
	—	—	10S	0.109	2.157	3.65	0.776	0.622	0.565	2.638	1.582	0.499	0.420	0.802
2	40	Std	40S	0.154	2.067	3.36	1.075	0.622	0.541	3.653	1.455	0.666	0.561	0.787
2.375	80	XS	80S	0.218	1.939	2.953	1.477	0.622	0.508	5.022	1.280	0.868	0.731	0.766
	160	—	—	0.343	1.689	2.240	2.190	0.622	0.442	7.444	0.971	1.163	0.979	0.729
	—	XXS	—	0.436	1.503	1.774	2.656	0.622	0.393	9.029	0.769	1.312	1.104	0.703
	—	—	—	0.562	1.251	1.229	3.199	0.622	0.328	10.882	0.533	1.442	1.2140	0.6710
	—	—	—	0.687	1.001	0.787	3.641	0.622	0.262	12.385	0.341	1.5130	1.2740	0.6440
	—	—	5S	0.083	2.709	5.76	0.728	0.753	0.709	2.475	2.499	0.710	0.494	0.988
	—	—	10S	0.120	2.635	5.45	1.039	0.753	0.690	3.531	2.361	0.988	0.687	0.975
	40	Std	40S	0.203	2.469	4.79	1.704	0.753	0.646	5.793	2.076	1.530	1.064	0.947
2½	80	XS	80S	0.276	2.323	4.24	2.254	0.753	0.608	7.661	1.837	1.925	1.339	0.924
2.875	160	—	—	0.375	2.125	3.55	2.945	0.753	0.556	10.01	1.535	2.353	1.637	0.894
	—	XXS	—	0.552	1.771	2.464	4.03	0.753	0.464	13.70	1.067	2.872	1.998	0.844
	—	—	—	0.675	1.525	1.826	4.663	0.753	0.399	15.860	0.792	3.0890	2.1490	0.8140
	—	—	—	0.800	1.275	1.276	5.212	0.753	0.334	17.729	0.554	3.2250	2.2430	0.7860
	—	—	5S	0.083	3.334	8.73	0.891	0.916	0.873	3.03	3.78	1.301	0.744	1.208
	—	—	10S	0.120	3.260	8.35	1.274	0.916	0.853	4.33	3.61	1.822	1.041	1.196
	40	Std	40S	0.216	3.068	7.39	2.228	0.916	0.803	7.58	3.20	3.02	1.724	1.164
3	80	XS	80S	0.300	2.900	6.61	3.02	0.916	0.759	10.25	2.864	3.90	2.226	1.136
3.500	160	—	—	0.437	2.626	5.42	4.21	0.916	0.687	14.32	2.348	5.03	2.876	1.094
	—	XXS	—	0.600	2.300	4.15	5.47	0.916	0.602	18.58	1.801	5.99	3.43	1.047
	—	—	—	0.725	2.050	3.299	6.317	0.916	0.537	21.487	1.431	6.5010	3.7150	1.0140
	—	—	—	0.850	1.800	2.543	7.073	0.916	0.471	24.057	1.103	6.8530	3.9160	0.9840
	—	—	5S	0.083	3.834	11.55	1.021	1.047	1.004	3.47	5.01	1.960	0.980	1.385
3½	—	—	10S	0.120	3.760	11.10	1.463	1.047	0.984	4.97	4.81	2.756	1.378	1.372
4.000	40	Std	40S	0.226	3.548	9.89	2.680	1.047	0.929	9.11	4.28	4.79	2.394	1.337
	80	XS	80S	0.318	3.364	8.89	3.68	1.047	0.881	12.51	3.85	6.28	3.14	1.307
	—	XXS	—	0.636	2.728	5.845	6.721	1.047	0.716	22.850	2.530	9.8480	4.9240	1.2100
	—	—	5S	0.083	4.334	14.75	1.152	1.178	1.135	3.92	6.40	2.811	1.249	1.562
	—	—	10S	0.120	4.260	14.25	1.651	1.178	1.115	5.61	6.17	3.96	1.762	1.549
	—	—	—	0.188	4.124	13.357	2.547	1.178	1.082	8.560	5.800	5.8500	2.6000	1.5250
	40	Std	40S	0.237	4.026	12.73	3.17	1.178	1.054	10.79	5.51	7.23	3.21	1.510
4	80	XS	80S	0.337	3.826	11.50	4.41	1.178	1.002	14.98	4.98	9.61	4.27	1.477
4.500	120	—	—	0.437	3.626	10.33	5.58	1.178	0.949	18.96	4.48	11.65	5.18	1.445
	—	—	—	0.500	3.500	9.621	6.283	1.178	0.916	21.360	4.160	12.7710	5.6760	1.4250
	160	—	—	0.531	3.438	9.28	6.62	1.178	0.900	22.51	4.02	13.27	5.90	1.416
	—	XXS	—	0.674	3.152	7.80	8.10	1.178	0.825	27.54	3.38	15.29	6.79	1.374
	—	—	—	0.800	2.900	6.602	9.294	1.178	0.759	31.613	2.864	16.6610	7.4050	1.3380
	—	—	—	0.925	2.650	5.513	10.384	1.178	0.694	35.318	2.391	17.7130	7.8720	1.3060
	—	—	5S	0.109	5.345	22.44	1.868	1.456	1.399	6.35	9.73	6.95	2.498	1.929
	—	—	10S	0.134	5.295	22.02	2.285	1.456	1.386	7.77	9.53	8.43	3.03	1.920
	40	Std	40S	0.258	5.047	20.01	4.30	1.456	1.321	14.62	8.66	15.17	5.45	1.878
5	80	XS	80S	0.375	4.813	18.19	6.11	1.456	1.260	20.78	7.89	20.68	7.43	1.839
5.563	120	—	—	0.500	4.563	16.35	7.95	1.456	1.195	27.04	7.09	25.74	9.25	1.799
	160	—	—	0.625	4.313	14.61	9.70	1.456	1.129	32.96	6.33	30.0	10.80	1.760
	—	XXS	—	0.750	4.063	12.97	11.34	1.456	1.064	38.55	5.62	33.6	12.10	1.722
	—	—	—	0.875	3.813	11.413	12.880	1.456	0.998	43.810	4.951	36.6450	13.1750	1.6860
	—	—	—	1.000	3.563	9.966	14.328	1.456	0.933	47.734	4.232	39.1110	14.0610	1.6520

PROPERTIES OF PIPE (CONTINUED)
(Courtesy of Anvil International)

nominal pipe size *outside diameter,* in.	schedule number*			wall thick-ness, in.	inside diam-eter, in.	inside area, sq. in.	metal area, sq. in.	sq ft outside surface, per ft	sq ft inside surface, per ft	weight per ft, lb†	weight of water per ft, lb	moment of inertia, in.⁴	section modul-lus, in.³	radius gyra-tion, in.
	a	b	c											
	—	—	5S	0.109	6.407	32.2	2.231	1.734	1.677	5.37	13.98	11.85	3.58	2.304
	—	—	10S	0.134	6.357	31.7	2.733	1.734	1.664	9.29	13.74	14.40	4.35	2.295
	—	—	—	0.129	6.187	30.100	4.410	1.734	1.620	15.020	13.100	22.6600	6.8400	2.2700
	40	Std	40S	0.280	6.065	28.89	5.58	1.734	1.588	18.97	12.51	28.14	8.50	2.245
6	80	XS	80S	0.432	5.761	26.07	8.40	1.734	1.508	28.57	11.29	40.5	12.23	2.195
6.625	120	—	—	0.562	5.501	23.77	10.70	1.734	1.440	36.39	10.30	49.6	14.98	2.153
	160	—	—	0.718	5.189	21.15	13.33	1.734	1.358	45.30	9.16	59.0	17.81	2.104
	—	XXS	—	0.864	4.897	18.83	15.64	1.734	1.282	53.16	8.17	66.3	20.03	2.060
	—	—	—	1.000	4.625	16.792	17.662	1.734	1.211	60.076	7.284	72.1190	21.7720	2.0200
	—	—	—	1.125	4.375	15.025	19.429	1.734	1.145	66.084	6.517	76.5970	23.1240	1.9850
	—	—	5S	0.109	8.407	55.5	2.916	2.258	2.201	9.91	24.07	26.45	6.13	3.01
	—	—	10S	0.148	8.329	54.5	3.94	2.258	2.180	13.40	23.59	35.4	8.21	3.00
	—	—	—	0.219	8.187	52.630	5.800	2.258	2.150	19.640	22.900	51.3200	11.9000	2.9700
8	20	—	—	0.250	8.125	51.8	6.58	2.258	2.127	22.36	22.48	57.7	13.39	2.962
8.625	30	—	—	0.277	8.071	51.2	7.26	2.258	2.113	24.70	22.18	63.4	14.69	2.953
	40	Std	40S	0.322	7.981	50.0	8.40	2.258	2.089	28.55	21.69	72.5	16.81	2.938
	60	—	—	0.406	7.813	47.9	10.48	2.258	2.045	35.64	20.79	88.8	20.58	2.909
	80	XS	80S	0.500	7.625	45.7	12.76	2.258	1.996	43.39	19.80	105.7	24.52	2.878
	100	—	—	0.593	7.439	43.5	14.96	2.258	1.948	50.87	18.84	121.4	28.14	2.847
	120	—	—	0.718	7.189	40.6	17.84	2.258	1.882	60.63	17.60	140.6	32.6	2.807
8	140	—	—	0.812	7.001	38.5	19.93	2.258	1.833	67.76	16.69	153.8	35.7	2.777
8.625	160	—	—	0.906	6.813	36.5	21.97	2.258	1.784	74.69	15.80	165.9	38.5	2.748
	—	—	—	1.000	6.625	34.454	23.942	2.258	1.734	81.437	14.945	177.1320	41.0740	2.7190
	—	—	—	1.125	6.375	31.903	26.494	2.258	1.669	90.114	13.838	190.6210	44.2020	2.6810
	—	—	5S	0.134	10.482	86.3	4.52	2.815	2.744	15.15	37.4	63.7	11.85	3.75
	—	—	10S	0.165	10.420	85.3	5.49	2.815	2.728	18.70	36.9	76.9	14.30	3.74
	—	—	—	0.219	10.312	83.52	7.24	2.815	2.70	24.63	36.2	100.46	18.69	3.72
	20	—	—	0.250	10.250	82.5	8.26	2.815	2.683	28.04	35.8	113.7	21.16	3.71
	30	—	—	0.307	10.136	80.7	10.07	2.815	2.654	34.24	35.0	137.5	25.57	3.69
	40	Std	40S	0.365	10.020	78.9	11.91	2.815	2.623	40.48	34.1	160.8	29.90	3.67
10	60	XS	80S	0.500	9.750	74.7	16.10	2.815	2.553	54.74	32.3	212.0	39.4	3.63
10.750	80	—	—	0.593	9.564	71.8	18.92	2.815	2.504	64.33	31.1	244.9	45.6	3.60
	100	—	—	0.718	9.314	68.1	22.63	2.815	2.438	76.93	29.5	286.2	53.2	3.56
	120	—	—	0.843	9.064	64.5	26.24	2.815	2.373	89.20	28.0	324	60.3	3.52
	—	—	—	0.875	9.000	63.62	27.14	2.815	2.36	92.28	27.6	333.46	62.04	3.50
	140	—	—	1.000	8.750	60.1	30.6	2.815	2.291	104.13	26.1	368	68.4	3.47
	160	—	—	1.125	8.500	56.7	34.0	2.815	2.225	115.65	24.6	399	74.3	3.43
	—	—	—	1.250	8.250	53.45	37.31	2.815	2.16	126.82	23.2	428.17	79.66	3.39
	—	—	—	1.500	7.750	47.15	43.57	2.815	2.03	148.19	20.5	478.59	89.04	3.31
	—	—	5S	0.156	12.438	121.4	6.17	3.34	3.26	20.99	52.7	122.2	19.20	4.45
	—	—	10S	0.180	12.390	120.6	7.11	3.34	3.24	24.20	52.2	140.5	22.03	4.44
	20	—	—	0.250	12.250	117.9	9.84	3.34	3.21	33.38	51.1	191.9	30.1	4.42
	30	—	—	0.330	12.090	114.8	12.88	3.34	3.17	43.77	49.7	248.5	39.0	4.39
	—	Std	40S	0.375	12.000	113.1	14.58	3.34	3.14	49.56	49.0	279.3	43.8	4.38
	40	—	—	0.406	11.938	111.9	15.74	3.34	3.13	53.53	48.5	300	47.1	4.37
	—	XS	80S	0.500	11.750	108.4	19.24	3.34	3.08	65.42	47.0	362	56.7	4.33
12	60	—	—	0.562	11.626	106.2	21.52	3.34	3.04	73 16	46.0	401	62.8	4.31
12.750	80	—	—	0.687	11.376	101.6	26.04	3.34	2.978	88.51	44.0	475	74.5	4.27
	—	—	—	0.750	11.250	99.40	28.27	3.34	2.94	96.2	43.1	510.7	80.1	4.25
	100	—	—	0.843	11.064	96.1	31.5	3.34	2.897	107.20	41.6	562	88.1	4.22
	—	—	—	0.875	11.000	95.00	32.64	3.34	2.88	110.9	41.1	578.5	90.7	4.21
	120	—	—	1.000	10.750	90.8	36.9	3.34	2.814	125.49	39.3	642	100.7	4.17
	140	—	—	1.125	10.500	86.6	41.1	3.34	2.749	139.68	37.5	701	109.9	4.13
	—	—	—	1.250	10.250	82.50	45.16	3.34	2.68	153.6	35.8	755.5	118.5	4.09
	160	—	—	1.312	10.126	80.5	47.1	3.34	2.651	160.27	34.9	781	122.6	4.07

PROPERTIES OF PIPE (CONTINUED)
(Courtesy of Anvil International)

nominal pipe size *outside diameter,* in.	schedule number* a	b	c	wall thick-ness, in.	inside diam-eter, in.	inside area, sq. in.	metal area, sq. in.	sq ft outside surface, per ft	sq ft inside surface, per ft	weight per ft, lb†	weight of water per ft, lb	moment of inertia, in.⁴	section modul-lus, in.³	radius gyra-tion, in.
	—	—	5S	0.156	13.688	147.20	6.78	3.67	3.58	23.0	63.7	162.6	23.2	4.90
	—	—	10S	0.188	13.624	145.80	8.16	3.67	3.57	27.7	63.1	194.6	27.8	4.88
	—	—	—	0.210	13.580	144.80	9.10	3.67	3.55	30.9	62.8	216.2	30.9	4.87
	—	—	—	0.219	13.562	144.50	9.48	3.67	3.55	32.2	62.6	225.1	32.2	4.87
	10	—	—	0.250	13.500	143.1	10.80	3.67	3.53	36.71	62.1	255.4	36.5	4.86
	—	—	—	0.281	13.438	141.80	12.11	3.67	3.52	41.2	61.5	285.2	40.7	4.85
	20	—	—	0.312	13.376	140.5	13.42	3.67	3.50	45.68	60.9	314	44.9	4.84
	—	—	—	0.344	13.312	139.20	14.76	3.67	3.48	50.2	60.3	344.3	49.2	4.83
14	30	Std	—	0.375	13.250	137.9	16.05	3.67	3.47	54.57	59.7	373	53.3	4.82
14.000	40	—	—	0.437	13.126	135.3	18.62	3.67	3.44	63.37	58.7	429	61.2	4.80
	—	—	—	0.469	13.062	134.00	19.94	3.67	3.42	67.8	58.0	456.8	65.3	4.79
	—	XS	—	0.500	13.000	132.7	21.21	3.67	3.40	72.09	57.5	484	69.1	4.78
	60	—	—	0.593	12.814	129.0	24.98	3.67	3.35	84.91	55.9	562	80.3	4.74
	—	—	—	0.625	12.750	127.7	26.26	3.67	3.34	89.28	55.3	589	84.1	4.73
	80	—	—	0.750	12.500	122.7	31.2	3.67	3.27	106.13	53.2	687	98.2	4.69
	100	—	—	0.937	12.126	115.5	38.5	3.67	3.17	130.73	50.0	825	117.8	4.63
	120	—	—	1.093	11.814	109.6	44.3	3.67	3.09	150.67	47.5	930	132.8	4.58
	140	—	—	1.250	11.500	103.9	50.1	3.67	3.01	170.22	45.0	1028	146.8	4.53
	160	—	—	1.406	11.188	98.3	55.6	3.67	2.929	189.12	42.6	1117	159.6	4.48
	—	—	5S	0.165	15.670	192.90	8.21	4.19	4.10	28	83.5	257	32.2	5.60
	—	—	10S	0.188	15.624	191.70	9.34	4.19	4.09	32	83.0	292	36.5	5.59
	10	—	—	0.250	15.500	188.7	12.37	4.19	4.06	42.05	81.8	384	48.0	5.57
	20	—	—	0.312	15.376	185.7	15.38	4.19	4.03	52.36	80.5	473	59.2	5.55
	30	Std	—	0.375	15.250	182.6	18.41	4.19	3.99	62.58	79.1	562	70.3	5.53
16	40	XS	—	0.500	15.000	176.7	24.35	4.19	3.93	82.77	76.5	732	91.5	5.48
16.000	60	—	—	0.656	14.688	169.4	31.6	4.19	3.85	107.50	73.4	933	116.6	5.43
	80	—	—	0.843	14.314	160.9	40.1	4.19	3.75	136.46	69.7	1157	144.6	5.37
	100	—	—	1.031	13.938	152.6	48.5	4.19	3.65	164.83	66.1	1365	170.6	5.30
	120	—	—	1.218	13.564	144.5	56.6	4.19	3.55	192.29	62.6	1556	194.5	5.24
	140	—	—	1.437	13.126	135.3	65.7	4.19	3.44	223.64	58.6	1760	220.0	5.17
	160	—	—	1.593	12.814	129.0	72.1	4.19	3.35	245.11	55.9	1894	236.7	5.12
	—	—	5S	0.165	17.670	245.20	9.24	4.71	4.63	31	106.2	368	40.8	6.31
	—	—	10S	0.188	17.624	243.90	10.52	4.71	4.61	36	105.7	417	46.4	6.30
	10	—	—	0.250	17.500	240.5	13.94	4.71	4.58	47.39	104.3	549	61.0	6.28
	20	—	—	0.312	17.376	237.1	17.34	4.71	4.55	59.03	102.8	678	75.5	6.25
	—	Std	—	0.375	17.250	233.7	20.76	4.71	4.52	70.59	101.2	807	89.6	6.23
18	30	—	—	0.437	17.126	230.4	24.11	4.71	4.48	82.06	99.9	931	103.4	6.21
18.000	—	XS	—	0.500	17.00	227.0	27.49	4.71	4.45	93.45	98.4	1053	117.0	6.19
	40	—	—	0.562	16.876	223.7	30.8	4.71	4.42	104.75	97.0	1172	130.2	6.17
	60	—	—	0.750	16.500	213.8	40.6	4.71	4.32	138.17	92.7	1515	168.3	6.10
	80	—	—	0.937	16.126	204.2	50.2	4.71	4.22	170.75	88.5	1834	203.8	6.04
	100	—	—	1.156	15.688	193.3	61.2	4.71	4.11	207.96	83.7	2180	242.2	5.97
	120	—	—	1.375	15.250	182.6	71.8	4.71	3.99	244.14	79.2	2499	277.6	5.90
	140	—	—	1.562	14.876	173.8	80.7	4.71	3.89	274.23	75.3	2750	306	5.84
	160	—	—	1.781	14.438	163.7	90.7	4.71	3.78	308.51	71.0	3020	336	5.77
	—	—	5S	0.188	19.634	302.40	11.70	5.24	5.14	40	131.0	574	57.4	7.00
	—	—	10S	0.218	19.564	300.60	13.55	5.24	5.12	46	130.2	663	66.3	6.99
	10	—	—	0.250	19.500	298.60	15.51	5.24	5.11	52.73	129.5	757	75.7	6.98
	20	Std	—	0.375	19.250	291.0	23.12	5.24	5.04	78.60	126.0	1114	111.4	6.94
20	30	XS	—	0.500	19.000	283.5	30.6	5.24	4.97	104.13	122.8	1457	145.7	6.90
20,000	40	—	—	0.593	18.814	278.0	36.2	5.24	4.93	122.91	120.4	1704	170.4	6.86
	60	—	—	0.812	18.376	265.2	48.9	5.24	4.81	166.40	115.0	2257	225.7	6.79
	—	—	—	0.875	18.250	261.6	52.6	5.24	4.78	178.73	113.4	2409	240.9	6.77
	80	—	—	1.031	17.938	252.7	61.4	5.24	4.70	208.87	109.4	2772	277.2	6.72
	100	—	—	1.281	17.438	238.8	75.3	5.24	4.57	256.10	103.4	3320	332	6.63

PROPERTIES OF PIPE (CONTINUED)
(Courtesy of Anvil International)

nominal pipe size *outside diameter,* in.	schedule number* a	b	c	wall thick-ness, in.	inside diam-eter, in.	inside area, sq. in.	metal area, sq. in.	sq ft outside surface, per ft	sq ft inside surface, per ft	weight per ft, lb†	weight of water per ft, lb	moment of inertia, in.⁴	section modul-lus, in.³	radius gyra-tion, in.
20 *20.000*	120	—	—	1.500	17.000	227.0	87.2	5.24	4.45	296.37	98.3	3760	376	6.56
	140	—	—	1.750	16.500	213.8	100.3	5.24	4.32	341.10	92.6	4220	422	6.48
	160	—	—	1.968	16.064	202.7	111.5	5.24	4.21	379.01	87.9	4590	459	6.41
22 *22.000*	—	—	5S	0.188	21.624	367.3	12.88	5.76	5.66	44	159.1	766	69.7	7.71
	—	—	10S	0.218	21.564	365.2	14.92	5.76	5.65	51	158.2	885	80.4	7.70
	10	—	—	0.250	21.500	363.1	17.18	5.76	5.63	58	157.4	1010	91.8	7.69
	20	Std	—	0.375	21.250	354.7	25.48	5.76	5.56	87	153.7	1490	135.4	7.65
	30	XS	—	0.500	21.000	346.4	33.77	5.76	5.50	115	150.2	1953	177.5	7.61
	—	—	—	0.625	20.750	338.2	41.97	5.76	5.43	143	146.6	2400	218.2	7.56
	—	—	—	0.750	20.500	330.1	50.07	5.76	5.37	170	143.1	2829	257.2	7.52
	60	—	—	0.875	20.250	322.1	58.07	5.76	5.30	197	139.6	3245	295.0	7.47
	80	—	—	1.125	19.750	306.4	73.78	5.76	5.17	251	132.8	4029	366.3	7.39
	100	—	—	1.375	19.250	291.0	89.09	5.76	5.04	303	126.2	4758	432.6	7.31
	120	—	—	1.625	18.750	276.1	104.02	5.76	4.91	354	119.6	5432	493.8	7.23
	140	—	—	1.875	18.250	261.6	118.55	5.76	4.78	403	113.3	6054	550.3	7.15
	160	—	—	2.125	17.750	247.4	132.68	5.76	4.65	451	107.2	6626	602.4	7.07
24 *24.000*	10	—	—	0.250	23.500	434	18.65	6.28	6.15	63.41	188.0	1316	109.6	8.40
	20	Std	—	0.375	23.250	425	27.83	6.28	6.09	94.62	183.8	1943	161.9	8.35
	—	XS	—	0.500	23.000	415	36.9	6.28	6.02	125.49	180.1	2550	212.5	8.31
	30	—	—	0.562	22.876	411	41.4	6.28	5.99	140.80	178.1	2840	237.0	8.29
	—	—	—	0.625	22.750	406	45.9	6.28	5.96	156.03	176.2	3140	261.4	8.27
	40	—	—	0.687	22.626	402	50.3	6.28	5.92	171.17	174.3	3420	285.2	8.25
	—	—	—	0.750	22.500	398	54.8	6.28	5.89	186.24	172.4	3710	309	8.22
	—	—	5S	0.218	23.564	436.1	16.29	6.28	6.17	55	188.9	1152	96.0	8.41
	—	—	—	0.875	22.250	388.6	63.54	6.28	5.83	216	168.6	4256	354.7	8.18
	60	—	—	0.968	22.064	382	70.0	6.28	5.78	238.11	165.8	4650	388	8.15
	80	—	—	1.218	21.564	365	87.2	6.28	5.65	296.36	158.3	5670	473	8.07
	100	—	—	1.531	20.938	344	108.1	6.28	5.48	367.40	149.3	6850	571	7.96
	120	—	—	1.812	20.376	326	126.3	6.28	5.33	429.39	141.4	7830	652	7.87
	140	—	—	2.062	19.876	310	142.1	6.28	5.20	483.13	134.5	8630	719	7.79
	160	—	—	2.343	19.314	293	159.4	6.28	5.06	541.94	127.0	9460	788	7.70
26 *26.000*	—	—	—	0.250	25.500	510.7	19.85	6.81	6.68	67	221.4	1646	126.6	9.10
	10	—	—	0.312	25.376	505.8	25.18	6.81	6.64	86	219.2	2076	159.7	9.08
	—	Std	—	0.375	25.250	500.7	30.19	6.81	6.61	103	217.1	2478	190.6	9.06
	20	XS	—	0.500	25.000	490.9	40.06	6.81	6.54	136	212.8	3259	250.7	9.02
	—	—	—	0.625	24.750	481.1	49.82	6.81	6.48	169	208.6	4013	308.7	8.98
	—	—	—	0.750	24.500	471.4	59.49	6.81	6.41	202	204.4	4744	364.9	8.93
	—	—	—	0.875	24.250	461.9	69.07	6.81	6.35	235	200.2	5458	419.9	8.89
	—	—	—	1.000	24.000	452.4	78.54	6.81	6.28	267	196.1	6149	473.0	8.85
	—	—	—	1.125	23.750	443.0	87.91	6.81	6.22	299	192.1	6813	524.1	8.80
28 *28.000*	—	—	—	0.250	27.500	594.0	21.80	7.33	7.20	74	257.3	2098	149.8	9.81
	10	—	—	0.312	27.376	588.6	27.14	7.33	7.17	92	255.0	2601	185.8	9.79
	—	Std	—	0.375	27.250	583.2	32.54	7.33	7.13	111	252.6	3105	221.8	9.77
	20	XS	—	0.500	27.000	572.6	43.20	7.33	7.07	147	248.0	4085	291.8	9.72
	30	—	—	0.625	26.750	562.0	53.75	7.33	7.00	183	243.4	5038	359.8	9.68
	—	—	—	0.750	26.500	551.6	64.21	7.33	6.94	218	238.9	5964	426.0	9.64
	—	—	—	0.875	26.250	541.2	74.56	7.38	6.87	253	234.4	6865	490.3	9.60
	—	—	—	1.000	26.000	530.9	84.82	7.33	6.81	288	230.0	7740	552.8	9.55
	—	—	—	1.125	25.750	520.8	94.98	7.33	6.74	323	225.6	8590	613.6	9.51
30 *30.000*	—	—	5S	0.250	29.500	683.4	23.37	7.85	7.72	79	296.3	2585	172.3	10.52
	10	—	10S	0.312	29.376	677.8	29.19	7.85	7.69	99	293.7	3201	213.4	10.50
	—	Std	—	0.375	29.250	672.0	34.90	7.85	7.66	119	291.2	3823	254.8	10.48
	20	XS	—	0.500	29.000	660.5	46.34	7.85	7.59	158	286.2	5033	335.5	10.43
	30	—	—	0.625	28.750	649.2	57.68	7.85	7.53	196	281.3	6213	414.2	10.39

PROPERTIES OF PIPE (CONTINUED)
(Courtesy of Anvil International)

nominal pipe size *outside diameter,* in.	schedule number* a	b	c	wall thick-ness, in.	inside diam-eter, in.	inside area, sq. in.	metal area, sq. in.	sq ft outside surface, per ft	sq ft inside surface, per ft	weight per ft, lb†	weight of water per ft, lb	moment of inertia, in.[4]	section modul-lus, in.[3]	radius gyra-tion, in.
	40	—	—	0.750	28.500	637.9	68.92	7.85	7.46	234	276.6	7371	491.4	10.34
30	—	—	—	0.875	28.250	620.7	80.06	7.85	7.39	272	271.8	8494	566.2	10.30
30.000	—	—	—	1.000	28.000	615.7	91.11	7.85	7.33	310	267.0	9591	639.4	10.26
	—	—	—	1.125	27.750	604.7	102.05	7.85	7.26	347	262.2	10653	710.2	10.22
	—	—	—	0.250	31.500	779.2	24.93	8.38	8.25	85	337.8	3141	196.3	11.22
	10	—	—	0.312	31.376	773.2	31.02	8.38	8.21	106	335.2	3891	243.2	11.20
	—	Std	—	0.375	31.250	766.9	37.25	8.38	8.18	127	332.5	4656	291.0	11.18
	20	XS	—	0.500	31.000	754.7	49.48	8.38	8.11	168	327.2	6140	383.8	11.14
32	30	—	—	0.625	30.750	742.5	61.59	8.38	8.05	209	321.9	7578	473.6	11.09
32.000	40	—	—	0.688	30.624	736.6	67.68	8.38	8.02	230	319.0	8298	518.6	11.07
	—	—	—	0.750	30.500	730.5	73.63	8.38	7.98	250	316.7	8990	561.9	11.05
	—	—	—	0.875	30.250	718.3	85.52	8.38	7.92	291	311.6	10372	648.2	11.01
	—	—	—	1.000	30.000	706.8	97.38	8.38	7.85	331	306.4	11680	730.0	10.95
	—	—	—	1.125	29.750	694.7	109.0	8.38	7.79	371	301.3	13023	814.0	10.92
	—	—	—	0.250	33.500	881.2	26.50	8.90	8.77	90	382.0	3773	221.9	11.93
	10	—	—	0.312	33.376	874.9	32.99	8.90	8.74	112	379.3	4680	275.3	11.91
	—	Std	—	0.375	33.250	867.8	39.61	8.90	8.70	135	376.2	5597	329.2	11.89
	20	XS	—	0.500	33.000	855.3	52.62	8.90	8.64	179	370.8	7385	434.4	11.85
34	30	—	—	0.625	32.750	841.9	65.53	8.90	8.57	223	365.0	9124	536.7	11.80
34.000	40		—	0.688	32.624	835.9	72.00	8.90	8.54	245	362.1	9992	587.8	11.78
	—	—	—	0.750	32.500	829.3	78.34	8.90	8.51	266	359.5	10829	637.0	11.76
	—	—	—	0.875	32.250	816.4	91.01	8.90	8.44	310	354.1	12501	735.4	11.72
	—	—	—	1.000	32.000	804.2	103.67	8.90	8.38	353	348.6	14114	830.2	11.67
	—	—	—	1.125	31.750	791.3	116.13	8.90	8.31	395	343.2	15719	924.7	11.63
	—	—	—	0.250	35.500	989.7	28.11	9.42	9.29	96	429.1	4491	249.5	12.64
	10	—	—	0.312	35.376	982.9	34.95	9.42	9.26	119	426.1	5565	309.1	12.62
	—	Std	—	0.375	35.250	975.8	42.01	9.42	9.23	143	423.1	6664	370.2	12.59
36	20	XS	—	0.500	35.000	962.1	55.76	9.42	9.16	190	417.1	8785	488.1	12.55
36.000	30	—	—	0.625	34.750	948.3	69.50	9.42	9.10	236	411.1	10872	604.0	12.51
	40	—	—	0.750	34.500	934.7	83.01	9.42	9.03	282	405.3	12898	716.5	12.46
	—	—	—	0.875	34.250	920.6	96.50	9.42	8.97	328	399.4	14903	827.9	12.42
	—	—	—	1.000	34.000	907.9	109.96	9.42	8.90	374	393.6	16851	936.2	12.38
	—	—	—	1.125	33.750	894.2	123.19	9.42	8.89	419	387.9	18763	1042.4	12.34
	—	—	—	0.250	41.500	1352.6	32.82	10.99	10.86	112	586.4	7126	339.3	14.73
	—	Std	—	0.375	41.250	1336.3	49.08	10.99	10.80	167	579.3	10627	506.1	14.71
	20	XS	—	0.500	41.000	1320.2	65.18	10.99	10.73	222	572.3	14037	668.4	14.67
42	30	—	—	0.625	40.750	1304.1	81.28	10.99	10.67	276	565.4	17373	827.3	14.62
42.000	40	—	—	0.750	40.500	1288.2	97.23	10.99	10.60	330	558.4	20689	985.2	14.59
	—	—	—	1.000	40.000	1256.6	128.81	10.99	10.47	438	544.8	27080	1289.5	14.50
	—	—	—	1.250	39.500	1225.3	160.03	10.99	10.34	544	531.2	33233	1582.5	14.41
	—	—	—	1.500	39.000	1194.5	190.85	10.99	10.21	649	517.9	39181	1865.7	14.33

1-3 PRESSURE RATINGS OF STANDARD COMPONENTS

Pressure ratings provided in listed standards (those listed in Table 326.1 of ASME B31.3) are provided below. They are grouped by Iron (1-3.1), Steel (1-3.2), and Copper and Copper Alloys (1-3.3).

1-3.1 Iron

ASME B16.1 covers **Gray Iron Pipe Flanges and Flanged Fittings, Classes 25,125, and 250.** It provides both dimensional information and pressure-temperature ratings. Sizes from NPS 1 to NPS 96 are included. Class A or B relates to class of iron. The pressure-temperature ratings are provided in the following table, Table 1-1 in ASME B 16.1.

ASME B16.1 PRESSURE-TEMPERATURE RATINGS

	Class 25 (1) ASTM A 126		Class 125 ASTM A 126				Class 250 (1) ASTM A 126			
	Class A		Class A	Class B			Class A	Class B		
Temperature (°F)	NPS 4-36	NPS 42-96	NPS 1-12	NPS 1-12	NPS 14-24	NPS 30-48	NPS 1-12	NPS 1-12	NPS 14-24	NPS 30-48
-20 to 150	45	25	175	200	150	150	400	500	300	300
200	40	25	165	190	135	115	370	460	280	250
225	35	25	155	180	130	100	355	440	270	225
250	30	25	150	175	125	85	340	415	260	200
275	25	25	145	170	120	65	325	395	250	175
300	-	-	140	165	110	50	310	375	240	150
325	-	-	130	155	105	-	295	355	230	125
353 (2)	-	-	125	150	100	-	280	335	220	100
375	-	-	-	145	-	-	265	315	210	-
406 (3)	-	-	-	140	-	-	250	290	200	-
425	-	-	-	130	-	-	-	270	-	-
450	-	-	-	125	-	-	-	250	-	-
				Hydrostatic Shell Test Pressures (4)						
100	70	40	270	300	230	230	600	750	450	450

GENERAL NOTES:

(a) Pressure is in lb/sq in. gage.
(b) NPS is nominal pipe size.
(1) Limitations:
  a. Class 25. When Class 25 cast iron flanges and flanged fittings are used for gaseous service, the maximum pressure shall be limited to 25 psig. Tabulated pressure-temperature ratings above 25 psig for Class 25 cast iron flanges and flanged fittings are applicable for nonshock hydraulic service only.
  b. Class 250. When used for liquid service, the tabulated pressure-temperature ratings in NPS 14 and larger are applicable to Class 250 flanges only and not to Class 250 fittings.
(2) 353°F (max.) to reflect the temperature of saturated steam at 125 psig.
(3) 406°F (max.) to reflect the temperature of saturated steam at 250 psig.
(4) Hydrostatic tests are not required unless specified by the user.

ASME B16.3 covers Malleable Iron Threaded Fittings, Classes 150 and 300. It provides both dimensional information and pressure-temperature ratings. It includes sizes from NPS 1/8 to NPS 6. The pressure-temperature ratings are provided in the following table, Table 1 in ASME B16.3.

ASME B16.3 PRESSURE-TEMPERATURE RATINGS

Temperature (°F)	Class 150 (psig)	Class 300 (psig)		
		Sizes (1/4-1)	Sizes (1 1/4-2)	Sizes (2 1/2-3)
-20 to 150	300	2000	1500	1000
200	265	1785	1350	910
250	225	1575	1200	825
300	185	1360	1050	735
350	150 (1)	1150	900	650
400	-	935	750	560
450	-	725	600	475
500	-	510	450	385
550	-	300	300	300

NOTE:

(1) Permissible for service temperature up to 366°F, reflecting the temperature of saturated steam at 150 psig.

ASME B16.4 covers **Gray Iron Threaded Fittings, Classes 125 and 250.** It provides both dimensional information and pressure-temperature ratings. It includes sizes from NPS 1/4 to NPS 12. The pressure-temperature ratings are provided in the following table, Table 1 in ASME B16.4

ASME B16.4 PRESSURE-TEMPERATURE RATINGS

Temperature (°F)	Class 125 (psi)	Class 250 (psi)
-20 to 150	175	400
200	165	370
250	150	340
300	140	310
350	125 (1)	300
400	-	250 (2)

NOTES:

1. Permissible for service temperature up to 353°F reflecting the temperature of saturated steam at 125 psig.
2. Permissible for service temperature up to 406°F reflecting the temperature of saturated steam at 250 psig.

ASME B16.39 covers **Malleable Iron Threaded Pipe Unions, Classes 150, 250, and 300.** It provides both dimensional information and pressure-temperature ratings. It covers sizes from NPS 1/8 to NPS 4. The pressure-temperature ratings are provided in the following table, Table 1 in ASME B16.39. Unions with copper or copper alloy seats are not to be used for temperatures in excess of 232°C (450°F).

ASME B 16.39 PRESSURE-TEMPERATURE RATINGS

	Pressure (psig)		
Pressure (psig)	**Class 150**	**Class 250**	**Class 300**
-20 to 150	300	500	600
200	265	455	550
250	225	405	505
300	185	360	460
350	150	315	415
400	110	270	370
450	75	225	325
500	-	180	280
550	-	130	230

ASME B16.42 covers **Ductile Iron Pipe Flanges and Flanged Fittings, Classes 150 and 300.** It provides both dimensional information and pressure-temperature ratings. It covers sizes from NPS 1 to NPS 24. The pressure-temperature ratings are provided in the following table, Table 1 in ASME B16.42.

ASME B16.42 PRESSURE-TEMPERATURE RATINGS

Temperature (°F)	Working Pressure (psi gage)	
	Class 150	**Class 300**
-20 to 100	250	640
200	235	600
300	215	565
400	200	525
500	170	495
600	140	465
650	125	450

1-3.2 Steel

ASME B16.5 covers **Pipe Flanges and Flanged Fittings, NPS 1/2 through NPS 24.** It provides both dimensional information and pressure-temperature ratings. The pressure temperature ratings provided in ASME B16.5 are the commonly used pressure classes. These include Classes 150, 300,400, 600, 900,1500, and 2500. Tables of pressure-temperature ratings are provided for groups of materials. Examples for commonly used carbon steel and austenitic stainless steel materials are provided in the following tables, Tables 2-1.1C and 2-2.1C in ASME B16.5.

Table 2-1.1C
PRESSURE-TEMPERATURE RATINGS FOR GROUP 1.1 MATERIALS (ASME B16.5-2020)

Table 2-1.1C Pressure–Temperature Ratings for Group 1.1 Materials

Nominal Designation	Forgings	Castings	Plates
C–Si	A105 (1)	A216 Gr. WCB (1)	A515 Gr. 70 (1)
C–Mn–Si	A350 Gr. LF2 (1)	...	A516 Gr. 70 (1), (2)
C–Mn–Si	...	...	A537 Cl. 1 (3)
C–Mn–Si–V	A350 Gr. LF6 Cl. 1 (4)	...	...
$3\frac{1}{2}$Ni	A350 Gr. LF3 (5)	...	...

	Working Pressures by Classes, psig						
	Class						
Temp., °F	150	300	400	600	900	1500	2500
−20 to 100	285	740	985	1,480	2,220	3,705	6,170
200	260	680	905	1,360	2,035	3,395	5,655
300	230	655	870	1,310	1,965	3,270	5,450
400	200	635	845	1,265	1,900	3,170	5,280
500	170	605	805	1,205	1,810	3,015	5,025
600	140	570	755	1,135	1,705	2,840	4,730
650	125	550	730	1,100	1,650	2,745	4,575
700	110	530	710	1,060	1,590	2,655	4,425
750	95	505	675	1,015	1,520	2,535	4,230
800	80	410	550	825	1,235	2,055	3,430
850	65	320	425	640	955	1,595	2,655
900	50	230	305	460	690	1,150	1,915
950	35	135	185	275	410	685	1,145
1,000	20	85	115	170	255	430	715

NOTES:
(1) Prolonged use above 800°F is permissible but not recommended. Prolonged exposure above 800°F may cause the carbide phase of steel to convert to graphite.
(2) Not to be used over 850°F.
(3) Not to be used over 700°F.
(4) Not to be used over 500°F.
(5) Not to be used over 650°F.

Table 2-1.1C
PRESSURE-TEMPERATURE RATINGS FOR GROUP 2.1 MATERIALS
(ASME B16.5-2020, Table 2-2.1C)

Table 2-2.1C Pressure–Temperature Ratings for Group 2.1 Materials

Nominal Designation	Forgings	Castings	Plates
18Cr–8Ni	A182 Gr. F304 (1)	A351 Gr. CF3 (2)	A240 Gr. 304 (1)
18Cr–8Ni	A182 Gr. F304H	A351 Gr. CF8 (1)	A240 Gr. 304H

	Working Pressures by Classes, psig						
	Class						
Temp., °F	150	300	400	600	900	1500	2500
−20 to 100	275	720	960	1,440	2,160	3,600	6,000
200	230	600	800	1,200	1,800	3,000	5,000
300	205	540	715	1,075	1,615	2,690	4,480
400	190	495	660	995	1,490	2,485	4,140
500	170	465	620	930	1,395	2,330	3,880
600	140	440	590	885	1,325	2,210	3,680
650	125	430	575	865	1,295	2,160	3,600
700	110	420	565	845	1,265	2,110	3,520
750	95	415	550	825	1,240	2,065	3,440
800	80	405	540	810	1,215	2,030	3,380
850	65	395	530	790	1,190	1,980	3,300
900	50	390	520	780	1,165	1,945	3,240
950	35	380	510	765	1,145	1,910	3,180
1,000	20	355	470	710	1,065	1,770	2,950
1,050	...	325	435	650	975	1,630	2,715
1,100	...	255	345	515	770	1,285	2,145
1,150	...	205	275	410	615	1,030	1,715
1,200	...	165	220	330	495	825	1,370
1,250	...	135	180	265	400	670	1,115
1,300	...	115	150	225	340	565	945
1,350	...	95	125	185	280	465	770
1,400	...	75	100	150	225	380	630
1,450	...	60	80	115	175	290	485
1,500	...	40	55	85	125	205	345

NOTES:

(1) At temperatures over 1,000°F, use only when the carbon content is 0.04% or higher.

(2) Not to be used over 800°F.

ASME B16.9 covers **Factory-Made Wrought Buttwelding Fittings.** It includes sizes up to NPS 24 for most fittings and up to NPS 48 for some. Pressure-temperature ratings are not provided; rather, the allowable pressure ratings for the fittings are stated to be the same as for straight seamless pipe made from the fitting material in accordance with the rules established in the applicable section of ASME B31. ASME B31.3 requires that the calculations be based on the nominal thickness corresponding to the wall thickness of the fitting, less applicable allowances (e.g., thread depth and corrosion), and considering the manufacturing under tolerance of the fitting. For B16.9, the under tolerance is 12.5% unless specified otherwise.

ASME **B16.ll** covers **Forged Fittings, Socket-Welding and Threaded.** It includes sizes from NPS 1/8 to NPS 4. Pressure-temperature ratings are not provided; rather, the allowable pressure ratings for the fittings are stated to be the same as for straight seamless pipe made from the fitting material in accordance with the rules established in the applicable section of ASME B31. The correlation between fitting class and matching pipe is provided in the following table, Table 2 in ASME B 16.11. ASME B31.3 requires that the calculations be based on the nominal thickness corresponding to the class designation of the fitting, less all applicable allowances (e.g., thread depth and corrosion allowance). The minimum wall thickness is equal to the nominal, so for these fittings, under tolerance need not be considered.

ASME B16.11 CORRELATION OF FITTINGS CLASS WITH SCHEDULE NUMBER OR WALL DESIGNATION OF PIPE FOR CALCULATION OF RATINGS

Class Designation of Fitting	Type of Fitting	Pipe Used For Rating Basis (1)	
		Schedule No.	Wall Designation
2000	Threaded	80	XS
3000	Threaded	160	-
6000	Threaded	-	XXS
3000	Socket-Welding	80	XS
6000	Socket-Welding	160	-
9000	Socket-Welding	-	XXS

NOTE:

1. This table is not intended to restrict the use of pipe of thinner or thicker wall with fittings. Pipe actually used may be thinner or thicker in nominal wall than that shown in this table. When thinner pipe is used, its strength may govern the rating. When thicker pipe is used (e.g., for mechanical strength), the strength of the fitting governs the rating.

ASME B16.34 covers **Valves—Flanged, Threaded and Welding End.** It provides both dimensional information and pressure-temperature ratings. Ratings for both Standard Class and Special Class valves are provided. Standard Class valves have the same pressure-temperature ratings as provided in ASME B 16.5 for flanges, with the addition of Class 4500. Special Class valves have a higher pressure-temperature rating. Since the

Table 2-1.1C

RATINGS FOR GROUP 1.1 MATERIALS (ASME B16.34-2020)

A105 (1), (2)	A350 Gr. LF3 (6)	A516 Gr. 70 (1), (4)	A672 Gr. C70 (1)
A216 Gr. WCB (1)	A350 Gr. LF6 Cl.1 (5)	A537 Cl. 1 (3)	A696 Gr. C (3)
A350 Gr. LF2 (1)	A515 Gr. 70 (1)	A672 Gr. B70 (1)	

A — Standard Class

Temperature, °F	Working Pressures by Class, psig						
	150	300	600	900	1500	2500	4500
-20 to 100	285	740	1,480	2,220	3,705	6,170	11,110
200	260	680	1,360	2,035	3,395	5,655	10,185
300	230	655	1,310	1,965	3,270	5,450	9,815
400	200	635	1,265	1,900	3,170	5,280	9,505
500	170	605	1,205	1,810	3,015	5,025	9,040
600	140	570	1,135	1,705	2,840	4,730	8,515
650	125	550	1,100	1,650	2,745	4,575	8,240
700	110	530	1,060	1,590	2,655	4,425	7,960
750	95	505	1,015	1,520	2,535	4,230	7,610
800	80	410	825	1,235	2,055	3,430	6,170
850	65	320	640	955	1,595	2,655	4,785
900	50	230	460	690	1,150	1,915	3,455
950	35	135	275	410	685	1,145	2,055
1,000	20	85	170	255	430	715	1,285

B — Special Class

Temperature, °F	Working Pressure by Class, psig						
	150	300	600	900	1500	2500	4500
-20 to 100	290	750	1,500	2,250	3,750	6,250	11,250
200	290	750	1,500	2,250	3,750	6,250	11,250
300	285	740	1,480	2,220	3,700	6,170	11,105
400	280	735	1,465	2,200	3,665	6,105	10,995
500	280	735	1,465	2,200	3,665	6,105	10,995
600	280	735	1,465	2,200	3,665	6,105	10,995
650	275	715	1,430	2,145	3,575	5,960	10,730
700	265	690	1,380	2,075	3,455	5,760	10,365
750	245	635	1,270	1,905	3,170	5,285	9,515
800	195	515	1,030	1,545	2,570	4,285	7,715
850	155	400	795	1,195	1,995	3,320	5,980
900	110	285	575	860	1,435	2,395	4,305
950	65	170	345	515	855	1,430	2,570
1,000	40	105	215	320	535	895	1,605

NOTES:

(1) Upon prolonged exposure to temperatures above 800°F, the carbide phase of steel may be converted to graphite. Permissible, but not recommended for prolonged use above 800°F.

(2) Only killed steel shall be used above 850°F.

(3) Not to be used over 700°F.

(4) Not to be used over 850°F.

(5) Not to be used over 500°F.

(6) Not to be used over 650°F.

flanges are essentially governed by ASME B16.5, flanged valves are limited to Standard Class, and to Class 2500 and less. Special Class valves are subjected to additional examination requirements. The pressure-temperature ratings provided in ASME B 16.34 are the commonly used pressure classes plus Class 4500. Tables of pressure-temperature ratings are provided for groups of materials. An example for commonly used carbon steel material is provided in the following table, Table 2-1.1C in ASME B 16.34.

ASME B16.47 covers **Large Diameter Steel Flanges, NPS 26 through NPS 60.** It provides both dimensional information and pressure-temperature ratings. The pressure-temperature ratings are the same as in ASME B16.5, with the addition of Class 75 for Series B flanges. Two flange dimensions are provided; Series A flange dimensions were based on MSS SP 44 and Series B were based on API 605.

API 600 covers **Steel Gate Valves—Flanged and Butt-Welding Ends.** The requirements are essentially supplementary to ASME B 16.34 and the pressure-temperature ratings are the same as in ASME B16.34.

API 602 covers Gate, Globe, and Check Valves for Sizes DN 100 (NPS 4) and Smaller. It provides requirements for valves with pressure-temperature ratings of Classes 150, 300, 600, 800, and 1500. All except Class 800 use the same pressure-temperature ratings as listed in ASME B 16.34. The pressure-temperature ratings for Class 800 is determined in accordance with the following equation

$$p_8 = 1/3/p_6 + 2/3/p9$$

where

p_8 = the pressure rating at the specified temperature for Class 800, rounded to the nearest 0.1 bar(10 kps)
p_6 - the pressure rating at the specified temperature for Class 600 (from ASME B 16.34)
p_9 = the pressure rating at the specified temperature for Class 900 (from ASME B16.34)

MSS SP-42 covers **Class 150 Corrosion-Resistant Gate, Globe, Angle and Check Valves With Flanged and Buttweld Ends.** The pressure-temperature ratings are provided in the following table, Table 4 in MSS SP-42.

MSS SP-42 PRESSURE-TEMPERATURE RATINGS

	Maximum Working Pressure (psig)				
Service Temperature(°F)	**CF8, CF3, 304**	**CF8M, CF3M, 316**	**304L, 316L**	**CF8C, 347**	**CN7M**
-20 to 100	275	275	230	275	230
150	253	255	213	265	215
200	230	235	195	255	200
250	218	225	185	243	195
300	205	215	175	230	190
350	198	205	168	215	190
400	190	195	160	200	190
450	180	183	153	185	180
500	170	170	145	170	170

GENERAL NOTE

This table gives data in U.S. customary units, the temperature being indicated in °F and the actual values being represented in metric units.

MSS SP-42 PRESSURE-TEMPERATURE RATINGS

Service Temperature(°F)	Maximum Working Pressure (bar)				
	CF8, CF3, 304	CF8M, CF3M, 316	304L, 316L	CF8C, 347	CN7M
-29 to 38	19.0	19.0	15.9	19.0	15.9
50	18.3	18.4	15.3	18.7	15.4
100	15.7	16.0	13.3	17.4	13.7
150	14.1	14.8	12.0	15.8	13.1
200	13.2	13.6	11.1	14.0	13.1
250	12.0	12.0	10.2	12.1	12.0
300	10.2	10.2	9.8	10.3	10.2
260	11.7	11.7	10.0	11.7	11.7

MSS SP-51 covers **Class 150LW Corrosion-Resistant Cast Flanges and Flanged Fittings.** The pressure-temperature ratings are provided in the following table, Table 1 in MSS SP-51.

MSS SP-51 PRESSURE-TEMPERATURE RATINGS

Temperature (°C)	Pressure Rating (Bar)	Temperature (°F)	Pressure Rating (psig)
-29 to 38	6.9	-20 to 100	100
50	6.8	200	90
100	6.1	300	80
150	5.5	400	70
200	4.9	500	60
260	4.1		

PRESSURE TEMPERATURE RATING
(ASME B16.22-2018)

Joining Material	Temperature, °C (°F)	Maximum Gage Pressure for Standard Water Tube Sizes, kPa (psi) [Note (1)] ⅛ Through 1	1¼ Through 2	2½ Through 4	5 Through 8
Alloy Sn50 50–50 tin–lead solder [Notes (2), (3)]	38 (100)	1 375 (200)	1 205 (175)	1 030 (150)	930 (135)
	66 (150)	1 030 (150)	860 (125)	685 (100)	620 (90)
	93 (200)	685 (100)	620 (90)	515 (75)	480 (70)
	121 (250)	585 (85)	515 (75)	340 (50)	310 (45)
Alloy Sb5 95–5 tin–antimony solder [Note (4)]	38 (100)	7 540 (1,090) [Note (5)]	5 880 (850) [Note (6)]	4 880 (705) [Note (6)]	4 555 (660) [Note (6)]
	66 (150)	4 315 (625) [Note (7)]	3 365 (485) [Note (7)]	2 790 (405) [Note (7)]	2 605 (375) [Note (7)]
	93 (200)	3 500 (505) [Note (8)]	2 730 (395) [Note (7)]	2 265 (325) [Note (7)]	2 115 (305) [Note (7)]
	121 (250)	1 885 (270)	1 475 (210)	1 220 (175)	1 135 (165)
Alloy E [Note (9)]	38 (100)	4 905 (710) [Note (7)]	3 825 (555) [Note (7)]	3 175 (460) [Note (7)]	2 965 (430) [Note (7)]
	66 (150)	3 275 (475) [Note (8)]	2 550 (370) [Note (7)]	2 115 (305) [Note (7)]	1 975 (285) [Note (8)]
	93 (200)	2 595 (375)	2 025 (290)	1 680 (240) [Note (8)]	1 570 (225) [Note (8)]
	121 (250)	2 230 (320)	1 735 (250)	1 440 (205)	1 340 (195)
Alloy HB [Note (10)]	38 (100)	7 135 (1,035) [Note (5)]	5 560 (805) [Note (6)]	4 615 (670) [Note (6)]	4 305 (625) [Note (5)]
	66 (150)	4 905 (710) [Note (7)]	3 825 (555) [Note (7)]	3 175 (460) [Note (7)]	2 965 (430) [Note (7)]
	93 (200)	3 045 (440) [Note (8)]	2 375 (345) [Note (8)]	1 970 (285) [Note (8)]	1 840 (265) [Note (8)]
	121 (250)	2 970 (430) [Note (8)]	2 315 (335) [Note (8)]	1 920 (275) [Note (8)]	1 800 (260) [Note (8)]
Joining materials melting at or above 593°C (1,100°F) [Note (11)]		Pressure–temperature ratings consistent with the materials and procedures employed			

GENERAL NOTE: For temperatures in the −18°C to −93°C (0°F to −200°F) range, it is recommended that a joint material melting at or above 593°C (1,100°F) be employed [see Note (9)].

NOTES:
(1) Standard water tube sizes per ASTM B88.
(2) ASTM B32 Alloy Grade Sn50.
(3) The Safe Drinking Water Act Amendments of 1986 prohibit the use of any solder having a lead content in excess of 0.2% in potable water systems.
(4) ASTM B32 Alloy Grade Sb5.
(5) The solder joint exceeds the strength of Types L and M tube in drawn temper and Type K tube in annealed temper.
(6) The solder joint exceeds the strength of Types K, L, and M tube in drawn and annealed tempers.
(7) The solder joint exceeds the strength of Type M tube in drawn temper and Types K and L in annealed temper.
(8) The solder joint exceeds the strength of Type L tube in annealed temper.
(9) ASTM B32 Alloy Grade E.
(10) ASTM B32 Alloy Grade HB.
(11) These joining materials are defined as "brazing alloys" by the American Welding Society.

1-3.3 COPPER AND COPPER ALLOYS

ASME B16.15 covers **Cast Bronze Threaded Fittings, Classes 125 and 250.** It includes sizes from NPS 1/8 to NPS 4. The pressure-temperature ratings for fittings per this standard are provided in the following table, Table 3.1-1 in ASME B16.15.

TABLE 3.1-1
PRESSURE-TEMPERATURE RATINGS

Table 3.1-1 Pressure–Temperature Ratings

Temperature, °C (°F)	Class 125, bar (psi)	Class 250, bar (psi)
−29 to 66 (−20 to 150)	13.8 (200)	27.6 (400)
100 (200)	12.9 (190)	26.2 (385)
125 (250)	12.3 (180)	24.9 (365)
150 (300)	11.3 (165)	23.0 (335)
175 (350)	10.4 (150)	20.8 (300)
200 (400)	8.9 (125)	17.8 (250)

GENERAL NOTES:
(a) 1 bar = 14.5 psi = 10^5 Pa.
(b) °C = 0.5556(°F − 32).

ASME B16.18 covers **Cast Copper Alloy Solder Joint Pressure Fittings.** It also includes solder x threaded fittings. It provides both dimensional information and information on the permissible pressure-temperature. Sizes that are included range from Standard Water Tube 1/4 to 12. The pressure ratings of these fittings are required to be the same as ASTM B88 Type L Copper Water Tube, annealed. However, the pressure rating cannot exceed the rating of the joint. For joints that are soldered, ratings are described in the provided in the following table, Table Al in ASME B16.18, Annex A. There is no rating reduction for the fittings when they are brazed.

ASME B16.22 covers **Wrought Copper and Copper Alloy Solder Joint Pressure Fittings.** It also includes solder x threaded fittings. It provides both dimensional information and pressure-temperature ratings. Sizes included range from Standard Water Tube 1/4 to 8. The pressure rating of the fittings is the that shown in the following table, Table 3.1.1-2 in B 16.22, For soldered joints, the rating cannot exceed that shown in the table on the following page from ASME B16.18.

INTERNAL WORKING PRESSURE (1) FOR COPPER FITTINGS (PSI) (asme B16.22-2018, Table 3.1.1-2)

Standard Water Tube Size [Note (1)]	−29°C to 38°C (−20°F to 100°F)	66°C (150°F)	93°C (200°F)	121°C (250°F)	149°C (300°F)	177°C (350°F)	204°C (400°F)
1/4	6 280 (910)	5 340 (770)	5 130 (740)	5 020 (725)	4 920 (710)	4 190 (605)	3 140 (455)
3/8	5 360 (775)	4 560 (660)	4 380 (635)	4 290 (620)	4 200 (610)	3 570 (515)	2 680 (385)
1/2	4 970 (720)	4 220 (610)	4 060 (585)	3 980 (575)	3 890 (565)	3 310 (480)	2 480 (360)
5/8	4 350 (630)	3 700 (535)	3 550 (515)	3 480 (505)	3 410 (490)	2 900 (420)	2 170 (315)
3/4	4 010 (580)	3 410 (490)	3 270 (475)	3 210 (465)	3 140 (455)	2 670 (385)	2 000 (290)
1	3 400 (490)	2 890 (420)	2 780 (400)	2 720 (395)	2 660 (385)	2 270 (325)	1 700 (245)
1 1/4	3 020 (435)	2 570 (370)	2 470 (355)	2 420 (350)	2 370 (340)	2 010 (290)	1 510 (215)
1 1/2	2 810 (405)	2 390 (345)	2 300 (330)	2 250 (325)	2 200 (315)	1 870 (270)	1 400 (200)
2	2 500 (360)	2 130 (305)	2 040 (295)	2 000 (290)	1 960 (280)	1 670 (240)	1 250 (180)
2 1/2	2 310 (335)	1 960 (285)	1 890 (270)	1 850 (265)	1 810 (260)	1 540 (220)	1 150 (165)
3	2 180 (315)	1 850 (265)	1 780 (255)	1 740 (250)	1 710 (245)	1 450 (210)	1 090 (155)
3 1/2	2 090 (300)	1 770 (255)	1 700 (245)	1 670 (240)	1 630 (235)	1 390 (200)	1 040 (150)
4	2 020 (290)	1 710 (245)	1 650 (235)	1 610 (230)	1 580 (225)	1 340 (195)	1 010 (145)
5	1 850 (265)	1 570 (225)	1 510 (215)	1 480 (215)	1 450 (210)	1 230 (175)	920 (130)
6	1 720 (250)	1 460 (210)	1 410 (200)	1 380 (200)	1 350 (195)	1 150 (165)	860 (125)
8	1 860 (270)	1 580 (225)	1 520 (220)	1 490 (215)	1 460 (210)	1 240 (180)	930 (135)

GENERAL NOTES:

(a) The fitting pressure–temperature rating applies to the largest opening of the fitting.

(b) The fitting pressure–temperature rating is calculated as shown in Nonmandatory Appendix A, then rounded down to the nearest unit of 10.

NOTE: (1) For size designation of fittings, see para. 4.1.

ASME B16.18 PRESSURE-TEMPERATURE RATINGS

Joining Material	Working Temperature		Maximum Working Gage Pressure, for Standard Water Tube Sizes (1)									
			1/4 Through 1		1 1/4 Through 2		2 1/2 Through 4		5 Through 8		10 Through 12	
	°F	°C	psi	kPa	psi	kPa	psi	kPa	psi	kPa	psi	kPa
Alloy Sn50	100	38	200	1375	175	1205	150	1030	135	930	100	685
50-50 tin-lead	150	66	150	1030	125	860	100	685	90	620	70	480
solder (2,3)	200	93	100	685	90	620	75	515	70	480	50	340
	250	120	85	585	75	515	50	340	45	310	40	275
Alloy Sb5 95-5	100	38	1090 (9)	7540 (9)	850 (8)	5880 (8)	705 (8)	4880 (8)	660 (8)	4555 (8)	500 (9)	3460 (9)
tin-antimony	150	66	625 (10)	4315 (10)	485 (10)	3365 (10)	405 (10)	2790 (10)	375 (10)	2605 (10)	285 (12)	1975 (11)
solder (4)	200	93	505 (11)	3500 (11)	395 (10)	2730 (10)	325 (10)	2265 (10)	305 (10)	2115 (10)	230 (11)	1605 (11)
	250	120	270	1885	210	1475	175	1220	165	1135	125	865
Alloy E (5)	100	38	710 (10)	4905 (10)	555 (10)	3825 (10)	460 (10)	3175 (10)	430 (10)	2965 (10)	325 (11)	2255 (11)
	150	66	475 (11)	3275 (11)	370(10)	2550 (10)	305 (10)	2115 (10)	285 (11)	1975 (11)	215 (11)	1500 (11)
	200	93	375	2595	290	2025	240 (11)	1680 (11)	225 (11)	1570 (11)	170	1190
	250	120	320	2230	250	1735	205	1440	195	1340	145	1020
Alloy HB (6)	100	38	1035 (9)	7135 (9)	805 (8)	5560 (8)	670 (8)	4615 (8)	625 (9)	4305 (9)	475 (9)	3275 (9)
	150	66	710 (10)	4905 (10)	555 (10)	3825 (10)	460 (10)	3175 (10)	430 (10)	2965 (10)	325 (10)	2255 (10)
	200	93	440 (11)	3045 (11)	345 (11)	2375 (11)	285 (11)	1970 (11)	265 (11)	1840 (11)	200	1400
	250	120	430 (11)	2970 (11)	335 (11)	2315 (11)	275 (11)	1920 (11)	260(11)	1800 (11)	195	1365
Joining materials melting at or above 1100°F (593°C) (7)			Pressure-temperature ratings consistent with the materials and procedures employed.									

(Continued)

INTERNAL WORKING PRESSURE (1) FOR COPPER FITTINGS (PSI) (asme B16.22-2018, Table 3.1.1-2)

GENERAL NOTE: For extremely low working temperatures in the 0°F to minus 200°F (minus 18°C to minus 93°C) range, it is recommended that a joint material melting at or above 1100°F (593°C) be employed [see Note (7)].

NOTES:

1. Standard water tube sizes per ASTM B 88.
2. ASTM B 32 Alloy Grade Sn50.
3. The Safe Drinking Water Act Amendment of 1986 prohibits the use in potable water systems of any solder having a lead content in excess of 0.2%.
4. ASTM B 32 Alloy Grade Sb5.
5. ASTM B 32 Alloy Grade E.
6. ASTM B 32 Alloy Grade HB.
7. These joining materials are defined as *brazing alloys* by the American Welding Society.
8. The solder joint exceeds the strength of Types K, L, and M tube in drawn and annealed tempers.
9. The solder joint exceeds the strength of Types L and M tube in drawn temper and Type K tube in annealed temper.
10. The solder joint exceeds the strength of Type M tube in drawn temper and Types K and L tube in annealed temper.
11. The solder joint exceeds the strength of Type L tube in annealed temper.

ASME B16.24 covers **Cast Copper Alloy Pipe Flanges and Flanged Fittings, Classes 150, 300, 400, 600, 900, 1500, and 2500.** It provides both dimensional information and pressure-temperature ratings. Sizes included range from NPS **1/2** to NPS 12. The pressure-temperature rating depends upon the alloy, and is provided in the following three tables, Tables I-1, I-6 and I-7 in ASME B 16.24.

ASME B16.26 covers **Cast Copper Alloy Fittings for Flared Copper Tubes.** They are rated for a maximum cold water service pressure of 175 psig.

PRESSURE–TEMPERATURE RATINGS FOR ASTM B61 ALLOY C92200 AND ASTM B62 ALLOY C83600 (ASME B16.24-2016, Table 1)

Service Temperature, °C	Pressure, bar			
	Class 150		Class 300	
	ASTM B62 C83600	ASTM B61 C92200	ASTM B62 C83600	ASTM B61 C92200
−29 to 66	15.5	15.5	34.5	34.5
100	14.3	14.6	31.4	32.4
125	13.4	14.1	29.1	30.9
150	12.4	13.4	26.8	29.3
175	11.4	12.4	24.3	27.6
200	. . .	11.9	. . .	26.1
208 [Note (1)]	10.3	. . .	21.4	. . .
225	. . .	11.5	. . .	24.6
232	9.3	. . .	19.3	. . .
250	. . .	10.5	. . .	23.0
275	. . .	10.0	. . .	21.6
289	. . .	9.7	. . .	20.7
Test pressure	24.1	24.1	51.7	51.7

GENERAL NOTE: 1 bar = 100 kPa.

NOTE:

(1) Some codes (e.g., ASME Boiler and Pressure Vessel Code, Section I; ASME B31.1; and ASME B31.5) limit the rating temperature of the indicated material to 208°C (406°F).

PRESSURE-TEMPERATURE RATINGS FOR ASTM B148 ALLOY C95200 FLANGES (ASME B16.24-2020, Table I-6)

Service Temperature, °F	Pressure, psi					
	Class 150	Class 300	Class 600	Class 900	Class 1500	Class 2500
−20 to 100	195	515	1,030	1,555	2,570	4,285
150	185	480	965	1,450	2,415	4,030
200	180	470	940	1,405	2,345	3,910
250	175	455	915	1,370	2,285	3,805
300	170	445	895	1,340	2,230	3,720
350	170	440	880	1,320	2,200	3,670
400	170	440	875	1,315	2,190	3,650
450	165	435	870	1,310	2,180	3,635
500	165	435	870	1,310	2,180	3,635
550	155	400	800	1,205	2,005	3,345
600	95	255	505	760	1,270	2,115

APPENDIX

II

GUIDELINES FOR COMPUTER FLEXIBILITY ANALYSIS

Prior to initiating a computer piping flexibility analysis, the analyst should gather as much information as they can to perform the job, in an organized manner. This information includes the following.

1. Piping system operating and design conditions. The analyst should have an understanding of the various potential combinations of operating conditions that may occur. There may be various combinations of lines and vessels at different temperature conditions. For example, with parallel heat exchangers, operating with one heat exchanger in operation and hot, and the other blocked off and at an ambient temperature condition, may be one of the potential modes of operation. Note that while it is not a Code requirement that the flexibility analysis be conducted using the design temperature, it may be the requirement of the owner that it be done in that manner.
2. Determine if there are steam out or other conditions that may cause different temperatures (e.g. steam tracing temperature).
3. Standard piping classes or other information indicating the materials used to construct the piping system.
4. Dimensional information on the piping system, in the form of isometric, orthographic, or other format.
5. Information on attached equipment. This typically includes outline drawings of vessels, rotating equipment, etc. Load limits for the equipment. Operating conditions /temperatures for attached equipment to consider thermal growth. Thermal movements of fired equipment are typically provided on the equipment drawings.
6. If supports and restraints have been designed, drawings or other information to determine the types of supports, to appropriately model them.
7. If the piping system is existing, photographs, walkdown, or other means to confirm the supports used in construction. It is not unusual for the supports found in the field to differ from those shown on the original design drawings.
8. Allowable stresses. Code of construction.
9. Weights of piping components.
10. Density of fluid.
11. Insulation thickness and type (including properties such as density, if standard insulation type densities are not provided by the flexibility analysis program).
12. Information on expansion joints, if present in the system, including pressure thrust area and stiffnesses.
13. Information on refractory lining, if present in the system, to determine the effect of the refractory on the stiffness of the piping. The increase in stiffness can effect equipment loads and load sensitive components in a system (e.g. a hot walled slide valve in a refractory lined system).
14. Wind, seismic (seismic may be governed by a building code, e.g. UBC, BOCA, IBC) and other occasional loads, such as waterhammer, that require consideration.
15. Any special requirements, such as natural frequency limitations.
16. For vapor lines, information on how the leak test is to be conducted. Must the piping be designed to carry the weight of water during a hydrotest?

The first step in conducting a flexibility analysis is to gather this information. It is then used to construct a pipe stress isometric. The stress isometric is used to organize the information for input into the computer program. The following information should generally be on the piping stress isometric. Note that while it is certainly possible to input a pipe stress model without preparing a piping stress isometric, the chance of making errors is greatly increased.

1. A global coordinate system with the positive direction indicated for the x, y and z reference axes used in the stress analysis. The positive y axis is typically assigned to the vertical up direction. The plant north direction shown with respect to this global reference frame. It may be helpful on a project to consistently model plant north as the positive x axis and east as the positive z axis.
2. Node points used in the analysis, labeled and located. Dimensions between node points, resolved into components parallel to each of the three global axes.
3. Locations, function, and lines of action of all supports and restraints. Spring hanger stiffnesses, hot and cold load settings, and design travel. When known, spring hanger model number and manufacturer.
4. Piping design parameters such as pipe size, thickness or schedule, piping material, corrosion/erosion allowance, mechanical allowances, insulation, refractory density, valve and flange (and other piping component) weights and locations.
5. The general arrangement of equipment to which the piping is attached.

The final stress report should also contain the following information. Having this information gathered together makes it possible to check the analysis with relative ease, and provides complete documentation of the basis for the work performed.

1. Equipment outline drawings for all vessels, heat exchangers, fired heaters, rotating equipment, tankage, etc. to which the piping is attached. These drawings should include, as a minimum, basic overall dimensions, material, nozzle locations and details where piping is attached, and equipment support locations. As a minimum, references to the appropriate drawings should be provided.
2. Pressures, temperatures and content specific gravities for all load cases considered. Data on any pipe sections with high temperature gradients, thermal bowing, or transients should be included as well.
3. Expansion joint stiffness and movements (axial, transverse, and rotational), bellows dimensional information and overall expansion joint assembly weight.
4. Any dynamic load data used in the analysis.
5. Tabulations showing compliance of loads on rotating and other load sensitive equipment with their allowables.
6. Program input and output showing all analysis results for the specified load cases as well as a record of the software used, its version number, and record of its setup or configuration. This may be an electronic record.
7. Program generated plots showing principal nodes (annotated as appropriate).
8. Assumptions/supporting calculations for stress intensification factors when values other than program defaults are used.
9. Support stiffness calculations, when appropriate.
10. If the analysis is also a pressure design check, appropriate calculations for wall thickness, area replacement, etc.

The following is a checklist of considerations.

1. Is thermal movement of attached equipment included?
2. If design temperature is being used for the flexibility analysis, is the thermal movement of the attached equipment consistent?
3. Have design temperatures of the equipment been improperly used in lieu of operating temperatures? This can occur when pressure vessels are given higher design temperatures when the allowable stress does not change up to that temperature (e.g. 650F) even though the expected metal temperature may be substantially less. In that case, the design temperature has nothing to do with the expected metal temperature for determining thermal movement.

4. Is the pressure thrust in the piping properly considered in loads on attached equipment (e.g. pressure thrust typically must be added to the reported force in the piping, for conducting a nozzle analysis)?
5. Are there conditions that may cause reversing moments in the system, such that the range between these conditions must be considered in determining the stress range?
6. Has friction been included, when significant?
7. Has the effect of friction on sliding support loads been considered?
8. If friction has been included in the piping stress analysis, has an analysis been run without friction? In general, friction is not something that is relied on. The harmful effects of friction are considered, but the benefit is not. Thus, in some cases analyses should be run with and without friction, to determine which is worse.
9. Are the loads on attached equipment within the allowable limits?
10. May the flexibility of the structural steel or other supporting elements significantly effect the results?
11. Has thermal expansion of the pipe support, or equipment support, been considered?
12. Has corrosion allowance been included?
13. Is the geometry correct? Check the plots.
14. Are stresses due to internal pressure included in the S_L calculation?
15. Are the correct valve and other component weights included?
16. Has the weight of valve actuators been included?
17. Does the flange weight include the weight of the bolting?
18. Has the effect of settlement (e.g. around tankage) been considered?
19. Have all combinations of operating conditions been considered?
20. Have environmental loads (e.g. wind, seismic, ice) been considered?
21. Has the correct minimum temperature for the location been considered in the analysis, in determining the range of temperatures the piping may be exposed to, or has a default value been used?
22. Are springs properly modeled?
23. If gaps at supports are included in the model, are adequate controls/documentation in place to assure that they will be installed in that manner in the field, and checked?
24. Has the effect of welding trunions on elbows (on their flexibility) been considered?
25. Is there a potential for elastic followup or other strain concentration condition?
26. Is radial thermal expansion of the line potentially significant (e.g. bottom supported large diameter hot lines)?
27. If the piping lifts off of supports modeled as nonlinear supports, has the lack of that support been considered in calculating the sustained stress for that operating condition?
28. Has pressure thrust of expansion joints been considered?
29. Has thermal expansion and the action of expansion joint hardware been considered?
30. Have axial load due to thermal expansion been considered in the thermal expansion stress range, if significant?
31. Have the effect of any flanges (or heavy walled components) welded adjacent to elbows been considered with respect to elbow stress intensification factors and flexibility?
32. Have the stresses in elbows and bends been considered at more than one location on the elbow or bend? These should typically be at least checked at the ends and the midpoint.
33. Has deflection between supports been considered, when necessary? Note that deflection is only calculated, reported, and plotted based on values at node points. Midspan nodes may be necessary.
34. On large diameter lines, has the effect of pipe radius on geometry been considered? Default modeling typically assumes the pipe spans between centerlines; however with short branch runs between large diameter lines, or between a large diameter line and some point of restraint, the assumption can be significantly unconservative.
35. Have sufficient means to direct the movements of a piping system been incorporated. Remember, ideal conditions do not typically exist, so floating systems without any intermediate points of fixity generally are not desirable.
36. Have any unlisted components been considered for pressure and fatigue?

37. On large D/t piping, have limitations to the SIF's been considered?
38. Should pressure stiffening of bends be considered in the analysis?
39. Are loads on the flanges high? Has the impact of loads on flange leakage been considered?
40. Should the change in pipe length due to internal pressure be considered? This depends upon the hoop stress, axial stress (which may be effected by the presence of expansion joints) and the material of construction (e.g. reinforced thermosetting resin piping may extend or contract due to internal pressure, depending upon the fiber wind angle).

APPENDIX

III

USEFUL INFORMATION FOR FLEXIBILITY ANALYSIS

Weights of Piping Materials
Weights of Flanged Valves
Weights of Flange Bolting
Dimensions of Fittings (ASME B16.9)
Dimensions of Welding Neck Flanges (ASME B16.5)
Dimensions of Valves (ASME B16.10)
Coefficient of Thermal Expansion (ASME B31.3)

WEIGHT OF PIPING MATERIALS
(Courtesy of Anvil International)

The tabulation of weights of standard piping materials has been arranged for convenience of selection of data that formerly consumed considerable time to develop. For special materials, the three formulae listed below for weights of tubes, weights of contents of tubes, and weights of piping insulation will be helpful.

Weight of tube = $F \times 10.68 \times T \times (D - T)$ lb/ft

T = wall thickness in inches

D = outside diameter in inches

F = relative weight factor

The weight of tube furnished in this piping data is based on low carbon steel weighing 0.2833 pounds per cubic inch.

Relative Weight Factor F

Aluminum	0.34
Brass	1.09
Cast Iron	0.92
Copper	1.14
Ferritic stainless steel	0.95
Austenitic stainless steel	1.02
Steel	1.00
Wrought iron	0.99

Weight of contents of a tube –

$G \times .3405 \times (D - 2T)^2$ lb/ft

G = specific gravity of contents

T = tube wall thickness in inches

D = tube outside diameter in inches

The weight per foot of steel pipe is subject to the following tolerances:

SPECIFICATION	TOLERANCE		
ASTM A-53 ASTM A-120	STD WT	+5%	-5%
	XS WT	+5%	-5%
	XXS WT	+10%	-10%
ASTM A-106	SCH 10-120	+6.5%	-3.5%
	SCH 140-160	+10%	-3.5%
ASTM A-335	12" and under	+6.5%	-3.5%
	over 12"	+10%	-5%
ASTM A-312 ASTM A-376	12" and under	+6.5%	-3.5%
API 5L	All sizes	+6.5%	-3.5%

The weight of welding tees and laterals are for full size fittings. The weights of reducing fittings are approximately the same as for full size fittings.

The weights of welding reducers are for one size reduction, and are approximately correct for other reductions.

Weights of valves of the same type may vary because of individual manufacturer's designs. Listed valve weights are approximate only. Specific valve weights should be used when available.

Where specific insulation thicknesses and densities differ from those shown, refer to "Weight of Piping Insulation" formula below.

Weight of piping insulation -

$I \times .0218 \times T \times (D + T)$ lb/ft

I = insulation density in pounds per cubic foot

T = insulation thickness in inches

D = outside diameter of pipe in inches

WEIGHT OF PIPING MATERIALS (Courtesy of Anvil International)

1.313" O.D. 1" PIPE

PIPE	Schedule No.	5S	10S	40	80	160	
	Wall Designation			Std.	XS		XXS
	Thickness — In.	0.065	0.109	.133	.179	.250	.358
	Pipe — Lbs/Ft	**0.868**	**1.404**	**1.68**	**2.17**	**2.84**	**3.66**
	Water — Lbs/Ft	**0.478**	**0.409**	**.37**	**.31**	**.23**	**.12**
WELDING FITTINGS	L.R. 90° Elbow	**.2** .3	**.4** .3	**.4** .3	**.4** .3	**.6** .3	**1.0** .3
	S.R. 90° Elbow			**.3** .2			
	L.R. 45° Elbow	**.1** .2	**.3** .2	**.3** .2	**.3** .2	**.4** .2	**.5** .2
	Tee	**.4** .4	**.6** .4	**.8** .4	**.9** .4	**1.1** .4	**1.3** .4
	Lateral	**.7** 1.1	**1.2** 1.1	**1.7** 1.1	**2.5** 1.1		
	Reducer	**.2** .2	**.4** .2	**.3** .2	**.4** .2	**.5** .2	**.5** .2
	Cap	**.1** .3	**.1** .3	**.3** .3	**.3** .3	**.4** .3	**.5** .3

	Temperature Range °F	100-199	200-299	300-399	400-499	500-599	600-699	700-799	800-899	900-999	1000-1099	1100-1200
INSULATION — 85% Magnesia Calcium Silicate	Nom. Thick., In.	1	1	1½	2	2						
	Lbs./Ft	**.57**	**.57**	**.97**	**1.54**	**1.54**						
INSULATION — Combination	Nom. Thick., In.						2½	2½	2½	3	3	3
	Lbs/Ft						**3.30**	**3.30**	**3.30**	**4.70**	**4.70**	**4.70**

	Pressure Rating psi	Cast Iron		Steel						
		125	250	150	300	400	600	900	1500	2500
FLANGES	Screwed or Slip-On	**2.3** 1.5	**4** 1.5	**2.5** 1.5	**4** 1.5	**5** 1.5	**5** 1.5	**12** 1.5	**12** 1.5	**15** 1.5
	Welding Neck			**3** 1.5	**5** 1.5	**7** 1.5	**7** 1.5	**12** 1.5	**12** 1.5	**16** 1.5
	Lap Joint			**2.5** 1.5	**4** 1.5	**5** 1.5	**5** 1.5	**12** 1.5	**12** 1.5	**15** 1.5
	Blind	**2.5** 1.5	**5** 1.5	**2.5** 1.5	**5** 1.5	**5** 1.5	**5** 1.5	**12** 1.5	**12** 1.5	**15** 1.5
FLANGED FITTINGS	S.R. 90° Elbow						**15** 3.7		**28** 3.8	
	L.R. 90° Elbow									
	45° Elbow						**14** 3.4		**26** 3.6	
	Tee						**20** 5.6		**39** 5.7	
VALVES	Flanged Bonnet Gate				**20** 1.2		**25** 1.5		**80** 4.3	
	Flanged Bonnet Globe or Angle								**84** 3.5	
	Flanged Bonnet Check									
	Pressure Seal Bonnet — Gate						**31** 1.7	**31** 1.7		
	Pressure Seal Bonnet — Globe									

Boldface type is weight in pounds. Lightface type beneath weight is weight factor for insulation.

Insulation thicknesses and weights are based on average conditions and do not constitute a recommendation for specific thicknesses of materials. Insulation weights are based on 85% magnesia and hydrous calcium silicate at 11 lbs/cubic foot. The listed thicknesses and weights of combination covering are the sums of the inner layer of diatomaceous earth at 21 lbs/cubic foot and the outer layer at 11 lbs/cubic foot.

Insulation weights include allowances for wire, cement, canvas, bands and paint, but not special surface finishes.

To find the weight of covering on flanges, valves or fittings, multiply the weight factor by the weight per foot of covering used on straight pipe.

Valve weights are approximate. When possible, obtain weights from the manufacturer.

Cast iron valve weights are for flanged end valves; steel weights for welding end valves.

All flanged fitting, flanged valve and flange weights include the proportional weight of bolts or studs to make up all joints.

WEIGHT OF PIPING MATERIALS
(Courtesy of Anvil International)

1.660" O.D. 1¼" PIPE

PIPE	Schedule No.	5S	10S	40	80	160	
	Wall Designation			Std.	XS		XXS
	Thickness — In.	.065	.109	.140	.191	.250	.382
	Pipe — Lbs/Ft	**1.11**	**1.81**	**2.27**	**3.00**	**3.77**	**5.22**
	Water — Lbs/Ft	**.80**	**.71**	**.65**	**.56**	**.46**	**.27**
WELDING FITTINGS	L.R. 90° Elbow	**.3** .3	**.5** .3	**.6** .3	**.8** .3	**1.0** .3	**1.3** .3
	S.R. 90° Elbow			**.4** .2			
	L.R. 45° Elbow	**.2** .2	**.3** .2	**.3** .2	**.5** .2	**.6** .2	**.7** .2
	Tee	**.7** .5	**1.1** .5	**1.6** .5	**1.6** .5	**1.9** .5	**2.4** .5
	Lateral	**1.1** 1.2	**1.9** 1.2	**2.4** 1.2	**3.8** 1.2		
	Reducer	**.3** .2	**.4** .2	**.5** .2	**.6** .2	**.7** .2	**.8** .2
	Cap	**.1** .3	**.1** .3	**.4** .3	**.4** .3	**.6** .3	**.6** .3

	Temperature Range °F	100-199	200-299	300-399	400-499	500-599	600-699	700-799	800-899	900-999	1000-1099	1100-1200
INSULATION 85% Magnesia Calcium Silicate	Nom. Thick., In.	1	1	1½	2	2	2½	2½	2½	3	3	3
	Lbs./Ft	**.85**	**.65**	**1.47**	**1.83**	**1.83**	**2.65**	**2.65**	**2.65**	**3.58**	**3.58**	**3.58**
Combination	Nom. Thick., In.						2½	2½	2½	3	3	3
	Lbs/Ft						**3.17**	**3.17**	**3.17**	**5.76**	**5.76**	**5.76**

	Pressure Rating psi	Cast Iron		Steel						
		125	250	150	300	400	600	900	1500	2500
FLANGES	Screwed or Slip-On	**2.5** 1.5	**4.8** 1.5	**3.5** 1.5	**5** 1.5	**7** 1.5	**7** 1.5	**13** 1.5	**13** 1.5	**23** 1.5
	Welding Neck			**3** 1.5	**7** 1.5	**8** 1.5	**8** 1.5	**13** 1.5	**13** 1.5	**25** 1.5
	Lap Joint			**3.5** 1.5	**5** 1.5	**7** 1.5	**7** 1.5	**13** 1.5	**13** 1.5	**22** 1.5
	Blind	**2.8** 1.5	**5.5** 1.5	**3.5** 1.5	**4** 1.5	**7** 1.5	**7** 1.5	**13** 1.5	**13** 1.5	**23** 1.5
FLANGED FITTINGS	S.R. 90° Elbow				**17** 3.7		**18** 3.8		**33** 3.9	
	L.R. 90° Elbow				**18** 3.9					
	45° Elbow				**15** 3.4		**16** 3.5		**31** 3.7	
	Tee				**23** 5.6		**28** 5.6		**49** 5.9	
VALVES	Flanged Bonnet Gate				**40** 4		**60** 4.2		**97** 4.6	
	Flanged Bonnet Globe or Angle									
	Flanged Bonnet Check				**21** 4					
	Pressure Seal Bonnet — Gate							**38** 1.8	**38** 1.1	
	Pressure Seal Bonnet — Globe									

Boldface type is weight in pounds. Lightface type beneath weight is weight factor for insulation.

Insulation thicknesses and weights are based on average conditions and do not constitute a recommendation for specific thicknesses of materials. Insulation weights are based on 85% magnesia and hydrous calcium silicate at 11 lbs/cubic foot. The listed thicknesses and weights of combination covering are the sums of the inner layer of diatomaceous earth at 21 lbs/cubic foot and the outer layer at 11 lbs/cubic foot.

Insulation weights include allowances for wire, cement, canvas, bands and paint, but not special surface finishes.

To find the weight of covering on flanges, valves or fittings, multiply the weight factor by the weight per foot of covering used on straight pipe.

Valve weights are approximate. When possible, obtain weights from the manufacturer.

Cast iron valve weights are for flanged end valves; steel weights for welding end valves.

All flanged fitting, flanged valve and flange weights include the proportional weight of bolts or studs to make up all joints.

WEIGHT OF PIPING MATERIALS
(Courtesy of Anvil International)

1.900" O.D. 1½" PIPE

PIPE	Schedule No.	5S	10S	40	80	160			
PIPE	Wall Designation			Std.	XS		XXS		
PIPE	Thickness — In.	.065	.109	.145	.200	.281	.400	.525	.650
PIPE	Pipe — Lbs/Ft	**1.27**	**2.09**	**2.72**	**3.63**	**4.86**	**6.41**	**7.71**	**8.68**
PIPE	Water — Lbs/Ft	**1.07**	**.96**	**.88**	**.77**	**.61**	**.41**	**.25**	**.12**
WELDING FITTINGS	L.R. 90° Elbow	**.4** .4	**.8** .4	**.9** .4	**1.2** .4	**1.5** .4	**2.0** .4		
WELDING FITTINGS	S.R. 90° Elbow			**.6** .3	**.8** .3				
WELDING FITTINGS	L.R. 45° Elbow	**.3** .2	**.5** .2	**.5** .2	**.7** .2	**.8** .2	**.1** .2		
WELDING FITTINGS	Tee	**.9** .6	**1.5** .6	**2** .6	**2.4** .6	**3.0** .6	**3.7** .6		
WELDING FITTINGS	Lateral	**1.3** 1.3	**2.1** 1.3	**3.3** 1.3	**5.5** 1.3				
WELDING FITTINGS	Reducer	**.3** .2	**.6** .2	**.6** .2	**.8** .2	**1.0** .2	**1.2** .2		
WELDING FITTINGS	Cap	**.1** .3	**.2** .3	**.4** .3	**.5** .3	**.7** .3	**.8** .3		

		Temperature Range °F	100-199	200-299	300-399	400-499	500-599	600-699	700-799	800-899	900-999	1000-1099	1100-1200
INSULATION	85% Magnesia Calcium Silicate	Nom. Thick., In.	1	1	1½	2	2	2½	2½	2½	3	3	3
INSULATION	85% Magnesia Calcium Silicate	Lbs./Ft	**.84**	**.84**	**1.35**	**2.52**	**2.52**	**3.47**	**3.47**	**3.47**	**4.52**	**4.52**	**4.52**
INSULATION	Combination	Nom. Thick., In.						2½	2½	2½	3	3	3
INSULATION	Combination	Lbs/Ft						**4.20**	**4.20**	**4.20**	**5.62**	**5.62**	**5.62**

	Pressure Rating psi	Cast Iron 125	Cast Iron 250	Steel 150	Steel 300	Steel 400	Steel 600	Steel 900	Steel 1500	Steel 2500
FLANGES	Screwed or Slip-On	**3** 1.5	**6** 1.5	**3.5** 1.5	**6** 1.5	**9** 1.5	**9** 1.5	**19** 1.5	**19** 1.5	**30** 1.5
FLANGES	Welding Neck			**4.5** 1.5	**8** 1.5	**12** 1.5	**12** 1.5	**19** 1.5	**19** 1.5	**34** 1.5
FLANGES	Lap Joint			**3.5** 1.5	**6** 1.5	**9** 1.5	**9** 1.5	**19** 1.5	**19** 1.5	**30** 1.5
FLANGES	Blind	**4** 1.5	**6** 1.5	**3.5** 1.5	**8** 1.5	**10** 1.5	**10** 1.5	**19** 1.5	**19** 1.5	**31** 1.5
FLANGED FITTINGS	S.R. 90° Elbow	**9** 3.7		**12** 3.7	**23** 3.8		**26** 3.9		**46** 4	
FLANGED FITTINGS	L.R. 90° Elbow	**12** 4		**13** 4	**24** 4					
FLANGED FITTINGS	45° Elbow	**8** 3.4		**11** 3.4	**21** 3.5		**23** 3.5		**39** 3.7	
FLANGED FITTINGS	Tee	**15** 5.6		**20** 5.6	**30** 5.7		**37** 5.8		**70** 6	
VALVES	Flanged Bonnet Gate	**27** 6.8			**55** 4.2		**70** 4.5		**125** 5	
VALVES	Flanged Bonnet Globe or Angle				**40** 4.2		**45** 4.2		**170** 5	
VALVES	Flanged Bonnet Check			**30** 4.1	**35** 4.1		**40** 4.2		**110** 4.5	
VALVES	Pressure Seal Bonnet — Gate							**42** 1.9	**42** 1.2	
VALVES	Pressure Seal Bonnet — Globe									

Boldface type is weight in pounds. Lightface type beneath weight is weight factor for insulation.

Insulation thicknesses and weights are based on average conditions and do not constitute a recommendation for specific thicknesses of materials. Insulation weights are based on 85% magnesia and hydrous calcium silicate at 11 lbs/cubic foot. The listed thicknesses and weights of combination covering are the sums of the inner layer of diatomaceous earth at 21 lbs/cubic foot and the outer layer at 11 lbs/cubic foot.

Insulation weights include allowances for wire, cement, canvas, bands and paint, but not special surface finishes.

To find the weight of covering on flanges, valves or fittings, multiply the weight factor by the weight per foot of covering used on straight pipe.

Valve weights are approximate. When possible, obtain weights from the manufacturer.

Cast iron valve weights are for flanged end valves; steel weights for welding end valves.

All flanged fitting, flanged valve and flange weights include the proportional weight of bolts or studs to make up all joints.

WEIGHT OF PIPING MATERIALS
(Courtesy of Anvil International)

2.375" O.D. 2" PIPE

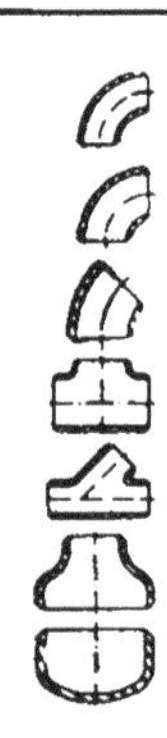

PIPE	Schedule No.	5S	10S	40	80	160							
	Wall Designation			Std.	XS		XXS						
	Thickness — In.	.065	.109	.154	.218	.343	.436	.562	.687				
	Pipe — Lbs/Ft	**1.60**	**2.64**	**3.65**	**5.02**	**7.44**	**9.03**	**10.88**	**12.39**				
	Water — Lbs/Ft	**1.72**	**1.58**	**1.46**	**1.28**	**.97**	**.77**	**.53**	**.34**				
WELDING FITTINGS	L.R. 90° Elbow	**.6** .5	**1.1** .5	**1.5** .5	**2.1** .5	**3.0** .5	**4.0** .5						
	S.R. 90° Elbow			**1** .3	**1.4** .3								
	L.R. 45° Elbow	**.4** .2	**.6** .2	**.9** .2	**1.1** .2	**1.6** .2	**2.0** .2						
	Tee	**1.1** .6	**1.8** .6	**2.9** .6	**3.7** .6	**4.9** .6	**5.7** .6						
	Lateral	**1.9** 1.4	**3.2** 1.4	**5** 1.4	**7.7** 1.4								
	Reducer	**.4** .3	**.9** .3	**.9** .3	**1.2** .3	**1.6** .3	**1.9** .3						
	Cap	**.2** .4	**.3** .4	**.6** .4	**.7** .4	**1.1** .4	**1.2** .4						

	Temperature Range °F	100-199	200-299	300-399	400-499	500-599	600-699	700-799	800-899	900-999	1000-1099	1100-1200
INSULATION — 85% Magnesia Calcium Silicate	Nom. Thick., In.	1	1	1½	2	2	2½	2½	2½	3	3	3
	Lbs./Ft	**1.01**	**1.01**	**1.71**	**2.53**	**2.53**	**3.48**	**3.48**	**4.42**	**4.42**	**4.42**	**5.59**
INSULATION — Combination	Nom. Thick., In.						2½	2½	2½	3	3	3
	Lbs/Ft						**4.26**	**4.28**	**5.93**	**5.93**	**5.93**	**7.80**

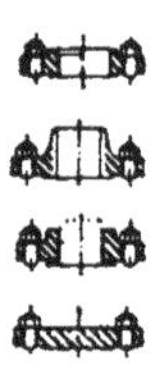

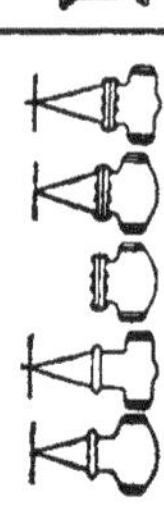

	Pressure Rating psi	Cast Iron		Steel						
		125	250	150	300	400	600	900	1500	2500
FLANGES	Screwed or Slip-On	**5** 1.5	**7** 1.5	**6** 1.5	**9** 1.5	**11** 1.5	**11** 1.5	**32** 1.5	**32** 1.5	**49** 1.5
	Welding Neck			**7** 1.5	**11** 1.5	**14** 1.5	**14** 1.5	**32** 1.5	**32** 1.5	**53** 1.5
	Lap Joint			**6** 1.5	**9** 1.5	**11** 1.5	**11** 1.5	**32** 1.5	**32** 1.5	**48** 1.5
	Blind	**5** 1.5	**8** 1.5	**5** 1.5	**10** 1.5	**12** 1.5	**12** 1.5	**32** 1.5	**32** 1.5	**50** 1.5
FLANGED FITTINGS	S.R. 90° Elbow	**14** 3.8	**20** 3.8	**19** 3.8	**29** 3.8		**35** 4		**83** 4.2	
	L.R. 90° Elbow	**16** 4.1	**27** 4.1	**22** 4.1	**31** 4.1					
	45° Elbow	**12** 3.4	**18** 3.5	**18** 3.4	**24** 3.5		**33** 3.7		**73** 3.9	
	Tee	**21** 5.7	**32** 5.7	**27** 5.7	**41** 5.7		**52** 6		**129** 6.3	
VALVES	Flanged Bonnet Gate	**37** 6.9	**52** 7.1	**40** 4	**65** 4.2		**80** 4.5		**190** 5	
	Flanged Bonnet Globe or Angle	**30** 7	**64** 7.3	**30** 3.8	**45** 4		**85** 4.5		**235** 5.5	
	Flanged Bonnet Check	**26** 7	**51** 7.3	**35** 3.8	**40** 4		**60** 4.2		**300** 5.8	
	Pressure Seal Bonnet — Gate								**150** 2.5	
	Pressure Seal Bonnet — Globe								**165** 3	

Boldface type is weight in pounds. Lightface type beneath weight is weight factor for insulation.

Insulation thicknesses and weights are based on average conditions and do not constitute a recommendation for specific thicknesses of materials. Insulation weights are based on 85% magnesia and hydrous calcium silicate at 11 lbs/cubic foot. The listed thicknesses and weights of combination covering are the sums of the inner layer of diatomaceous earth at 21 lbs/cubic foot and the outer layer at 11 lbs/cubic foot.

Insulation weights include allowances for wire, cement, canvas, bands and paint, but not special surface finishes.

To find the weight of covering on flanges, valves or fittings, multiply the weight factor by the weight per foot of covering used on straight pipe.

Valve weights are approximate. When possible, obtain weights from the manufacturer.

Cast iron valve weights are for flanged end valves; steel weights for welding end valves.

All flanged fitting, flanged valve and flange weights include the proportional weight of bolts or studs to make up all joints.

WEIGHT OF PIPING MATERIALS
(Courtesy of Anvil International)

2.875" O.D. 2½" PIPE

PIPE	Schedule No.	5S	10S	40	80	160							
	Wall Designation			Std.	XS		XXS						
	Thickness — In.	.083	.120	.203	.276	.375	.553	.675	.800				
	Pipe — Lbs/Ft	**2.48**	**3.53**	**5.79**	**7.66**	**10.01**	**13.70**	**15.86**	**17.73**				
	Water — Lbs/Ft	**2.50**	**2.36**	**2.08**	**1.84**	**1.54**	**1.07**	**.79**	**.55**				
WELDING FITTINGS	L.R. 90° Elbow	**1.2** .6	**1.8** .6	**3.0** .6	**3.8** .6	**5.0** .6	**7.0** .6						
	S.R. 90° Elbow			**2.2** .4	**2.5** .4								
	L.R. 45° Elbow	**.7** .3	**1.0** .3	**1.6** .3	**2.1** .3	**3.0** .3	**3.5** .3						
	Tee	**2.1** .8	**3.0** .8	**5.2** .8	**6.4** .8	**7.8** .8	**9.8** .8						
	Lateral	**3.5** 1.5	**4.9** 1.5	**9.0** 1.5	**13** 1.5								
	Reducer	**.6** .3	**1.2** .3	**1.6** .3	**2.0** .3	**2.7** .3	**3.3** .3						
	Cap	**.3** .4	**.4** .4	**.9** .4	**.1** .4	**1.9** .4	**2.0** .4						

	Temperature Range °F	100-199	200-299	300-399	400-499	500-599	600-699	700-799	800-899	900-999	1000-1099	1100-1200
INSULATION — 85% Magnesia Calcium Silicate	Nom. Thick., In.	1	1	1½	2	2	2½	2½	3	3	3½	3½
	Lbs./Ft	**1.14**	**1.14**	**2.29**	**3.23**	**3.23**	**4.28**	**4.28**	**5.46**	**5.46**	**6.86**	**6.86**
INSULATION — Combination	Nom. Thick., In.						2½	2½	3	3	3½	3½
	Lbs/Ft						**5.20**	**5.20**	**7.36**	**7.36**	**9.58**	**9.58**

	Pressure Rating psi	Cast Iron		Steel						
		125	250	150	300	400	600	900	1500	2500
FLANGES	Screwed or Slip-On	**7** 1.5	**12.5** 1.5	**8** 1.5	**14** 1.5	**17** 1.5	**17** 1.5	**46** 1.5	**46** 1.5	**69** 1.5
	Welding Neck			**11** 1.5	**16** 1.5	**22** 1.5	**22** 1.5	**46** 1.5	**46** 1.5	**66** 1.5
	Lap Joint			**8** 1.5	**14** 1.5	**16** 1.5	**16** 1.5	**45** 1.5	**45** 1.5	**67** 1.5
	Blind	**7.8** 1.5	**10** 1.5	**8** 1.5	**16** 1.5	**19** 1.5	**19** 1.5	**45** 1.5	**45** 1.5	**70** 1.5
FLANGED FITTINGS	S.R. 90° Elbow	**20** 3.8	**33** 3.9	**27** 3.8	**42** 3.9		**50** 4.1		**114** 4.4	
	L.R. 90° Elbow	**24** 4.2		**30** 4.2	**47** 4.2					
	45° Elbow	**18** 3.5	**31** 3.6	**22** 3.5	**35** 3.6		**46** 3.8		**99** 3.9	
	Tee	**31** 5.7	**49** 5.8	**42** 5.7	**61** 5.9		**77** 6.2		**169** 6.6	
VALVES	Flanged Bonnet Gate	**50** 7	**82** 7.1	**60** 4	**100** 4.2		**105** 4.6		**275** 5.2	
	Flanged Bonnet Globe or Angle	**43** 7.1	**87** 7.4	**50** 4	**70** 4.1		**120** 4.6		**325** 5.5	
	Flanged Bonnet Check	**36** 7.1	**71** 7.4	**40** 4	**50** 4		**105** 4.6		**320** 5.5	
	Pressure Seal Bonnet — Gate								**215** 2.5	
	Pressure Seal Bonnet — Globe								**230** 2.8	

Boldface type is weight in pounds. Lightface type beneath weight is weight factor for insulation.

Insulation thicknesses and weights are based on average conditions and do not constitute a recommendation for specific thicknesses of materials. Insulation weights are based on 85% magnesia and hydrous calcium silicate at 11 lbs/cubic foot. The listed thicknesses and weights of combination covering are the sums of the inner layer of diatomaceous earth at 21 lbs/cubic foot and the outer layer at 11 lbs/cubic foot.

Insulation weights include allowances for wire, cement, canvas, bands and paint, but not special surface finishes.

To find the weight of covering on flanges, valves or fittings, multiply the weight factor by the weight per foot of covering used on straight pipe.

Valve weights are approximate. When possible, obtain weights from the manufacturer.

Cast iron valve weights are for flanged end valves; steel weights for welding end valves.

All flanged fitting, flanged valve and flange weights include the proportional weight of bolts or studs to make up all joints.

WEIGHT OF PIPING MATERIALS
(Courtesy of Anvil International)

3.500" O.D. 3" PIPE

PIPE	Schedule No.	5S	10S	40	80	160			
	Wall Designation			Std.	XS		XXS		
	Thickness — In.	.083	.120	.216	.300	.438	.600	.725	.850
	Pipe — Lbs/Ft	**3.03**	**4.33**	**7.58**	**10.25**	**14.32**	**18.58**	**21.49**	**24.06**
	Water — Lbs/Ft	**3.78**	**3.61**	**3.20**	**2.86**	**2.35**	**1.80**	**1.43**	**1.10**
WELDING FITTINGS	L.R. 90° Elbow	**1.7** .8	**2.5** .8	**4.7** .8	**6.0** .8	**8.5** .8	**11** .8		
	S.R. 90° Elbow			**3.3** .5	**4.1** .5				
	L.R. 45° Elbow	**.9** .3	**1.3** .3	**2.5** .3	**3.3** .3	**4.5** .3	**5.5** .3		
	Tee	**2.7** .8	**3.9** .8	**7.0** .8	**10** .8	**12.2** .8	**14.8** .8		
	Lateral	**4.5** 1.8	**6.4** 1.8	**12.5** 1.8	**18** 1.8				
	Reducer	**.8** .3	**1.5** .3	**2.1** .3	**2.6** .3	**3.7** .3	**4.6** .3		
	Cap	**.5** .5	**.7** .5	**1.4** .5	**1.8** .5	**3.5** .5	**3.6** .5		

INSULATION	Temperature Range °F	100-199	200-299	300-399	400-499	500-599	600-699	700-799	800-899	900-999	1000-1099	1100-1200
85% Magnesia Calcium Silicate	Nom. Thick., In.	1	1	1½	2	2	2½	3	3	3	3½	3½
	Lbs./Ft	**1.25**	**1.25**	**2.08**	**3.01**	**3.01**	**4.07**	**5.24**	**5.24**	**5.24**	**6.65**	**6.65**
Combination	Nom. Thick., In.						2½	3	3	3	3½	3½
	Lbs/Ft						**5.07**	**6.94**	**6.94**	**6.94**	**9.17**	**9.17**

	Pressure Rating psi	Cast Iron		Steel						
		125	250	150	300	400	600	900	1500	2500
FLANGES	Screwed or Slip-On	**8.6** 1.5	**15.8** 1.5	**9** 1.5	**17** 1.5	**20** 1.5	**20** 1.5	**37** 1.5	**61** 1.5	**102** 1.5
	Welding Neck			**12** 1.5	**19** 1.5	**27** 1.5	**27** 1.5	**38** 1.5	**61** 1.5	**113** 1.5
	Lap Joint			**9** 1.5	**17** 1.5	**19** 1.5	**19** 1.5	**36** 1.5	**60** 1.5	**99** 1.5
	Blind	**9** 1.5	**17.5** 1.5	**10** 1.5	**20** 1.5	**24** 1.5	**24** 1.5	**38** 1.5	**61** 1.5	**105** 1.5
FLANGED FITTINGS	S.R. 90° Elbow	**25** 3.9	**44** 4	**32** 3.9	**53** 4		**67** 4.1	**98** 4.3	**150** 4.6	
	L.R. 90° Elbow	**29** 4.3		**40** 4.3	**63** 4.3					
	45° Elbow	**21** 3.5	**39** 3.6	**28** 3.5	**46** 3.6		**60** 3.8	**93** 3.9	**135** 4	
	Tee	**38** 5.9	**62** 6	**52** 5.9	**81** 6		**102** 6.2	**151** 6.5	**238** 6.9	
VALVES	Flanged Bonnet Gate	**66** 7	**112** 7.4	**70** 4	**125** 4.4		**155** 4.8	**260** 5	**410** 5.5	
	Flanged Bonnet Globe or Angle	**56** 7.2	**87** 7.6	**60** 4.3	**95** 4.5		**155** 4.8	**225** 5	**495** 5.5	
	Flanged Bonnet Check	**46** 7.2	**100** 7.6	**60** 4.3	**70** 4.4		**120** 4.8	**150** 4.9	**440** 5.8	
	Pressure Seal Bonnet — Gate							**208** 3	**235** 3.2	
	Pressure Seal Bonnet — Globe							**135** 2.5	**180** 3	

Boldface type is weight in pounds. Lightface type beneath weight is weight factor for insulation.

Insulation thicknesses and weights are based on average conditions and do not constitute a recommendation for specific thicknesses of materials. Insulation weights are based on 85% magnesia and hydrous calcium silicate at 11 lbs/cubic foot. The listed thicknesses and weights of combination covering are the sums of the inner layer of diatomaceous earth at 21 lbs/cubic foot and the outer layer at 11 lbs/cubic foot.

Insulation weights include allowances for wire, cement, canvas, bands and paint, but not special surface finishes.

To find the weight of covering on flanges, valves or fittings, multiply the weight factor by the weight per foot of covering used on straight pipe.

Valve weights are approximate. When possible, obtain weights from the manufacturer.

Cast iron valve weights are for flanged end valves; steel weights for welding end valves.

All flanged fitting, flanged valve and flange weights include the proportional weight of bolts or studs to make up all joints.

WEIGHT OF PIPING MATERIALS
(Courtesy of Anvil International)

4.000" O.D. 3½" PIPE

PIPE	Schedule No.	5S	10S	40	80	160							
	Wall Designation			Std.	XS		XXS						
	Thickness — In.	.083	.120	.226	.318	.636							
	Pipe — Lbs/Ft	**3.47**	**4.97**	**9.11**	**12.51**	**22.85**							
	Water — Lbs/Ft	**5.01**	**4.81**	**4.28**	**3.85**	**2.53**							
WELDING FITTINGS	L.R. 90° Elbow	**2.4** .9	**3.4** .9	**6.7** .9	**8.7** .9	**15** .9							
	S.R. 90° Elbow			**4.2** .6	**5.75** .6								
	L.R. 45° Elbow	**1.2** .4	**1.7** .4	**3.3** .4	**4.4** .4	**8** .4							
	Tee	**3.4** .9	**4.9** .9	**10.3** .9	**13.8** .9	**20.2** .9							
	Lateral	**6.2** 1.8	**8.9** 1.8	**17.2** 1.8	**25** 1.8								
	Reducer	**1.2** .3	**2.1** .3	**3.0** .3	**4.0** .3	**6.8** .3							
	Cap	**.6** .6	**.8** .6	**2.1** .6	**2.8** .6	**5.5** .6							

	Temperature Range °F		100-199	200-299	300-399	400-499	500-599	600-699	700-799	800-899	900-999	1000-1099	1100-1200
INSULATION	85% Magnesia Calcium Silicate	Nom. Thick., In.	1	1	1½	2	2½	2½	3	3	3½	3½	3½
		Lbs./Ft	**1.83**	**1.83**	**2.77**	**3.71**	**4.88**	**4.88**	**6.39**	**6.39**	**7.80**	**7.80**	**7.80**
	Combination	Nom. Thick., In.						2½	3	3	3½	3½	3½
		Lbs/Ft						**6.49**	**8.71**	**8.71**	**10.8**	**10.8**	**10.8**

		Cast Iron		Steel						
	Pressure Rating psi	125	250	150	300	400	600	900	1500	2500
FLANGES	Screwed or Slip-On	**11** 1.5	**20** 1.5	**13** 1.5	**21** 1.5	**27** 1.5	**27** 1.5			
	Welding Neck			**14** 1.5	**22** 1.5	**32** 1.5	**32** 1.5			
	Lap Joint			**13** 1.5	**21** 1.5	**26** 1.5	**26** 1.5			
	Blind	**13** 1.5	**23** 1.5	**15** 1.5	**25** 1.5	**35** 1.5	**35** 1.5			
FLANGED FITTINGS	S.R. 90° Elbow	**33** 4		**49** 4			**82** 4.3			
	L.R. 90° Elbow			**54** 4.4						
	45° Elbow	**29** 3.6		**39** 3.6			**75** 3.6			
	Tee	**51** 6	**103** 6.2	**70** 6			**133** 6.4			
VALVES	Flanged Bonnet Gate	**82** 7.1	**143** 7.5	**90** 4.1	**155** 4.5		**180** 4.8	**360** 5	**510** 5.5	
	Flanged Bonnet Globe or Angle	**74** 7.3	**137** 7.7				**160** 4.7			
	Flanged Bonnet Check	**71** 7.3	**125** 7.7				**125** 4.7			
	Pressure Seal Bonnet — Gate						**140** 2.5	**295** 2.8	**380** 3	
	Pressure Seal Bonnet — Globe									

Boldface type is weight in pounds. Lightface type beneath weight is weight factor for insulation.

Insulation thicknesses and weights are based on average conditions and do not constitute a recommendation for specific thicknesses of materials. Insulation weights are based on 85% magnesia and hydrous calcium silicate at 11 lbs/cubic foot. The listed thicknesses and weights of combination covering are the sums of the inner layer of diatomaceous earth at 21 lbs/cubic foot and the outer layer at 11 lbs/cubic foot.

Insulation weights include allowances for wire, cement, canvas, bands and paint, but not special surface finishes.

To find the weight of covering on flanges, valves or fittings, multiply the weight factor by the weight per foot of covering used on straight pipe.

Valve weights are approximate. When possible, obtain weights from the manufacturer.

Cast iron valve weights are for flanged end valves; steel weights for welding end valves.

All flanged fitting, flanged valve and flange weights include the proportional weight of bolts or studs to make up all joints.

WEIGHT OF PIPING MATERIALS
(Courtesy of Anvil International)

4.500" O.D. 4" PIPE

PIPE	Schedule No.	5S	10S		40	80	120		160				
	Wall Designation				Std.	XS				XXS			
	Thickness — In.	.083	.120	.188	.237	.337	.438	.500	.531	.674	.800	.925	
	Pipe — Lbs/Ft	**3.92**	**5.61**	**8.56**	**10.79**	**14.98**	**18.96**	**21.36**	**22.51**	**27.54**	**31.61**	**35.32**	
	Water — Lbs/Ft	**6.40**	**6.17**	**5.80**	**5.51**	**4.98**	**4.48**	**4.16**	**4.02**	**3.38**	**2.86**	**2.39**	
WELDING FITTINGS	L.R. 90° Elbow	**3.0** 1	**4.3** 1		**8.7** 1	**12.0** 1			**18.0** 1	**20.5** 1			
	S.R. 90° Elbow				**6.7** .7	**8.3** .7							
	L.R. 45° Elbow	**1.5** .4	**2.2** .4		**4.3** .4	**5.9** .4			**8.5** .4	**10** .4			
	Tee	**3.9** 1	**5.7** 1		**13.5** 1	**16.4** 1			**22.8** 1	**26.6** 1			
	Lateral	**6.6** 2.1	**10.0** 2.1		**20.5** 2.1	**32** 2.1							
	Reducer	**1.2** .3	**2.4** .3		**3.6** .3	**4.8** .3			**6.6** .3	**8.2** .3			
	Cap	**.8** .3	**1.2** .3		**2.5** .5	**3.4** .5			**6.5** 6.5	**6.6** 6.6			

INSULATION	Temperature Range °F	100-199	200-299	300-399	400-499	500-599	600-699	700-799	800-899	900-999	1000-1099	1100-1200
85% Magnesia Calcium Silicate	Nom. Thick., In.	1	1	1½	2	2½	2½	3	3	3½	3½	4
	Lbs./Ft	**1.62**	**1.62**	**2.55**	**3.61**	**4.66**	**4.66**	**6.07**	**6.07**	**7.48**	**7.48**	**9.10**
Combination	Nom. Thick., In.						2½	3	3	3½	3½	3½
	Lbs/Ft						**6.07**	**8.30**	**8.30**	**10.6**	**10.6**	**10.6**

	Pressure Rating psi	Cast Iron 125	Cast Iron 250	Steel 150	Steel 300	Steel 400	Steel 600	Steel 900	Steel 1500	Steel 2500
FLANGES	Screwed or Slip-On	**14** 1.5	**24** 1.5	**15** 1.5	**26** 1.5	**32** 1.5	**43** 1.5	**66** 1.5	**90** 1.5	**158** 1.5
	Welding Neck			**17** 1.5	**29** 1.5	**41** 1.5	**48** 1.5	**648** 1.5	**90** 1.5	**177** 1.5
	Lap Joint			**15** 1.5	**26** 1.5	**31** 1.5	**42** 1.5	**64** 1.5	**92** 1.5	**153** 1.5
	Blind	**16** 1.5	**27** 1.5	**19** 1.5	**31** 1.5	**39** 1.5	**47** 1.5	**67** 1.5	**90** 1.5	**164** 1.5
FLANGED FITTINGS	S.R. 90° Elbow	**43** 4.1	**69** 4.2	**59** 4.1	**85** 4.2	**99** 4.3	**128** 4.4	**185** 4.5	**254** 4.8	
	L.R. 90° Elbow	**50** 4.5		**72** 4.5	**98** 4.5					
	45° Elbow	**38** 3.7	**62** 3.8	**51** 3.7	**78** 3.8	**82** 3.9	**119** 4	**170** 4.1	**214** 4.2	
	Tee	**66** 6.1	**103** 6.3	**86** 6.1	**121** 6.3	**153** 6.4	**187** 6.6	**262** 6.8	**386** 7.2	
VALVES	Flanged Bonnet Gate	**109** 7.2	**188** 7.5	**100** 4.2	**175** 4.5	**195** 5	**255** 5.1	**455** 5.4	**735** 6	
	Flanged Bonnet Globe or Angle	**97** 7.4	**177** 7.8	**95** 4.3	**145** 4.8	**215** 5	**230** 5.1	**415** 5.5	**800** 6	
	Flanged Bonnet Check	**80** 7.4	**146** 7.8	**80** 4.3	**105** 4.5	**160** 4.8	**195** 5	**320** 5.6	**780** 6	
	Pressure Seal Bonnet — Gate						**215** 2.8	**380** 3	**520** 4	
	Pressure Seal Bonnet — Globe							**240** 2.7	**290** 3	

Boldface type is weight in pounds. Lightface type beneath weight is weight factor for insulation.

Insulation thicknesses and weights are based on average conditions and do not constitute a recommendation for specific thicknesses of materials. Insulation weights are based on 85% magnesia and hydrous calcium silicate at 11 lbs/cubic foot. The listed thicknesses and weights of combination covering are the sums of the inner layer of diatomaceous earth at 21 lbs/cubic foot and the outer layer at 11 lbs/cubic foot.

Insulation weights include allowances for wire, cement, canvas, bands and paint, but not special surface finishes.

To find the weight of covering on flanges, valves or fittings, multiply the weight factor by the weight per foot of covering used on straight pipe.

Valve weights are approximate. When possible, obtain weights from the manufacturer.

Cast iron valve weights are for flanged end valves; steel weights for welding end valves.

All flanged fitting, flanged valve and flange weights include the proportional weight of bolts or studs to make up all joints.

WEIGHT OF PIPING MATERIALS
(Courtesy of Anvil International)

5.563″ O.D. 5″ PIPE

PIPE	Schedule No.	5S	10S	40	80	120	160						
	Wall Designation			Std.	XS			XXS					
	Thickness — In.	.109	.134	.258	.375	.500	.625	.750	.875	1.000			
	Pipe — Lbs/Ft	**6.35**	**7.77**	**14.62**	**20.78**	**27.04**	**32.96**	**38.55**	**43.81**	**47.73**			
	Water — Lbs/Ft	**9.73**	**9.53**	**8.66**	**7.89**	**7.09**	**6.33**	**5.62**	**4.95**	**4.23**			
WELDING FITTINGS	L.R. 90° Elbow	**6.0** 1.3	**7.4** 1.3	**16.0** 1.3	**21.4** 1.3		**33** 1.3	**34** 1.3					
	S.R. 90° Elbow	**4.2** .8	**5.2** .8	**10.4** .8	**14.5** .8								
	L.R. 45° Elbow	**3.1** .5	**3.8** .5	**8.3** .5	**10.5** .5		**14** .5	**18** .5					
	Tee	**9.8** 1.2	**12.0** 1.2	**19.8** 1.2	**26.9** 1.2		**38.5** 1.2	**43.4** 1.2					
	Lateral	**15.3** 2.5	**18.4** 2.5	**31** 2.5	**49** 2.5								
	Reducer	**2.5** .4	**4.3** .4	**5.9** .4	**8.3** .4		**12.4** .4	**14.2** .4					
	Cap	**1.3** .7	**1.6** .7	**4.2** .7	**5.7** .7		**11** .7	**11** .7					

Temperature Range °F		100-199	200-299	300-399	400-499	500-599	600-699	700-799	800-899	900-999	1000-1099	1100-1200
INSULATION — 85% Magnesia Calcium Silicate	Nom. Thick., In.	1	1½	1½	2	2½	2½	3	3½	3½	4	4
	Lbs./Ft	**1.86**	**2.92**	**2.92**	**4.08**	**5.38**	**5.38**	**6.90**	**8.41**	**8.41**	**10.4**	**10.4**
INSULATION — Combination	Nom. Thick., In.						2½	3	3½	3½	4	4
	Lbs/Ft						**7.01**	**9.30**	**11.8**	**11.8**	**14.9**	**14.9**

	Pressure Rating psi	Cast Iron		Steel						
		125	250	150	300	400	600	900	1500	2500
FLANGES	Screwed or Slip-On	**17** 1.5	**28** 1.5	**18** 1.5	**32** 1.5	**37** 1.5	**73** 1.5	**100** 1.5	**162** 1.5	**259** 1.5
	Welding Neck			**22** 1.5	**36** 1.5	**49** 1.5	**78** 1.5	**103** 1.5	**162** 1.5	**293** 1.5
	Lap Joint			**18** 1.5	**32** 1.5	**35** 1.5	**71** 1.5	**98** 1.5	**179** 1.5	**253** 1.5
	Blind	**21** 1.5	**35** 1.5	**23** 1.5	**39** 1.5	**50** 1.5	**78** 1.5	**104** 1.5	**172** 1.5	**272** 1.5
FLANGED FITTINGS	S.R. 90° Elbow	**55** 4.3	**91** 4.3	**80** 4.3	**113** 4.3	**123** 4.5	**205** 4.7	**268** 4.8	**435** 5.2	
	L.R. 90° Elbow	**65** 4.7		**91** 4.7	**128** 4.7					
	45° Elbow	**48** 3.8	**80** 3.8	**66** 3.8	**98** 3.8	**123** 4	**180** 4.2	**239** 4.3	**350** 4.5	
	Tee	**84** 6.4	**139** 6.5	**119** 6.4	**172** 6.4	**179** 6.8	**304** 7	**415** 7.2	**665** 7.8	
VALVES	Flanged Bonnet Gate	**138** 7.3	**264** 7.9	**150** 4.3	**265** 4.9	**310** 5.3	**455** 5.5	**615** 6	**1340** 7	
	Flanged Bonnet Globe or Angle	**138** 7.6	**247** 8	**155** 4.3	**215** 5	**355** 5.2	**515** 5.8	**555** 5.8	**950** 6	
	Flanged Bonnet Check	**118** 7.6	**210** 8	**110** 4.3	**165** 5	**185** 5	**350** 5.8	**570** 6	**1150** 7	
	Pressure Seal Bonnet — Gate						**350** 3.1	**520** 3.8	**865** 4.5	
	Pressure Seal Bonnet — Globe							**280** 4	**450** 4.5	

Boldface type is weight in pounds. Lightface type beneath weight is weight factor for insulation.

Insulation thicknesses and weights are based on average conditions and do not constitute a recommendation for specific thicknesses of materials. Insulation weights are based on 85% magnesia and hydrous calcium silicate at 11 lbs/cubic foot. The listed thicknesses and weights of combination covering are the sums of the inner layer of diatomaceous earth at 21 lbs/cubic foot and the outer layer at 11 lbs/cubic foot.

Insulation weights include allowances for wire, cement, canvas, bands and paint, but not special surface finishes.

To find the weight of covering on flanges, valves or fittings, multiply the weight factor by the weight per foot of covering used on straight pipe.

Valve weights are approximate. When possible, obtain weights from the manufacturer.

Cast iron valve weights are for flanged end valves; steel weights for welding end valves.

All flanged fitting, flanged valve and flange weights include the proportional weight of bolts or studs to make up all joints.

WEIGHT OF PIPING MATERIALS
(Courtesy of Anvil International)

6.625" O.D. 6" PIPE

PIPE	Schedule No.	5S	10S		40	80	120	160					
	Wall Designation				Std.	XS			XXS				
	Thickness — In.	.109	.134	.219	.280	.432	.562	.718	.864	1.000	1.125		
	Pipe — Lbs/Ft	**5.37**	**9.29**	**15.02**	**18.97**	**28.57**	**36.39**	**45.3**	**53.2**	**60.01**	**66.08**		
	Water — Lbs/Ft	**13.98**	**13.74**	**13.10**	**12.51**	**11.29**	**10.30**	**9.2**	**8.2**	**7.28**	**6.52**		
WELDING FITTINGS	L.R. 90° Elbow	**8.9** 1.5	**11.0** 1.5		**22.8** 1.5	**32.2** 1.5	**43** 1.5	**55** 1.6	**62** 1.5				
	S.R. 90° Elbow	**6.1** 1	**7.5** 1		**16.6** 1	**22.9** 1	**30** 1						
	L.R. 45° Elbow	**4.5** .6	**5.5** .6		**11.3** .6	**16.4** .6	**21** .6	**26** .6	**30** .6				
	Tee	**13.8** 1.4	**17.0** 1.4		**31.3** 1.4	**39.5** 1.4		**59** 1.4	**68** 1.4				
	Lateral	**16.7** 2.9	**20.5** 2.9		**42** 2.9	**78** 2.9							
	Reducer	**3.3** .5	**5.8** .5		**8.6** .6	**12.6** .5		**18.8** .5	**21.4** .5				
	Cap	**1.6** .9	**1.9** .9		**6.4** .9	**9.2** .9	**13.3** .9	**17.5** .9	**17.5** .9				

INSULATION												
Temperature Range °F		100-199	200-299	300-399	400-499	500-599	600-699	700-799	800-899	900-999	1000-1099	1100-1200
85% Magnesia Calcium Silicate	Nom. Thick., In.	1	1½	2	2	2½	3	3	3½	3½	4	4
	Lbs./Ft	**2.11**	**3.28**	**4.57**	**4.57**	**6.09**	**7.60**	**7.60**	**9.82**	**9.82**	**11.5**	**11.4**
Combination	Nom. Thick., In.						3	3	3½	3½	4	4
	Lbs/Ft						**10.3**	**10.3**	**13.4**	**13.4**	**16.6**	**16.6**

	Pressure Rating psi	Cast Iron		Steel						
		125	250	150	300	400	600	900	1500	2500
FLANGES	Screwed or Slip-On	**20** 1.5	**38** 1.5	**22** 1.5	**45** 1.5	**54** 1.5	**95** 1.5	**128** 1.5	**202** 1.5	**396** 1.5
	Welding Neck			**27** 1.5	**48** 1.5	**67** 1.5	**96** 1.5	**130** 1.5	**202** 1.5	**451** 1.5
	Lap Joint			**22** 1.5	**45** 1.5	**52** 1.5	**93** 1.5	**125** 1.5	**208** 1.5	**387** 1.5
	Blind	**26** 1.5	**48** 1.5	**29** 1.5	**56** 1.5	**71** 1.5	**101** 1.5	**133** 1.5	**197** 1.5	**418** 1.5
FLANGED FITTINGS	S.R. 90° Elbow	**71** 4.3	**121** 4.4	**90** 4.3	**147** 4.4	**184** 4.6	**275** 4.8	**375** 5	**566** 5.3	
	L.R. 90° Elbow	**88** 4.9		**126** 4.9	**182** 4.9					
	45° Elbow	**63** 3.8	**111** 3.9	**82** 3.8	**132** 3.9	**149** 4.1	**240** 4.3	**320** 4.3	**487** 4.6	
	Tee	**108** 6.5	**186** 6.6	**149** 6.5	**218** 6.6	**280** 6.9	**400** 7.2	**585** 7.5	**839** 8	

		Cast Iron		Steel						
	Pressure Rating psi	125	250	150	300	400	600	900	1500	2500
VALVES	Flanged Bonnet Gate	**172** 7.3	**359** 8	**190** 4.3	**360** 5	**435** 5.5	**620** 5.8	**835** 6	**1595** 7	
	Flanged Bonnet Globe or Angle	**184** 7.8	**345** 8.2	**185** 4.4	**275** 5	**415** 5.3	**645** 5.8	**765** 6	**1800** 7	
	Flanged Bonnet Check	**154** 7.8	**286** 8.2	**150** 4.8	**200** 5	**360** 5.4	**445** 6	**800** 6.4	**1730** 7	
	Pressure Seal Bonnet — Gate						**580** 3.5	**750** 4	**1215** 5	
	Pressure Seal Bonnet — Globe							**730** 4	**780** 5	

Boldface type is weight in pounds. Lightface type beneath weight is weight factor for insulation.

Insulation thicknesses and weights are based on average conditions and do not constitute a recommendation for specific thicknesses of materials. Insulation weights are based on 85% magnesia and hydrous calcium silicate at 11 lbs/cubic foot. The listed thicknesses and weights of combination covering are the sums of the inner layer of diatomaceous earth at 21 lbs/cubic foot and the outer layer at 11 lbs/cubic foot.

Insulation weights include allowances for wire, cement, canvas, bands and paint, but not special surface finishes.

To find the weight of covering on flanges, valves or fittings, multiply the weight factor by the weight per foot of covering used on straight pipe.

Valve weights are approximate. When possible, obtain weights from the manufacturer.

Cast iron valve weights are for flanged end valves; steel weights for welding end valves.

All flanged fitting, flanged valve and flange weights include the proportional weight of bolts or studs to make up all joints.

WEIGHT OF PIPING MATERIALS
(Courtesy of Anvil International)

8.625" O.D. 8" PIPE

PIPE	Schedule No.	5S	10S		20	30	40	60	80	100	120	140	160
	Wall Designation						Std.		XS				
	Thickness — In.	.109	.148	.219	.250	.277	.322	.406	.500	.593	.718	.812	.906
	Pipe — Lbs/Ft	**9.91**	**13.40**	**19.64**	**22.36**	**24.70**	**28.55**	**35.64**	**43.4**	**50.9**	**60.6**	**67.8**	**74.7**
	Water — Lbs/Ft	**24.07**	**23.59**	**22.90**	**22.48**	**22.18**	**21.69**	**20.79**	**19.8**	**18.8**	**17.6**	**16.7**	**15.8**
WELDING FITTINGS	L.R. 90° Elbow	**15.4** 2	**21.0** 2				**44.9** 2		**70.3** 2				**120** 2
	S.R. 90° Elbow	**6.6** 1.3	**14.3** 1.3				**34.5** 1.3		**50.2** 1.3				
	L.R. 45° Elbow	**8.1** .8	**11.0** .8				**22.8** .8		**32.8** .8				**56** .8
	Tee	**18.4** 1.8	**25.0** 1.8				**60.2** 1.8		**78** 1.8				**120** 1.8
	Lateral	**25.3** 3.8	**41.1** 3.8				**76** 3.8		**140** 3.8				
	Reducer	**4.5** .5	**7.8** .5				**13.9** .5		**20.4** .5				**32.1** .5
	Cap	**2.1** 1	**2.8** 1				**11.3** 1		**16.3** 1				**32** 1

Temperature Range °F		100-199	200-299	300-399	400-499	500-599	600-699	700-799	800-899	900-999	1000-1099	1100-1200
INSULATION 85% Magnesia Calcium Silicate	Nom. Thick., In.	1½	1½	2	2	2½	3	3½	3½	4	4	4½
	Lbs./Ft	**4.13**	**4.13**	**5.64**	**5.64**	**7.85**	**9.48**	**11.5**	**11.5**	**13.8**	**13.8**	**16.0**
Combination	Nom. Thick., In.						3	3½	3½	4	4	4½
	Lbs/Ft						**12.9**	**16.2**	**16.2**	**20.4**	**20.4**	**23.8**

		Cast Iron		Steel						
	Pressure Rating psi	125	250	150	300	400	600	900	1500	2500
FLANGES	Screwed or Slip-On	**29** 1.5	**60** 1.5	**33** 1.5	**67** 1.5	**82** 1.5	**135** 1.5	**207** 1.5	**319** 1.5	**601** 1.5
	Welding Neck			**42** 1.5	**76** 1.5	**104** 1.5	**137** 1.5	**222** 1.5	**334** 1.5	**692** 1.5
	Lap Joint			**33** 1.5	**77** 1.5	**79** 1.5	**132** 1.5	**223** 1.5	**347** 1.5	**587** 1.5
	Blind	**43** 1.5	**79** 1.5	**48** 1.5	**90** 1.5	**115** 1.5	**159** 1.5	**232** 1.5	**363** 1.5	**649** 1.5
FLANGED FITTINGS	S.R. 90° Elbow	**113** 4.5	**194** 4.7	**157** 4.5	**238** 4.7	**310** 5	**435** 5.2	**639** 5.4	**995** 5.7	
	L.R. 90° Elbow	**148** 5.3		**202** 5.3	**283** 5.3					
	45° Elbow	**97** 3.9	**164** 4	**127** 3.9	**203** 4	**215** 4.1	**360** 4.4	**507** 4.5	**870** 4.8	
	Tee	**168** 6.8	**289** 7.1	**230** 6.8	**337** 7.1	**445** 7.5	**610** 7.8	**978** 8.1	**1465** 8.6	
VALVES	Flanged Bonnet Gate	**251** 7.5	**583** 8.1	**305** 4.5	**505** 5.1	**703** 6	**960** 6.3	**1180** 6.6	**2740** 7	
	Flanged Bonnet Globe or Angle	**317** 8.4	**554** 8.6	**475** 5.4	**505** 5.5	**610** 5.9	**1130** 6.3	**1160** 6.3	**2865** 7	
	Flanged Bonnet Check	**302** 8.4	**454** 8.6	**235** 5.2	**310** 5.3	**475** 5.6	**725** 6	**1140** 6.4	**2075** 7	
	Pressure Seal Bonnet — Gate						**925** 4.5	**1185** 4.7	**2345** 5.5	
	Pressure Seal Bonnet — Globe							**1550** 4	**1680** 5	

Boldface type is weight in pounds. Lightface type beneath weight is weight factor for insulation.

Insulation thicknesses and weights are based on average conditions and do not constitute a recommendation for specific thicknesses of materials. Insulation weights are based on 85% magnesia and hydrous calcium silicate at 11 lbs/cubic foot. The listed thicknesses and weights of combination covering are the sums of the inner layer of diatomaceous earth at 21 lbs/cubic foot and the outer layer at 11 lbs/cubic foot.

Insulation weights include allowances for wire, cement, canvas, bands and paint, but not special surface finishes.

To find the weight of covering on flanges, valves or fittings, multiply the weight factor by the weight per foot of covering used on straight pipe.

Valve weights are approximate. When possible, obtain weights from the manufacturer.

Cast iron valve weights are for flanged end valves; steel weights for welding end valves.

All flanged fitting, flanged valve and flange weights include the proportional weight of bolts or studs to make up all joints.

WEIGHT OF PIPING MATERIALS
(Courtesy of Anvil International)

10.750" O.D. 10" PIPE

PIPE	Schedule No.	5S	10S		20	30	40	60	80	100	120	140	160
	Wall Designation						Std.		XS				
	Thickness — In.	.134	.165	.219	.250	.307	.365	.500	.593	.718	.843	1.000	1.125
	Pipe — Lbs/Ft	**15.15**	**18.70**	**24.63**	**28.04**	**34.24**	**40.5**	**54.7**	**64.3**	**76.9**	**89.2**	**104.1**	**115.7**
	Water — Lbs/Ft	**37.4**	**36.9**	**36.2**	**35.77**	**34.98**	**34.1**	**32.3**	**31.1**	**29.5**	**28.0**	**26.1**	**24.6**
WELDING FITTINGS	L.R. 90° Elbow	**29.2** 2.5	**36.0** 2.5				**84** 2.5	**112** 2.5					**230** 2.5
	S.R. 90° Elbow	**20.3** 1.7	**24.9** 1.7				**62.2** 1.7	**74** 1.7					
	L.R. 45° Elbow	**14.6** 1	**18.0** 1				**42.4** 1	**53.8** 1					**109** 1
	Tee	**30.0** 2.1	**37.0** 2.1				**104** 2.1	**132** 2.1					**222** 2.1
	Lateral	**47.5** 4.4	**70.0** 4.4				**124** 4.4	**200** 4.4					
	Reducer	**8.1** .6	**14.0** .6				**23.2** .6	**31.4** .6					**58** .6
	Cap	**3.8** 1.3	**4.7** 1.3				**20** 1.3	**26.3** 1.3					**59** 1.3

Temperature Range °F		100-199	200-299	300-399	400-499	500-599	600-699	700-799	800-899	900-999	1000-1099	1100-1200
INSULATION: 85% Magnesia Calcium Silicate	Nom. Thick., In.	1½	1½	2	2½	2½	3	3½	3½	4	4	4½
	Lbs./Ft	**5.20**	**5.20**	**7.07**	**8.93**	**8.93**	**11.0**	**13.2**	**13.2**	**15.5**	**15.5**	**18.1**
INSULATION: Combination	Nom. Thick., In.						3	3½	3½	4	4	4½
	Lbs/Ft						**15.4**	**19.3**	**19.3**	**23**	**23**	**27.2**

	Pressure Rating psi	Cast Iron		Steel						
		125	250	150	300	400	600	900	1500	2500
FLANGES	Screwed or Slip-On	**45** 1.5	**93** 1.5	**50** 1.5	**100** 1.5	**117** 1.5	**213** 1.5	**293** 1.5	**528** 1.5	**1148** 1.5
	Welding Neck			**59** 1.5	**110** 1.5	**152** 1.5	**225** 1.5	**316** 1.5	**546** 1.5	**1291** 1.5
	Lap Joint			**50** 1.5	**110** 1.5	**138** 1.5	**231** 1.5	**325** 1.5	**577** 1.5	**1120** 1.5
	Blind	**66** 1.5	**120** 1.5	**77** 1.5	**146** 1.5	**181** 1.5	**267** 1.5	**338** 1.5	**599** 1.5	**1248** 1.5
FLANGED FITTINGS	S.R. 90° Elbow	**182** 4.8	**306** 4.9	**240** 4.8	**343** 4.9	**462** 5.2	**747** 5.6	**995** 5.8		
	L.R. 90° Elbow	**237** 5.8		**290** 5.8	**438** 5.8					
	45° Elbow	**152** 4.1	**256** 4.2	**185** 4.1	**288** 4.2	**332** 4.3	**572** 4.6	**732** 4.7		
	Tee	**277** 7.2	**446** 7.4	**353** 7.2	**527** 7.4	**578** 7.8	**1007** 8.4	**1417** 8.7		
VALVES	Flanged Bonnet Gate	**471** 7.7	**899** 8.3	**455** 4.5	**750** 5	**1035** 6	**1575** 6.9	**2140** 7.1	**3690** 8	
	Flanged Bonnet Globe or Angle	**541** 9.1	**943** 9.1	**485** 4.5	**855** 5.5	**1070** 6	**1500** 6.3	**2500** 6.8	**4160** 8	
	Flanged Bonnet Check	**453** 9.1	**751** 9.1	**370** 6	**485** 6.1	**605** 6.3	**1030** 6.8	**1350** 7	**2280** 7.5	
	Pressure Seal Bonnet — Gate						**1450** 4.9	**1860** 5.5	**3150** 6	
	Pressure Seal Bonnet — Globe							**1800** 5	**1910** 6	

Boldface type is weight in pounds. Lightface type beneath weight is weight factor for insulation.

Insulation thicknesses and weights are based on average conditions and do not constitute a recommendation for specific thicknesses of materials. Insulation weights are based on 85% magnesia and hydrous calcium silicate at 11 lbs/cubic foot. The listed thicknesses and weights of combination covering are the sums of the inner layer of diatomaceous earth at 21 lbs/cubic foot and the outer layer at 11 lbs/cubic foot.

Insulation weights include allowances for wire, cement, canvas, bands and paint, but not special surface finishes.

To find the weight of covering on flanges, valves or fittings, multiply the weight factor by the weight per foot of covering used on straight pipe.

Valve weights are approximate. When possible, obtain weights from the manufacturer.

Cast iron valve weights are for flanged end valves; steel weights for welding end valves.

All flanged fitting, flanged valve and flange weights include the proportional weight of bolts or studs to make up all joints.

WEIGHT OF PIPING MATERIALS
(Courtesy of Anvil International)

12.750" O.D. 12" PIPE

PIPE												
Schedule No.	5S	10S	20	30		40		60	80	120	140	160
Wall Designation					Std.		XS					
Thickness — In.	.156	.180	.250	.330	.375	.406	.500	.562	.687	1.000	1.125	1.312
Pipe — Lbs/Ft	**20.99**	**24.20**	**33.38**	**43.8**	**49.6**	**53.5**	**65.4**	**73.2**	**88.5**	**125.5**	**139.7**	**160.3**
Water — Lbs/Ft	**52.7**	**52.2**	**51.10**	**49.7**	**49.0**	**48.5**	**47.0**	**46.0**	**44.0**	**39.3**	**37.5**	**34.9**

WELDING FITTINGS

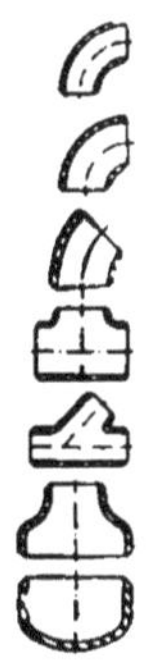

WELDING FITTINGS	5S	10S	20	30	Std.	40	XS	60	80	120	140	160
L.R. 90° Elbow	**51.2** 3	**57.0** 3			**122** 3		**156** 3					**375** 3
S.R. 90° Elbow	**33.6** 2	**38.1** 2			**82** 2		**104** 2					
L.R. 45° Elbow	**25.5** 1.3	**29.0** 1.3			**60.3** 1.3		**78** 1.3					**182** 1.3
Tee	**46.7** 2.5	**54.0** 2.5			**162** 2.5		**180** 2.5					**360** 2.5
Lateral	**74.7** 5.4	**86.2** 5.4			**180** 5.4		**273** 5.4					
Reducer	**14.1** .7	**20.9** .7			**33.4** .7		**43.6** .7					**94** .7
Cap	**6.2** 1.5	**7.1** 1.5			**29.5** 1.5		**38.1** 1.5					**95** 1.5

INSULATION

Temperature Range °F		100-199	200-299	300-399	400-499	500-599	600-699	700-799	800-899	900-999	1000-1099	1100-1200
85% Magnesia Calcium Silicate	Nom. Thick., In.	1½	1½	2	2½	3	3	3½	4	4	4½	4½
	Lbs./Ft	**6.04**	**6.04**	**8.13**	**10.5**	**12.7**	**12.7**	**15.1**	**17.9**	**17.9**	**20.4**	**20.4**
Combination	Nom. Thick., In.						3	3½	4	4	4½	4½
	Lbs/Ft						**17.7**	**21.9**	**26.7**	**26.7**	**31.1**	**31.1**

FLANGES

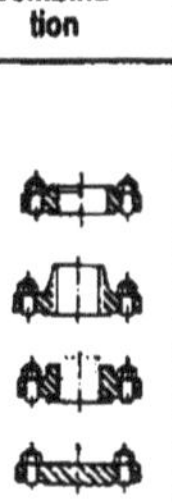

Pressure Rating psi	Cast Iron		Steel						
	125	250	150	300	400	600	900	1500	2500
Screwed or Slip-On	**58** 1.5	**123** 1.5	**71** 1.5	**140** 1.5	**164** 1.5	**261** 1.5	**388** 1.5	**820** 1.5	**1611** 1.5
Welding Neck			**87** 1.5	**163** 1.5	**212** 1.5	**272** 1.5	**434** 1.5	**843** 1.5	**1919** 1.5
Lap Joint			**71** 1.5	**164** 1.5	**187** 1.5	**286** 1.5	**433** 1.5	**902** 1.5	**1583** 1.5
Blind	**95** 1.5	**165** 1.5	**117** 1.5	**209** 1.5	**261** 1.5	**341** 1.5	**475** 1.5	**928** 1.5	**1775** 1.5
FLANGED FITTINGS									
S.R. 90° Elbow	**257** 5	**430** 5.2	**345** 5	**509** 5.2	**669** 5.5	**815** 5.8	**1474** 6.2		
L.R. 90° Elbow	**357** 6.2		**485** 6.2	**624** 6.2			**1598** 6.2		
45° Elbow	**227** 4.3	**360** 4.3	**282** 4.3	**414** 4.3	**469** 4.5	**705** 4.7	**1124** 4.8		
Tee	**387** 7.5	**640** 7.8	**513** 7.5	**754** 7.8	**943** 8.3	**1361** 8.7	**1928** 9.3		
VALVES									
Flanged Bonnet Gate	**687** 7.8	**1298** 8.5	**635** 4	**1015** 5	**1420** 5.5	**2155** 7	**2770** 7.2	**4650** 8	
Flanged Bonnet Globe or Angle	**808** 9.4	**1200** 9.5	**710** 5	**1410** 5.5					
Flanged Bonnet Check	**674** 9.4	**1160** 9.5	**560** 6	**720** 6.5		**1410** 7.2	**2600** 8	**3370** 8	
Pressure Seal Bonnet — Gate						**1975** 5.5	**2560** 6	**4515** 7	
Pressure Seal Bonnet — Globe									

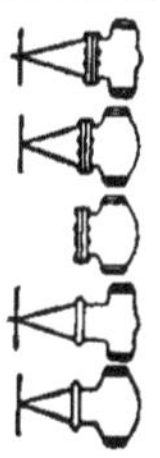

Boldface type is weight in pounds. Lightface type beneath weight is weight factor for insulation.

Insulation thicknesses and weights are based on average conditions and do not constitute a recommendation for specific thicknesses of materials. Insulation weights are based on 85% magnesia and hydrous calcium silicate at 11 lbs/cubic foot. The listed thicknesses and weights of combination covering are the sums of the inner layer of diatomaceous earth at 21 lbs/cubic foot and the outer layer at 11 lbs/cubic foot.

Insulation weights include allowances for wire, cement, canvas, bands and paint, but not special surface finishes.

To find the weight of covering on flanges, valves or fittings, multiply the weight factor by the weight per foot of covering used on straight pipe.

Valve weights are approximate. When possible, obtain weights from the manufacturer.

Cast iron valve weights are for flanged end valves; steel weights for welding end valves.

All flanged fitting, flanged valve and flange weights include the proportional weight of bolts or studs to make up all joints.

WEIGHT OF PIPING MATERIALS
(Courtesy of Anvil International)

14" O.D. 14" PIPE

PIPE	Schedule No.	5S	10S	10	20	30	40		60	80	120	140	160
	Wall Designation					Std.		XS					
	Thickness — In.	.156	.188	.250	.312	.375	.438	.500	.593	.750	1.093	1.250	1.406
	Pipe — Lbs/Ft	**23.0**	**27.7**	**36.71**	**45.7**	**54.6**	**63.4**	**72.1**	**84.9**	**106.1**	**150.7**	**170.2**	**189.1**
	Water — Lbs/Ft	**63.7**	**63.1**	**62.06**	**60.92**	**59.7**	**58.7**	**57.5**	**55.9**	**53.2**	**47.5**	**45.0**	**42.6**
WELDING FITTINGS	L.R. 90° Elbow	**65.6** 3.5	**78.0** 3.5			**157** 3.5		**200** 3.5					
	S.R. 90° Elbow	**43.1** 2.3	**51.7** 2.3			**108** 2.3		**135** 2.3					
	L.R. 45° Elbow	**32.5** 1.5	**39.4** 1.5			**80** 1.5		**98** 1.5					
	Tee	**49.4** 2.8	**59.6** 2.8			**196** 2.8		**220** 2.8					
	Lateral	**94.4** 5.8	**113** 5.8			**218** 5.8		**340** 5.8					
	Reducer	**25.0** 1.1	**31.2** 1.1			**63** 1.1		**83** 1.1					
	Cap	**7.6** 1.7	**9.2** 1.7			**35.3** 1.7		**45.9** 1.7					

INSULATION	Temperature Range °F	100-199	200-299	300-399	400-499	500-599	600-699	700-799	800-899	900-999	1000-1099	1100-1200
85% Magnesia Calcium Silicate	Nom. Thick., In.	1½	1½	2	2½	3	3	3½	4	4	4½	4½
	Lbs./Ft	**6.16**	**6.16**	**8.38**	**10.7**	**13.1**	**13.1**	**15.8**	**18.5**	**18.5**	**21.3**	**21.3**
Combination	Nom. Thick., In.						3	3½	4	4	4½	4½
	Lbs/Ft						**18.2**	**22.8**	**27.5**	**27.5**	**32.4**	**32.4**

	Pressure Rating psi	Cast Iron		Steel						
		125	250	150	300	400	600	900	1500	2500
FLANGES	Screwed or Slip-On	**90** 1.5	**184** 1.5	**95** 1.5	**195** 1.5	**235** 1.5	**316** 1.5	**460** 1.5	**1016** 1.5	
	Welding Neck			**130** 1.5	**217** 1.5	**277** 1.5	**406** 1.5	**642** 1.5	**1241** 1.5	
	Lap Joint			**119** 1.5	**220** 1.5	**254** 1.5	**349** 1.5	**477** 1.5	**1076** 1.5	
	Blind	**125** 1.5	**239** 1.5	**141** 1.5	**267** 1.5	**354** 1.5	**437** 1.5	**574** 1.5		
FLANGED FITTINGS	S.R. 90° Elbow	**360** 5.3	**617** 5.5	**497** 5.3	**632** 5.5	**664** 5.7	**916** 5.9	**1549** 6.4		
	L.R. 90° Elbow	**480** 6.6	**767** 6.6	**622** 6.6	**772** 6.6					
	45° Elbow	**280** 4.3	**497** 4.4	**377** 4.3	**587** 4.4	**638** 4.6	**883** 4.8	**1246** 4.9		
	Tee	**540** 8	**956** 8.4	**683** 8	**968** 8.3	**1131** 8.6	**1652** 8.9	**2318** 9.6		
VALVES	Flanged Bonnet Gate	**921** 7.9	**1762** 8.8	**905** 4.9	**1525** 6	**1920** 6.3	**2960** 7	**4170** 8	**6425** 8.8	
	Flanged Bonnet Globe or Angle	**1171** 9.9								
	Flanged Bonnet Check	**885** 9.9		**1010** 5	**1155** 5.2					
	Pressure Seal Bonnet — Gate						**2620** 6	**3475** 6.5	**6380** 7.5	
	Pressure Seal Bonnet — Globe									

Boldface type is weight in pounds. Lightface type beneath weight is weight factor for insulation.

Insulation thicknesses and weights are based on average conditions and do not constitute a recommendation for specific thicknesses of materials. Insulation weights are based on 85% magnesia and hydrous calcium silicate at 11 lbs/cubic foot. The listed thicknesses and weights of combination covering are the sums of the inner layer of diatomaceous earth at 21 lbs/cubic foot and the outer layer at 11 lbs/cubic foot.

Insulation weights include allowances for wire, cement, canvas, bands and paint, but not special surface finishes.

To find the weight of covering on flanges, valves or fittings, multiply the weight factor by the weight per foot of covering used on straight pipe.

Valve weights are approximate. When possible, obtain weights from the manufacturer.

Cast iron valve weights are for flanged end valves; steel weights for welding end valves.

All flanged fitting, flanged valve and flange weights include the proportional weight of bolts or studs to make up all joints.

WEIGHT OF PIPING MATERIALS
(Courtesy of Anvil International)

16" O.D. 16" PIPE

PIPE	Schedule No.	5S	10S	10	20	30	40	60	80	100	120	140	160
	Wall Designation					Std.		XS					
	Thickness — In.	.165	.188	.250	.312	.375	.500	.656	.843	1.031	1.218	1.438	1.593
	Pipe — Lbs/Ft	**28.0**	**32.0**	**42.1**	**52.4**	**62.6**	**82.8**	**107.5**	**136.5**	**164.8**	**192.3**	**223.6**	**245.1**
	Water — Lbs/Ft	**83.5**	**83.0**	**81.8**	**80.5**	**79.1**	**76.5**	**73.4**	**69.7**	**66.1**	**62.6**	**58.6**	**55.9**
WELDING FITTINGS	L.R. 90° Elbow	**89.8** 4	**102.0** 4			**208** 4	**270** 4						
	S.R. 90° Elbow	**59.7** 2.5	**67.7** 2.5			**135** 2.5	**177** 2.5						
	L.R. 45° Elbow	**44.9** 1.7	**51.0** 1.7			**104** 1.7	**136** 1.7						
	Tee	**66.8** 3.2	**75.9** 3.2			**250** 3.2	**278** 3.2						
	Lateral	**127.0** 6.7	**144.0** 6.7			**275** 6.7	**431** 6.7						
	Reducer	**31.3** 1.2	**35.7** 1.2			**77** 1.2	**102** 1.2						
	Cap	**10.1** 1.8	**11.5** 1.8			**44.3** 1.8	**57** 1.8						

	Temperature Range °F	100-199	200-299	300-399	400-499	500-599	600-699	700-799	800-899	900-999	1000-1099	1100-1200
INSULATION — 85% Magnesia Calcium Silicate	Nom. Thick., In.	1½	1½	2	2½	3	3	3½	4	4	4½	4½
	Lbs./Ft	**6.90**	**6.90**	**9.33**	**12.0**	**14.6**	**14.6**	**17.5**	**20.5**	**20.5**	**23.6**	**23.6**
Combination	Nom. Thick., In.						3	3½	4	4	4½	4½
	Lbs/Ft						**20.3**	**25.2**	**30.7**	**30.7**	**36.0**	**36.0**

	Pressure Rating psi	Cast Iron		Steel						
		125	250	150	300	400	600	900	1500	2500
FLANGES	Screwed or Slip-On	**114** 1.5	**233** 1.5	**107** 1.5	**262** 1.5	**310** 1.5	**442** 1.5	**559** 1.5	**1297** 1.5	
	Welding Neck			**141** 1.5	**288** 1.5	**351** 1.5	**577** 1.5	**785** 1.5	**1597** 1.5	
	Lap Joint			**142** 1.5	**282** 1.5	**337** 1.5	**476** 1.5	**588** 1.5	**1372** 1.5	
	Blind	**174** 1.5	**308** 1.5	**184** 1.5	**349** 1.5	**455** 1.5	**603** 1.5	**719** 1.5		
FLANGED FITTINGS	S.R. 90° Elbow	**484** 5.5	**826** 5.8	**656** 5.5	**956** 5.8	**1014** 6	**1402** 6.3	**1886** 6.7		
	L.R. 90° Elbow	**684** 7	**1036** 7	**781** 7	**1058** 7					
	45° Elbow	**374** 4.3	**696** 4.6	**481** 4.3	**708** 4.6	**839** 4.7	**1212** 5	**1586** 5		
	Tee	**714** 8.3	**1263** 8.7	**961** 8.3	**1404** 8.6	**1671** 9	**2128** 9.4	**3054** 10		
VALVES	Flanged Bonnet Gate	**1254** 8	**2321** 9	**1190** 5	**2015** 7	**2300** 7.2	**3675** 7.9	**4950** 8.2	**7875** 9	
	Flanged Bonnet Globe or Angle									
	Flanged Bonnet Check	**1166** 10.5			**1225** 6					
	Pressure Seal Bonnet — Gate						**3230** 7		**8130** 8	
	Pressure Seal Bonnet — Globe									

Boldface type is weight in pounds. Lightface type beneath weight is weight factor for insulation.

Insulation thicknesses and weights are based on average conditions and do not constitute a recommendation for specific thicknesses of materials. Insulation weights are based on 85% magnesia and hydrous calcium silicate at 11 lbs/cubic foot. The listed thicknesses and weights of combination covering are the sums of the inner layer of diatomaceous earth at 21 lbs/cubic foot and the outer layer at 11 lbs/cubic foot.

Insulation weights include allowances for wire, cement, canvas, bands and paint, but not special surface finishes.

To find the weight of covering on flanges, valves or fittings, multiply the weight factor by the weight per foot of covering used on straight pipe.

Valve weights are approximate. When possible, obtain weights from the manufacturer.

Cast iron valve weights are for flanged end valves; steel weights for welding end valves.

All flanged fitting, flanged valve and flange weights include the proportional weight of bolts or studs to make up all joints.

WEIGHT OF PIPING MATERIALS
(Courtesy of Anvil International)

18″ O.D. 18″ PIPE

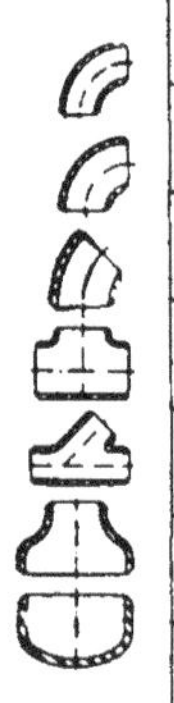

PIPE	Schedule No.	5S	10S	10	20		30		40	60	80	120	160
	Wall Designation					Std.		XS					
	Thickness — In.	.165	.188	.250	.312	.375	.438	.500	.562	750	937	1.375	1.781
	Pipe — Lbs/Ft	**31.0**	**36.0**	**47.4**	**59.0**	**70.6**	**82.1**	**93.5**	**104.8**	**138.2**	**170.8**	**244.1**	**308.5**
	Water — Lbs/Ft	**106.2**	**105.7**	**104.3**	**102.8**	**101.2**	**99.9**	**98.4**	**97**	**92.7**	**88.5**	**79.2**	**71.0**
WELDING FITTINGS	L.R. 90° Elbow	**114.0** 4.5	**129.0** 4.5			**256** 4.5		**332** 4.5					
	S.R. 90° Elbow	**75.7** 2.8	**85.7** 2.8			**176** 2.8		**225** 2.8					
	L.R. 45° Elbow	**57.2** 1.9	**64.5** 1.9			**132** 1.9		**168** 1.9					
	Tee	**83.2** 3.6	**94.7** 3.6			**282** 3.6		**351** 3.6					
	Lateral	**157.0** 7.5	**179.0** 7.5			**326** 7.5		**525** 7.5					
	Reducer	**42.6** 1.3	**48.5** 1.3			**94** 1.3		**123** 1.3					
	Cap	**12.7** 2.1	**14.5** 2.1			**57** 2.1		**75** 2.1					

	Temperature Range °F	100-199	200-299	300-399	400-499	500-599	600-699	700-799	800-899	900-999	1000-1099	1100-1200
INSULATION: **85% Magnesia Calcium Silicate**	Nom. Thick., In.	1½	1½	2	2½	3	3	3½	4	4	4½	4½
	Lbs./Ft	**7.73**	**7.73**	**10.4**	**13.3**	**16.3**	**16.3**	**19.3**	**22.6**	**22.6**	**25.9**	**25.9**
INSULATION: **Combination**	Nom. Thick., In.						3	3½	4	4	4½	4½
	Lbs/Ft						**22.7**	**28.0**	**33.8**	**33.8**	**39.5**	**39.5**

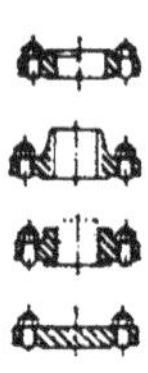

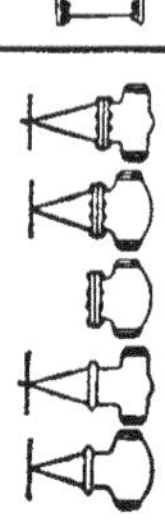

	Pressure Rating psi	Cast Iron		Steel						
		125	250	150	300	400	600	900	1500	2500
FLANGES	Screwed or Slip-On	**125** 1.5		**139** 1.5	**331** 1.5	**380** 1.5	**573** 1.5	**797** 1.5	**1694** 1.5	
	Welding Neck			**159** 1.5	**355** 1.5	**430** 1.5	**652** 1.5	**1074** 1.5	**2069** 1.5	
	Lap Joint			**165** 1.5	**355** 1.5	**415** 1.5	**566** 1.5	**820** 1.5	**1769** 1.5	
	Blind	**209** 1.5	**396** 1.5	**228** 1.5	**440** 1.5	**572** 1.5	**762** 1.5	**1030** 1.5		
FLANGED FITTINGS	S.R. 90° Elbow	**599** 5.8	**1060** 6	**711** 5.8	**1126** 6	**1340** 6.2	**1793** 6.6	**2817** 7		
	L.R. 90° Elbow		**1350** 7.4	**941** 7.4	**1426** 7.4					
	45° Elbow	**439** 4.4	**870** 4.7	**521** 4.4	**901** 4.7	**1040** 4.8	**1543** 5	**2252** 5.2		
	Tee	**879** 8.6	**1625** 9	**1010** 8.6	**1602** 9	**1909** 9.3	**2690** 9.9	**4327** 10.5		
VALVES	Flanged Bonnet Gate	**1629** 8.2	**2578** 9.3	**1510** 6	**2505** 6.5	**3765** 7	**4460** 7.8	**6675** 8.5		
	Flanged Bonnet Globe or Angle									
	Flanged Bonnet Check	**1371** 10.5								
	Pressure Seal Bonnet — Gate						**3100** 5.5	**3400** 5.6	**4200** 6	
	Pressure Seal Bonnet — Globe									

Boldface type is weight in pounds. Lightface type beneath weight is weight factor for insulation.

Insulation thicknesses and weights are based on average conditions and do not constitute a recommendation for specific thicknesses of materials. Insulation weights are based on 85% magnesia and hydrous calcium silicate at 11 lbs/cubic foot. The listed thicknesses and weights of combination covering are the sums of the inner layer of diatomaceous earth at 21 lbs/cubic foot and the outer layer at 11 lbs/cubic foot.

Insulation weights include allowances for wire, cement, canvas, bands and paint, but not special surface finishes.

To find the weight of covering on flanges, valves or fittings, multiply the weight factor by the weight per foot of covering used on straight pipe.

Valve weights are approximate. When possible, obtain weights from the manufacturer.

Cast iron valve weights are for flanged end valves; steel weights for welding end valves.

All flanged fitting, flanged valve and flange weights include the proportional weight of bolts or studs to make up all joints.

WEIGHT OF PIPING MATERIALS
(Courtesy of Anvil International)

20" O.D. 20" PIPE

PIPE	Schedule No.	5S	10S	10	20	30	40	60	80	100	120	140	160
	Wall Designation				Std.	XS							
	Thickness — In.	.188	.218	.250	.375	.500	.593	.812	1.031	1.281	1.500	1.750	1.968
	Pipe — Lbs/Ft	**40.0**	**46.0**	**52.7**	**78.6**	**104.1**	**122.9**	**166.4**	**208.9**	**256.1**	**296.4**	**341.1**	**379.0**
	Water — Lbs/Ft	**131.0**	**130.2**	**129.5**	**126.0**	**122.8**	**120.4**	**115.0**	**109.4**	**103.4**	**98.3**	**92.6**	**87.9**
WELDING FITTINGS	L.R. 90° Elbow	**160.0** 5	**185.0** 5		**322** 5	**438** 5							
	S.R. 90° Elbow	**106.0** 3.4	**122.0** 3.4		**238** 3.4	**278** 3.4							
	L.R. 45° Elbow	**80.3** 2.1	**92.5** 2.1		**160** 2.1	**228** 2.1							
	Tee	**112.0** 4	**130.0** 4		**378** 4	**490** 4							
	Lateral	**228.0** 8.3	**265.0** 8.3		**396** 8.3	**625** 8.3							
	Reducer	**71.6** 1.7	**87.6** 1.7		**142** 1.7	**186** 1.7							
	Cap	**17.7** 2.3	**20.5** 2.3		**71** 2.3	**93** 2.3							

INSULATION		100-199	200-299	300-399	400-499	500-599	600-699	700-799	800-899	900-999	1000-1099	1100-1200
	Temperature Range °F	100-199	200-299	300-399	400-499	500-599	600-699	700-799	800-899	900-999	1000-1099	1100-1200
85% Magnesia Calcium Silicate	Nom. Thick., In.	1½	1½	2	2½	3	3	3½	4	4	4½	4½
	Lbs./Ft	**8.45**	**8.45**	**11.6**	**14.6**	**17.7**	**17.7**	**21.1**	**24.6**	**24.6**	**28.1**	**28.1**
Combination	Nom. Thick., In.						3	3½	4	4	4½	4½
	Lbs/Ft						**24.7**	**30.7**	**37.0**	**37.0**	**43.1**	**43.1**

	Pressure Rating psi	Cast Iron		Steel						
		125	250	150	300	400	600	900	1500	2500
FLANGES	Screwed or Slip-On	**153** 1.5		**180** 1.5	**378** 1.5	**468** 1.5	**733** 1.5	**972** 1.5	**2114** 1.5	
	Welding Neck			**195** 1.5	**431** 1.5	**535** 1.5	**811** 1.5	**1344** 1.5	**2614** 1.5	
	Lap Joint			**210** 1.5	**428** 1.5	**510** 1.5	**725** 1.5	**1048** 1.5	**2189** 1.5	
	Blind	**275** 1.5	**487** 1.5	**297** 1.5	**545** 1.5	**711** 1.5	**976** 1.5	**1287** 1.5		
FLANGED FITTINGS	S.R. 90° Elbow	**792** 6	**1315** 6.3	**922** 6	**1375** 6.3	**1680** 6.5	**2314** 6.9	**3610** 7.3		
	L.R. 90° Elbow	**1132** 7.8	**1725** 7.8	**1352** 7.8	**1705** 7.8					
	45° Elbow	**592** 4.6	**1055** 4.8	**652** 4.6	**1105** 4.8	**1330** 4.9	**1917** 5.2	**2848** 5.4		
	Tee	**1178** 9	**2022** 9.5	**1378** 9	**1908** 9.5	**2370** 9.7	**3463** 10.1	**5520** 11		
VALVES	Flanged Bonnet Gate	**1934** 8.3	**3823** 9.5	**1855** 6	**3370** 7	**5700** 8	**5755** 8			
	Flanged Bonnet Globe or Angle									
	Flanged Bonnet Check	**1772** 11								
	Pressure Seal Bonnet — Gate									
	Pressure Seal Bonnet — Globe									

Boldface type is weight in pounds. Lightface type beneath weight is weight factor for insulation.

Insulation thicknesses and weights are based on average conditions and do not constitute a recommendation for specific thicknesses of materials. Insulation weights are based on 85% magnesia and hydrous calcium silicate at 11 lbs/cubic foot. The listed thicknesses and weights of combination covering are the sums of the inner layer of diatomaceous earth at 21 lbs/cubic foot and the outer layer at 11 lbs/cubic foot.

Insulation weights include allowances for wire, cement, canvas, bands and paint, but not special surface finishes.

To find the weight of covering on flanges, valves or fittings, multiply the weight factor by the weight per foot of covering used on straight pipe.

Valve weights are approximate. When possible, obtain weights from the manufacturer.

Cast iron valve weights are for flanged end valves; steel weights for welding end valves.

All flanged fitting, flanged valve and flange weights include the proportional weight of bolts or studs to make up all joints.

WEIGHT OF PIPING MATERIALS
(Courtesy of Anvil International)

24" O.D. 24" PIPE

PIPE	Schedule No.	5S	10	20		30	40	60	80	120	140	160	
	Wall Designation			Std.	XS								
	Thickness — In.	.218	.250	.375	.500	.562	.687	.968	1.218	1.812	2.062	2.343	
	Pipe — Lbs/Ft	**55.0**	**63.4**	**94.6**	**125.5**	**140.8**	**171.2**	**238.1**	**296.4**	**429.4**	**483.1**	**541.9**	
	Water — Lbs/Ft	**188.9**	**188.0**	**183.8**	**180.1**	**178.1**	**174.3**	**165.8**	**158.3**	**141.4**	**134.5**	**127.0**	
WELDING FITTINGS	L.R. 90° Elbow	**260.0** 6		**500** 6	**578** 6								
	S.R. 90° Elbow	**178.0** 3.7		**305** 3.7	**404** 3.7								
	L.R. 45° Elbow	**130.0** 2.5		**252** 2.5	**292** 2.5								
	Tee	**174.0** 4.9		**544** 4.9	**607** 4.9								
	Lateral	**361.0** 10		**544** 10	**875** 10								
	Reducer	**107.0** 1.7		**167** 1.7	**220** 1.7								
	Cap	**28.6** 2.8		**102** 2.8	**134** 2.8								

INSULATION													
Temperature Range °F		100-199	200-299	300-399	400-499	500-599	600-699	700-799	800-899	900-999	1000-1099	1100-1200
85% Magnesia Calcium Silicate	Nom. Thick., In.	1½	1½	2	2½	3	3	3½	4	4	4½	4½
	Lbs./Ft	**10.0**	**10.0**	**13.4**	**17.0**	**21.0**	**21.0**	**24.8**	**28.7**	**28.7**	**32.9**	**32.9**
Combination	Nom. Thick., In.						3	3½	4	4	4½	4½
	Lbs/Ft						**29.2**	**36.0**	**43.1**	**43.1**	**50.6**	**50.6**

	Pressure Rating psi	Cast Iron		Steel						
		125	250	150	300	400	600	900	1500	2500
FLANGES	Screwed or Slip-On	**236** 1.5		**245** 1.5	**577** 1.5	**676** 1.5	**1056** 1.5	**1823** 1.5	**3378** 1.5	
	Welding Neck			**295** 1.5	**632** 1.5	**777** 1.5	**1157** 1.5	**2450** 1.5	**4153** 1.5	
	Lap Joint			**295** 1.5	**617** 1.5	**752** 1.5	**1046** 1.5	**2002** 1.5	**3478** 1.5	
	Blind	**404** 1.5	**757** 1.5	**446** 1.5	**841** 1.5	**1073** 1.5	**1355** 1.5	**2442** 1.5		
FLANGED FITTINGS	S.R. 90° Elbow	**1231** 6.7	**2014** 6.8	**1671** 6.7	**2174** 6.8	**2474** 7.1	**3506** 7.6	**6155** 8.1		
	L.R. 90° Elbow	**1711** 8.7	**2644** 8.7	**1821** 8.7	**2874** 8.7					
	45° Elbow	**871** 4.8	**1604** 5	**1121** 4.8	**1634** 5	**1974** 5.1	**2831** 5.5	**5124** 6		
	Tee	**1836** 10	**3061** 10.2	**2276** 10	**3161** 10.2	**3811** 10.6	**5184** 11.4	**9387** 12.1		

VALVES	Flanged Bonnet Gate	**3062** 8.5	**6484** 9.8	**2500** 5	**4675** 7	**6995** 8.7	**8020** 9.5			
	Flanged Bonnet Globe or Angle									
	Flanged Bonnet Check	**2956** 12								
	Pressure Seal Bonnet — Gate									
	Pressure Seal Bonnet — Globe									

Boldface type is weight in pounds. Lightface type beneath weight is weight factor for insulation.

Insulation thicknesses and weights are based on average conditions and do not constitute a recommendation for specific thicknesses of materials. Insulation weights are based on 85% magnesia and hydrous calcium silicate at 11 lbs/cubic foot. The listed thicknesses and weights of combination covering are the sums of the inner layer of diatomaceous earth at 21 lbs/cubic foot and the outer layer at 11 lbs/cubic foot.

Insulation weights include allowances for wire, cement, canvas, bands and paint, but not special surface finishes.

To find the weight of covering on flanges, valves or fittings, multiply the weight factor by the weight per foot of covering used on straight pipe.

Valve weights are approximate. When possible, obtain weights from the manufacturer.

Cast iron valve weights are for flanged end valves; steel weights for welding end valves.

All flanged fitting, flanged valve and flange weights include the proportional weight of bolts or studs to make up all joints.

WEIGHT OF PIPING MATERIALS
(Courtesy of Anvil International)

26" O.D. 26" PIPE

PIPE	Schedule No.		10		20								
	Wall Designation			Std.	XS								
	Thickness — In.	.250	.312	.375	.500	.625	.750	.875	1.000	1.125			
	Pipe — Lbs/Ft	**67.0**	**85.7**	**102.6**	**136.2**	**169.0**	**202.0**	**235.0**	**267.0**	**299.0**			
	Water — Lbs/Ft	**221.4**	**219.2**	**216.8**	**212.5**	**208.6**	**204.4**	**200.2**	**196.1**	**192.1**			
WELDING FITTINGS 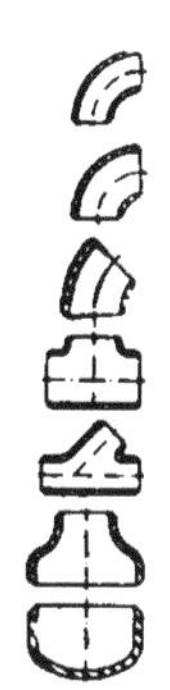	L.R. 90° Elbow			**502** 8.5	**713** 8.5								
	S.R. 90° Elbow			**359** 5	**474** 5								
	L.R. 45° Elbow			**269** 3.5	**355** 3.5								
	Tee			**634** 6.8	**794** 6.8								
	Lateral												
	Reducer			**200** 4.3	**272** 4.3								
	Cap			**110** 4.3	**145** 4.3								

INSULATION	Temperature Range °F	100-199	200-299	300-399	400-499	500-599	600-699	700-799	800-899	900-999	1000-1099	1100-1200
85% Magnesia Calcium Silicate	Nom. Thick., In.	1½	1½	2	2½	3	3½	4	4½	5	5	6
	Lbs./Ft	**10.4**	**10.4**	**14.1**	**18.0**	**21.9**	**26.0**	**30.2**	**34.6**	**39.1**	**39.1**	**48.4**
Combination	Nom. Thick., In.						3½	4½	5½	6	6½	7
	Lbs/Ft						**37.0**	**51.9**	**67.8**	**76.0**	**84.5**	**93.2**

	Pressure Rating psi	Cast Iron		Steel						
		125	250	150	300	400	600	900	1500	2500
FLANGES 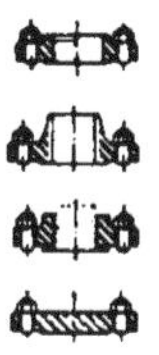	Screwed or Slip-On			**292** 1.5	**699** 1.5	**650** 1.5	**950** 1.5	**1525** 1.5		
	Welding Neck			**342** 1.5	**799** 1.5	**750** 1.5	**1025** 1.5	**1575** 1.5		
	Lap Joint									
	Blind			**567** 1.5	**1179** 1.5	**1125** 1.5	**1525** 1.5	**2200** 1.5		
FLANGED FITTINGS 	S.R. 90° Elbow									
	L.R. 90° Elbow									
	45° Elbow									
	Tee									
VALVES 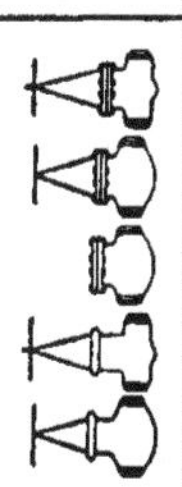	Flanged Bonnet Gate									
	Flanged Bonnet Globe or Angle									
	Flanged Bonnet Check									
	Pressure Seal Bonnet — Gate									
	Pressure Seal Bonnet — Globe									

Boldface type is weight in pounds. Lightface type beneath weight is weight factor for insulation.

Insulation thicknesses and weights are based on average conditions and do not constitute a recommendation for specific thicknesses of materials. Insulation weights are based on 85% magnesia and hydrous calcium silicate at 11 lbs/cubic foot. The listed thicknesses and weights of combination covering are the sums of the inner layer of diatomaceous earth at 21 lbs/cubic foot and the outer layer at 11 lbs/cubic foot.

Insulation weights include allowances for wire, cement, canvas, bands and paint, but not special surface finishes.

To find the weight of covering on flanges, valves or fittings, multiply the weight factor by the weight per foot of covering used on straight pipe.

Valve weights are approximate. When possible, obtain weights from the manufacturer.

Cast iron valve weights are for flanged end valves; steel weights for welding end valves.

All flanged fitting, flanged valve and flange weights include the proportional weight of bolts or studs to make up all joints.

WEIGHT OF PIPING MATERIALS
(Courtesy of Anvil International)

28″ O.D. 28″ PIPE

PIPE

Schedule No.		10		20	30							
Wall Designation			Std.	XS								
Thickness — In.	.250	.312	.375	.500	.625	.750	.875	1.000	1.125			
Pipe — Lbs/Ft	**74.0**	**92.4**	**110.6**	**146.9**	**182.7**	**218.0**	**253.0**	**288.0**	**323.0**			
Water — Lbs/Ft	**257.3**	**255.0**	**252.7**	**248.1**	**243.6**	**238.9**	**234.4**	**230.0**	**225.6**			

WELDING FITTINGS

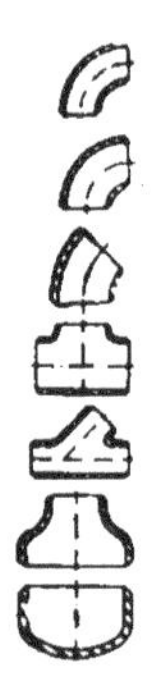

L.R. 90° Elbow			**626** 9	**829** 9								
S.R. 90° Elbow			**415** 5.4	**551** 5.4								
L.R. 45° Elbow			**312** 3.6	**413** 3.6								
Tee			**729** 7	**910** 7								
Lateral												
Reducer			**210** 2.7	**290** 2.7								
Cap			**120** 4.5	**160** 4.5								

INSULATION

Temperature Range °F		100-199	200-299	300-399	400-499	500-599	600-699	700-799	800-899	900-999	1000-1099	1100-1200
85% Magnesia Calcium Silicate	Nom. Thick., In.	1½	1½	2	2½	3	3½	4	4½	5	5	6
	Lbs./Ft	**11.2**	**11.2**	**15.1**	**19.2**	**23.4**	**27.8**	**32.3**	**36.9**	**41.6**	**41.6**	**51.4**
Combination	Nom. Thick., In.						3½	4½	5½	6	6½	7
	Lbs/Ft						**39.5**	**55.4**	**72.2**	**80.9**	**89.8**	**99.0**

FLANGES

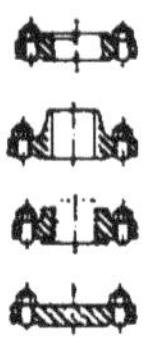

Pressure Rating psi	Cast Iron 125	Cast Iron 250	Steel 150	Steel 300	Steel 400	Steel 600	Steel 900	Steel 1500	Steel 2500
Screwed or Slip-On			**334** 1.5	**853** 1.5	**780** 1.5	**1075** 1.5	**1800** 1.5		
Welding Neck			**364** 1.5	**943** 1.5	**880** 1.5	**1175** 1.5	**1850** 1.5		
Lap Joint									
Blind			**669** 1.5	**1408** 1.5	**1425** 1.5	**1750** 1.5	**2575** 1.5		

FLANGED FITTINGS

S.R. 90° Elbow									
L.R. 90° Elbow									
45° Elbow									
Tee									

VALVES

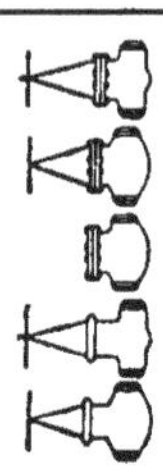

Flanged Bonnet Gate									
Flanged Bonnet Globe or Angle									
Flanged Bonnet Check									
Pressure Seal Bonnet — Gate									
Pressure Seal Bonnet — Globe									

Boldface type is weight in pounds. Lightface type beneath weight is weight factor for insulation.

Insulation thicknesses and weights are based on average conditions and do not constitute a recommendation for specific thicknesses of materials. Insulation weights are based on 85% magnesia and hydrous calcium silicate at 11 lbs/cubic foot. The listed thicknesses and weights of combination covering are the sums of the inner layer of diatomaceous earth at 21 lbs/cubic foot and the outer layer at 11 lbs/cubic foot.

Insulation weights include allowances for wire, cement, canvas, bands and paint, but not special surface finishes.

To find the weight of covering on flanges, valves or fittings, multiply the weight factor by the weight per foot of covering used on straight pipe.

Valve weights are approximate. When possible, obtain weights from the manufacturer.

Cast iron valve weights are for flanged end valves; steel weights for welding end valves.

All flanged fitting, flanged valve and flange weights include the proportional weight of bolts or studs to make up all joints.

WEIGHT OF PIPING MATERIALS
(Courtesy of Anvil International)

30″ O.D. 30″ PIPE

PIPE	Schedule No.	5S	10 & 10S		20	30							
	Wall Designation			Std.	XS								
	Thickness — In.	.250	.312	.375	.500	.625	.750	.875	1.000	1.125			
	Pipe — Lbs/Ft	**79.0**	**98.9**	**118.7**	**157.6**	**196.1**	**234.0**	**272.0**	**310.0**	**347.0**			
	Water — Lbs/Ft	**296.3**	**293.5**	**291.0**	**286.0**	**281.1**	**276.6**	**271.8**	**267.0**	**262.2**			
WELDING FITTINGS	L.R. 90° Elbow	**478.0** 10		**775** 10	**953** 10		**596.0** 10						
	S.R. 90° Elbow	**319.0** 5.9		**470** 5.9	**644** 5.9		**388.0** 5.9						
	L.R. 45° Elbow	**239.0** 3.9		**358** 3.9	**475** 3.9		**298.0** 3.9						
	Tee			**855** 7.8	**1065** 7.8								
	Lateral												
	Reducer			**220** 3.9	**315** 3.9								
	Cap			**125** 4.8	**175** 4.8								

INSULATION												
Temperature Range °F		100-199	200-299	300-399	400-499	500-599	600-699	700-799	800-899	900-999	1000-1099	1100-1200
85% Magnesia Calcium Silicate	Nom. Thick., In.	1½	1½	2	2½	3	3½	4	4½	5	5	6
	Lbs./Ft	**11.9**	**11.9**	**16.1**	**20.5**	**25.0**	**29.5**	**34.3**	**39.1**	**44.1**	**44.1**	**54.4**
Combination	Nom. Thick., In.						3½	4½	5½	6	6½	7
	Lbs/Ft						**42.1**	**58.9**	**76.5**	**85.7**	**95.1**	**104.7**

	Pressure Rating psi	Cast Iron		Steel						
		125	250	150	300	400	600	900	1500	2500
FLANGES	Screwed or Slip-On			**365** 1.5	**975** 1.5	**900** 1.5	**1175** 1.5	**2075** 1.5		
	Welding Neck			**410** 1.5	**1095** 1.5	**1000** 1.5	**1300** 1.5	**2150** 1.5		
	Lap Joint									
	Blind			**770** 1.5	**1665** 1.5	**1675** 1.5	**2000** 1.5	**3025** 1.5		
FLANGED FITTINGS	S.R. 90° Elbow									
	L.R. 90° Elbow									
	45° Elbow									
	Tee									
VALVES	Flanged Bonnet Gate									
	Flanged Bonnet Globe or Angle									
	Flanged Bonnet Check									
	Pressure Seal Bonnet — Gate									
	Pressure Seal Bonnet — Globe									

Boldface type is weight in pounds. Lightface type beneath weight is weight factor for insulation.

Insulation thicknesses and weights are based on average conditions and do not constitute a recommendation for specific thicknesses of materials. Insulation weights are based on 85% magnesia and hydrous calcium silicate at 11 lbs/cubic foot. The listed thicknesses and weights of combination covering are the sums of the inner layer of diatomaceous earth at 21 lbs/cubic foot and the outer layer at 11 lbs/cubic foot.

Insulation weights include allowances for wire, cement, canvas, bands and paint, but not special surface finishes.

To find the weight of covering on flanges, valves or fittings, multiply the weight factor by the weight per foot of covering used on straight pipe.

Valve weights are approximate. When possible, obtain weights from the manufacturer.

Cast iron valve weights are for flanged end valves; steel weights for welding end valves.

All flanged fitting, flanged valve and flange weights include the proportional weight of bolts or studs to make up all joints.

WEIGHT OF PIPING MATERIALS
(Courtesy of Anvil International)

32″ O.D. 32″ PIPE

PIPE	Schedule No.		10		20	30	40						
	Wall Designation			Std.	XS								
	Thickness — In.	.250	.312	.375	.500	.625	.688	.750	.875	1.000	1.125		
	Pipe — Lbs/Ft	**85.0**	**105.8**	**126.7**	**168.2**	**209.4**	**229.9**	**250.0**	**291.0**	**331.0**	**371.0**		
	Water — Lbs/Ft	**337.8**	**335.0**	**323.3**	**327.0**	**321.8**	**319.2**	**316.7**	**311.6**	**306.4**	**301.3**		
WELDING FITTINGS	L.R. 90° Elbow			**818** 10.5	**1090** 10.5								
	S.R. 90° Elbow			**546** 6.3	**722** 6.3								
	L.R. 45° Elbow			**408** 4.2	**541** 4.2								
	Tee			**991** 8.4	**1230** 8.4								
	Lateral												
	Reducer			**255** 3.1	**335** 3.1								
	Cap			**145** 5.2	**190** 5.2								

INSULATION		Temperature Range °F	100-199	200-299	300-399	400-499	500-599	600-699	700-799	800-899	900-999	1000-1099	1100-1200
	85% Magnesia Calcium Silicate	Nom. Thick., In.	1½	1½	2	2½	3	3½	4	4½	5	5	6
		Lbs./Ft	**12.7**	**12.7**	**17.1**	**21.7**	**26.5**	**31.3**	**36.3**	**41.4**	**46.6**	**46.6**	**57.5**
	Combination	Nom. Thick., In.						3½	4½	5½	6	6½	7
		Lbs/Ft						**44.7**	**62.3**	**80.9**	**90.5**	**100.4**	**110.5**

	Pressure Rating psi	Cast Iron		Steel						
		125	250	150	300	400	600	900	1500	2500
FLANGES	Screwed or Slip-On			**476** 1.5	**1093** 1.5	**1025** 1.5	**1375** 1.5	**2500** 1.5		
	Welding Neck			**516** 1.5	**1228** 1.5	**1150** 1.5	**1500** 1.5	**2575** 1.5		
	Lap Joint									
	Blind			**951** 1.5	**1978** 1.5	**1975** 1.5	**2300** 1.5	**3650** 1.5		
FLANGED FITTINGS	S.R. 90° Elbow									
	L.R. 90° Elbow									
	45° Elbow									
	Tee									
VALVES	Flanged Bonnet Gate									
	Flanged Bonnet Globe or Angle									
	Flanged Bonnet Check									
	Pressure Seal Bonnet — Gate									
	Pressure Seal Bonnet — Globe									

Boldface type is weight in pounds. Lightface type beneath weight is weight factor for insulation.

Insulation thicknesses and weights are based on average conditions and do not constitute a recommendation for specific thicknesses of materials. Insulation weights are based on 85% magnesia and hydrous calcium silicate at 11 lbs/cubic foot. The listed thicknesses and weights of combination covering are the sums of the inner layer of diatomaceous earth at 21 lbs/cubic foot and the outer layer at 11 lbs/cubic foot.

Insulation weights include allowances for wire, cement, canvas, bands and paint, but not special surface finishes.

To find the weight of covering on flanges, valves or fittings, multiply the weight factor by the weight per foot of covering used on straight pipe.

Valve weights are approximate. When possible, obtain weights from the manufacturer.

Cast iron valve weights are for flanged end valves; steel weights for welding end valves.

All flanged fitting, flanged valve and flange weights include the proportional weight of bolts or studs to make up all joints.

WEIGHT OF PIPING MATERIALS
(Courtesy of Anvil International)

34" O.D. 34" PIPE

PIPE	Schedule No.		10		20	30	40						
	Wall Designation			Std.	XS								
	Thickness — In.	.250	.312	.375	.500	.625	.688	.750	.875	1.000	1.125		
	Pipe — Lbs/Ft	**90.0**	**112.4**	**134.7**	**178.9**	**222.8**	**244.6**	**266.0**	**310.0**	**353.0**	**395.0**		
	Water — Lbs/Ft	**382.0**	**379.1**	**376.0**	**370.3**	**365.0**	**362.2**	**359.5**	**354.1**	**348.6**	**343.2**		
WELDING FITTINGS	L.R. 90° Elbow			**926** 11	**1230** 11								
	S.R. 90° Elbow			**617** 5.5	**817** 5.5								
	L.R. 45° Elbow			**463** 4.4	**615** 4.4								
	Tee			**1136** 8.9	**1420** 8.9								
	Lateral												
	Reducer			**270** 3.3	**355** 3.3								
	Cap			**160** 5.6	**210** 5.6								

INSULATION	Temperature Range °F	100-199	200-299	300-399	400-499	500-599	600-699	700-799	800-899	900-999	1000-1099	1100-1200
85% Magnesia Calcium Silicate	Nom. Thick., In.	1½	1½	2	2½	3	3½	4	4½	5	5	6
	Lbs./Ft	**13.4**	**13.4**	**18.2**	**23.0**	**28.0**	**33.1**	**38.3**	**43.7**	**49.1**	**49.1**	**60.5**
Combination	Nom. Thick., In.						3½	4½	5½	6	6½	7
	Lbs/Ft						**47.2**	**65.8**	**85.3**	**95.4**	**105.7**	**116.3**

	Pressure Rating psi	Cast Iron		Steel						
		125	250	150	300	400	600	900	1500	2500
FLANGES	Screwed or Slip-On			**515** 1.5	**1281** 1.5	**1150** 1.5	**1500** 1.5	**2950** 1.5		
	Welding Neck			**560** 1.5	**1406** 1.5	**1300** 1.5	**1650** 1.5	**3025** 1.5		
	Lap Joint									
	Blind			**1085** 1.5	**2231** 1.5	**2250** 1.5	**2575** 1.5	**4275** 1.5		
FLANGED FITTINGS	S.R. 90° Elbow									
	L.R. 90° Elbow									
	45° Elbow									
	Tee									
VALVES	Flanged Bonnet Gate									
	Flanged Bonnet Globe or Angle									
	Flanged Bonnet Check									
	Pressure Seal Bonnet — Gate									
	Pressure Seal Bonnet — Globe									

Boldface type is weight in pounds. Lightface type beneath weight is weight factor for insulation.

Insulation thicknesses and weights are based on average conditions and do not constitute a recommendation for specific thicknesses of materials. Insulation weights are based on 85% magnesia and hydrous calcium silicate at 11 lbs/cubic foot. The listed thicknesses and weights of combination covering are the sums of the inner layer of diatomaceous earth at 21 lbs/cubic foot and the outer layer at 11 lbs/cubic foot.

Insulation weights include allowances for wire, cement, canvas, bands and paint, but not special surface finishes.

To find the weight of covering on flanges, valves or fittings, multiply the weight factor by the weight per foot of covering used on straight pipe.

Valve weights are approximate. When possible, obtain weights from the manufacturer.

Cast iron valve weights are for flanged end valves; steel weights for welding end valves.

All flanged fitting, flanged valve and flange weights include the proportional weight of bolts or studs to make up all joints.

WEIGHT OF PIPING MATERIALS
(Courtesy of Anvil International)

36″ O.D. 36″ PIPE

PIPE												
Schedule No.		10		20	30	40						
Wall Designation			Std.	XS								
Thickness — In.	.250	.312	.375	.500	.625	.750	.875	1.000	1.125			
Pipe — Lbs/Ft	**96.0**	**119.1**	**142.7**	**189.6**	**236.1**	**282.4**	**328.0**	**374.0**	**419.0**			
Water — Lbs/Ft	**429.1**	**425.9**	**422.6**	**416.6**	**411.0**	**405.1**	**399.4**	**393.6**	**387.9**			

WELDING FITTINGS												
L.R. 90° Elbow			**1040** 12	**1380** 12								
S.R. 90° Elbow			**692** 5	**913** 5								
L.R. 45° Elbow			**518** 4.8	**686** 4.8								
Tee			**1294** 9.5	**1610** 9.5								
Lateral												
Reducer			**340** 3.6	**360** 3.6								
Cap			**175** 6	**235** 6								

INSULATION	Temperature Range °F	100-199	200-299	300-399	400-499	500-599	600-699	700-799	800-899	900-999	1000-1099	1100-1200
85% Magnesia Calcium Silicate	Nom. Thick., In.	1½	1½	2	2½	3	3½	4	4½	5	5	6
	Lbs./Ft	**14.2**	**14.2**	**19.2**	**24.2**	**29.5**	**34.8**	**40.3**	**45.9**	**51.7**	**51.7**	**63.5**
Combination	Nom. Thick., In.						3½	4½	5½	6	6½	7
	Lbs/Ft						**49.8**	**69.3**	**89.7**	**100.2**	**111.0**	**122.0**

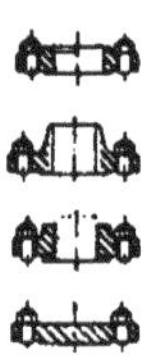

FLANGES	Cast Iron		Steel						
Pressure Rating psi	125	250	150	300	400	600	900	1500	2500
Screwed or Slip-On			**588** 1.5	**1485** 1.5	**1325** 1.5	**1600** 1.5	**3350** 1.5		
Welding Neck			**628** 1.5	**1585** 1.5	**1475** 1.5	**1750** 1.5	**3450** 1.5		
Lap Joint									
Blind			**1233** 1.5	**2560** 1.5	**2525** 1.5	**2950** 1.5	**4900** 1.5		

FLANGED FITTINGS									
S.R. 90° Elbow									
L.R. 90° Elbow									
45° Elbow									
Tee									

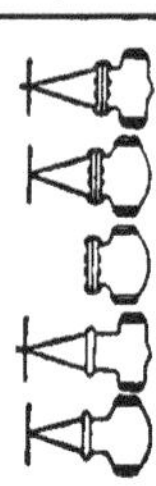

VALVES									
Flanged Bonnet Gate									
Flanged Bonnet Globe or Angle									
Flanged Bonnet Check									
Pressure Seal Bonnet — Gate									
Pressure Seal Bonnet — Globe									

Boldface type is weight in pounds. Lightface type beneath weight is weight factor for insulation.

Insulation thicknesses and weights are based on average conditions and do not constitute a recommendation for specific thicknesses of materials. Insulation weights are based on 85% magnesia and hydrous calcium silicate at 11 lbs/cubic foot. The listed thicknesses and weights of combination covering are the sums of the inner layer of diatomaceous earth at 21 lbs/cubic foot and the outer layer at 11 lbs/cubic foot.

Insulation weights include allowances for wire, cement, canvas, bands and paint, but not special surface finishes.

To find the weight of covering on flanges, valves or fittings, multiply the weight factor by the weight per foot of covering used on straight pipe.

Valve weights are approximate. When possible, obtain weights from the manufacturer.

Cast iron valve weights are for flanged end valves; steel weights for welding end valves.

All flanged fitting, flanged valve and flange weights include the proportional weight of bolts or studs to make up all joints.

WEIGHT OF PIPING MATERIALS
(Courtesy of Anvil International)

42" O.D. 42" PIPE

PIPE	Schedule No.			20	30	40							
	Wall Designation		Std.	XS									
	Thickness — In.	.250	.375	.500	.625	.750	1.000	1.250	1.500				
	Pipe — Lbs/Ft	**112.0**	**166.7**	**221.6**	**276.0**	**330.0**	**438.0**	**544.0**	**649.0**				
	Water — Lbs/Ft	**586.4**	**578.7**	**571.7**	**565.4**	**558.4**	**544.8**	**531.2**	**517.9**				
WELDING FITTINGS	L.R. 90° Elbow		**1420** 15	**1880** 15									
	S.R. 90° Elbow		**1079** 9	**1430** 9									
	L.R. 45° Elbow		**707** 6	**937** 6									
	Tee		**1870**	**2415**									
	Lateral												
	Reducer		**310** 4.5	**410** 4.5									
	Cap		**230** 7.5	**300** 7.5									

INSULATION	Temperature Range °F		100-199	200-299	300-399	400-499	500-599	600-699	700-799	800-899	900-999	1000-1099	1100-1200
	85% Magnesia Calcium Silicate	Nom. Thick., In.	1½	1½	2	2½	3	3½	4	4½	5	5	6
		Lbs./Ft	**16.5**	**16.5**	**22.2**	**28.0**	**34.0**	**40.1**	**46.4**	**52.7**	**59.2**	**59.2**	**72.6**
	Combination	Nom. Thick., In.						3½	4½	5½	6	6½	7
		Lbs/Ft						**57.4**	**79.7**	**102.8**	**114.8**	**126.9**	**139.3**

	Pressure Rating psi	Cast Iron		Steel						
		125	250	150	300	400	600	900	1500	2500
FLANGES	Screwed or Slip-On			**792** 1.5	**1895** 1.5	**1759** 1.5	**2320** 1.5			
	Welding Neck			**862** 1.5	**2024** 1.5	**1879** 1.5	**2414** 1.5			
	Lap Joint									
	Blind			**1733** 1.5	**3449** 1.5	**3576** 1.5	**4419** 1.5			
FLANGED FITTINGS	S.R. 90° Elbow									
	L.R. 90° Elbow									
	45° Elbow									
	Tee									
VALVES	Flanged Bonnet Gate									
	Flanged Bonnet Globe or Angle									
	Flanged Bonnet Check									
	Pressure Seal Bonnet — Gate									
	Pressure Seal Bonnet — Globe									

Boldface type is weight in pounds. Lightface type beneath weight is weight factor for insulation.

Insulation thicknesses and weights are based on average conditions and do not constitute a recommendation for specific thicknesses of materials. Insulation weights are based on 85% magnesia and hydrous calcium silicate at 11 lbs/cubic foot. The listed thicknesses and weights of combination covering are the sums of the inner layer of diatomaceous earth at 21 lbs/cubic foot and the outer layer at 11 lbs/cubic foot.

Insulation weights include allowances for wire, cement, canvas, bands and paint, but not special surface finishes.

To find the weight of covering on flanges, valves or fittings, multiply the weight factor by the weight per foot of covering used on straight pipe.

Valve weights are approximate. When possible, obtain weights from the manufacturer.

Cast iron valve weights are for flanged end valves; steel weights for welding end valves.

All flanged fitting, flanged valve and flange weights include the proportional weight of bolts or studs to make up all joints.

WEIGHT OF FLANGED GATE VALVES (lb)

Size (in.)	150#	300#	600#
2	46	74	84
3	76	108	160
4	110	165	300
5	155	235	
6	175	320	640
8	310	500	1080
10	455	760	1550
12	650	1020	2100
14	860	1380	
16	1120	1960	
18	1400	2450	
20	2125	3890	
24	3120	5955	

Courtesy of Crane Valves

WEIGHT OF FLANGED CHECK (SWING) VALVES (lb)

Size (in.)	150#	300#	600#
2	33	46	62
3	59	86	115
4	93	154	192
5	152	255	
6	165	276	495
8	275	420	780
10	440	640	1400
12	680	1000	1750
14	950	1550	
16	1225	1700	
18	1700	2200	
20	1850	2800	
24	2900	3650	

Courtesy of Crane Valves

WEIGHT OF FLANGED GLOBE VALVES (lb)

Size (in.)	150#	300#	600#
2	47	60	88
3	82	117	160
4	134	176	270
5	199	290	
6	240	340	550
8	370	530	1000
10	525	750	
12	900	1100	
14	1000		

Courtesy of Crane Valves

Bolt Weights (lb) (1)

Flange Size (in.)	Flange Schedule								
	Gray Iron		Steel						
	125	250	150	300	400	600	900	1500	2500
1/2	1.0		1.0	1.0		1.0		3.2	3.4
3/4	1.0		1.0	2.0		2.0		3.3	3.6
1	1.0	2.0	1.0	2.0		2.0		6.0	6.0
1 1/4	1.0	2.0	1.0	2.0		2.0		6.0	9.0
1 1/2	1.0	2.5	1.0	3.5		3.5		9.0	12.0
2	1.5	3.5	1.5	4.0		4.5		12.5	21.0
2 1/2	1.5	6.0	1.5	7.0		8.0	19.0	19.0	27.0
3	1.5	6.0	1.5	7.5		8.0	12.5	25.0	37.0
3 1/2	3.5	6.5	3.5	7.5		12.0			
4	4.0	6.5	4.0	7.5	12.0	12.5	25.0	34.0	61.0
5	6.0	6.5	6.0	8.0	12.5	19.5	33.0	60.0	98.0
6	6.0	10.0	6.0	11.5	19.0	30.0	40.0	76.0	145.0
8	6.5	16.0	6.5	18.0	30.0	40.0	69.0	121.0	232.0
10	15.0	33.0	15.0	38.0	52.0	72.0	95.0	184.0	445.0
12	15.0	44.0	15.0	49.0	69.0	91.0	124.0	306.0	622.0
14	22.0	57.0	22.0	62.0	88.0	118.0	159.0	425.0	
16	31.0	76.0	31.0	83.0	114.0	152.0	199.0	570.0	
18	41.0	93.0	41.0	101.0	139.0	193.0	299.0	770.0	
20	52.0	95.0	52.0	105.0	180.0	242.0	361.0	1010.0	
24	71.0	174.0	71.0	174.0	274.0	360.0	687.0	1560.0	

(1) *Weights are for one complete flanged joint and include bolts and nuts.*

Steel Butt-Welding Fitting Dimensions

These dimensions are per ASME B16.9 – 2001 edition.

(all dimensions in inches)

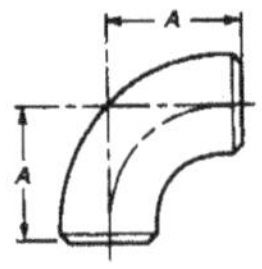

90° Long Radius Elbow Straight or Reducing

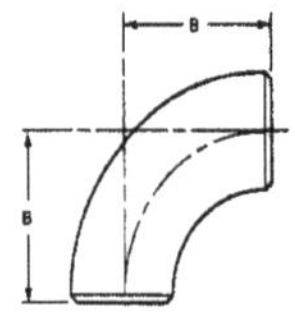

90° Short Radius Elbow

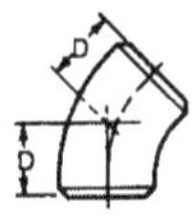

45° Long Radius

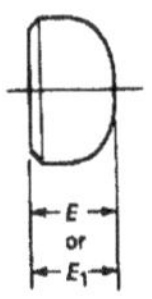

Cap

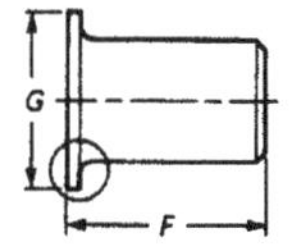

Lap Joint Stub End Long Pattern

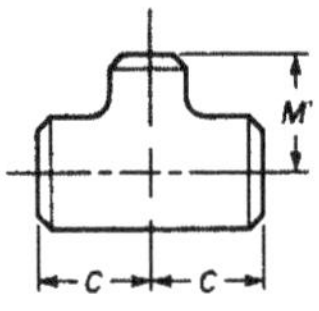

Reducing Tee

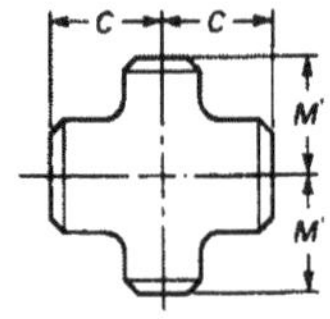

Reducing Cross

Dimension "M" is shown in table below; refer to large table for dimension "C"

Size a	Size b	M'	Size a	Size b	M'
1	x 3/4	1.50	6	x 3 1/2	5.00
1 1/4	x 1	1.88		x 3	4.88
	x 3/4	1.88		x 2 1/2	4.75
1 1/2	x 1 1/4	2.25	8	x 6	6.62
	x 1	2.25		x 5	6.38
	x 3/4	2.25		x 4	6.12
2	x 1 1/2	2.38		x 3 1/2	6.00
	x 1 1/4	2.25	10	x 8	8.00
	x 1	2.00		x 6	7.62
	x 3/4	1.75		x 5	7.50
2 1/2	x 2	2.75		x 4	7.25
	x 1 1/2	2.62	12	x 10	9.50
	x 1 1/4	2.50		x 8	9.00
	x 1	2.25		x 6	8.62
3	x 2 1/2	3.25		x 5	8.50
	x 2	3.00	14	x 12	10.62
	x 1 1/2	2.88		x 10	10.12
	x 1 1/4	2.75		x 8	9.75
3 1/2	x 3	3.62		x 6	9.38
	x 2 1/2	3.50	16	x 14	12.00
	x 2	3.25		x 12	11.62
	x 1 1/2	3.12		x 10	11.12
4	x 3 1/2	4.00		x 8	10.75
	x 3	3.88	18	x 16	13.00
	x 2 1/2	3.75		x 14	13.00
	x 2	3.50		x 12	12.62
	x 1 1/2	3.38		x 10	12.12
5	x 4	4.62	20	x 18	14.50
	x 3 1/2	4.50		x 16	14.00
	x 3	4.38		x 14	14.00
	x 2 1/2	4.25		x 12	13.62
	x 2	4.12	24	x 20	17.00
6	x 5	5.38		x 18	16.50
	x 4	5.12		x 16	16.00

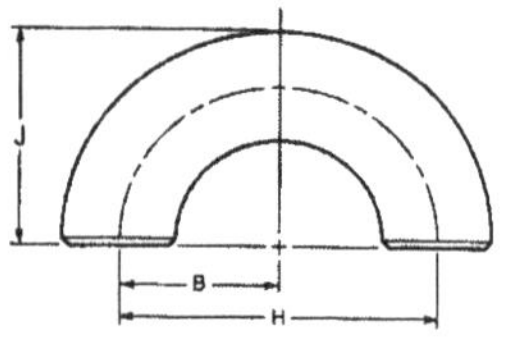

Short Radius Return Bend

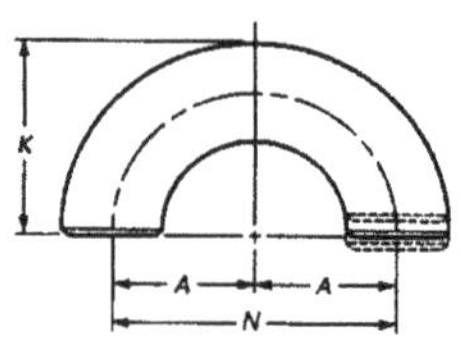

Long Radius Return Bend

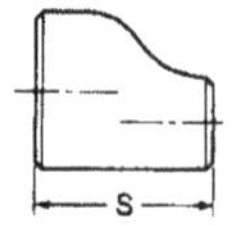

Eccentric Reducer

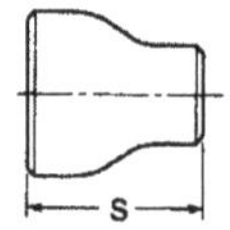

Concentric Reducer

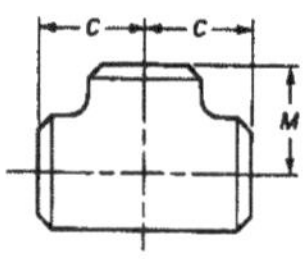

Straight Tee

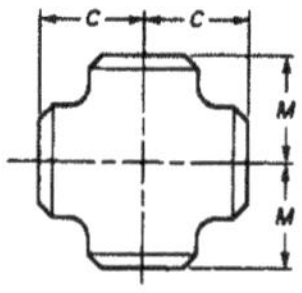

Straight Cross

Size	A	B	D	E	E[1]	F	G	H	J	K	M	N	S*	Pipe Schedule Numbers for: Std. Ftgs.	Pipe Schedule Numbers for: Extra Strong
1/2	1.50		0.62	1.00	1.00	3.00	1.38			1.88	1.00	3.00		40	80
3/4	1.50		0.75	1.00	1.00	3.00	1.38			2.00	1.12	3.00	1.50	40	80
1	1.50	1.00	0.88	1.50	1.50	4.00	2.00	2.00	1.62	2.19	1.50	3.00	2.00	40	80
1 1/4	1.88	1.25	1.00	1.50	1.50	4.00	2.50	2.50	2.06	2.75	1.88	3.75	2.00	40	80
1 1/2	2.25	1.50	1.12	1.50	1.50	4.00	2.88	3.00	2.44	3.25	2.25	4.50	2.50	40	80
2	3.00	2.00	1.38	1.50	1.75	6.00	3.62	4.00	3.19	4.19	2.50	6.00	3.00	40	80
2 1/2	3.75	2.50	1.75	1.50	2.00	6.00	4.12	5.00	3.94	5.19	3.00	7.50	3.50	40	80
3	4.50	3.00	2.00	2.00	2.50	6.00	5.00	6.00	4.75	6.25	3.38	9.00	3.50	40	80
3 1/2	5.25	3.40	2.25	2.50	3.00	6.00	5.50	7.00	5.50	7.25	3.75	10.50	4.00	40	80
4	6.00	4.00	2.50	2.50	3.00	6.00	5.19	8.00	6.25	8.25	4.12	12.00	4.00	40	80
5	7.50	5.00	3.12	3.00	3.50	8.00	7.31	10.00	7.75	10.31	4.88	15.00	5.00	40	80
6	9.00	6.00	3.75	3.50	4.00	8.00	8.50	12.00	9.31	12.31	5.62	18.00	5.50	40	80
8	12.00	8.00	5.00	4.00	5.00	8.00	10.62	16.00	12.31	16.31	7.00	24.00	6.00	40	80
10	15.00	10.00	6.25	5.00	6.00	10.00	12.75	20.00	15.38	20.38	8.50	30.00	7.00	40	60
12	18.00	12.00	7.50	6.00	7.00	10.00	15.00	24.00	18.38	24.38	10.00	36.00	8.00		
14	21.00	14.00	8.75	6.50	7.50	12.00	16.25	28.00	21.00	28.00	11.00	42.00	13.00	30	
16	24.00	16.00	10.00	7.00	8.00	12.00	18.50	32.00	24.00	32.00	12.00	48.00	14.00	30	40
18	27.00	18.00	11.25	8.00	9.00	12.00	21.00	36.00	27.00	36.00	13.50	54.00	15.00		
20	30.00	20.00	12.50	9.00	10.00	12.00	23.00	40.00	30.00	40.00	15.00	60.00	20.00	20	30
22	33.00	22.00	13.50	10.00	10.00	12.00	25.25	44.00	33.00	44.00	16.50	66.00	20.00		
24	36.00	24.00	15.00	10.50	12.00	12.00	27.25	48.00	36.00	48.00	17.00	72.00	20.00	20	

[1] Size is that of large end.

Welding Neck Flange Length Through Hub, Y (in.) (ASME B16.5, 2017 Edition)

	Flange Class						
NPS	150	300	400	600	900	1500	2500
½	1.81	2.00		2.06		2.38	2.88
¾	2.00	2.19		2.25		2.75	3.12
1	2.12	2.38		2.44		2.88	3.50
1¼	2.19	2.50		2.62		2.88	3.75
1½	2.38	2.62		2.75		3.25	4.38
2	2.44	2.69		2.88		4.00	5.00
2½	2.69	2.94		3.12		4.12	5.62
3	2.69	3.06		3.25	4.00	4.62	6.62
3½	2.75	3.12		3.38			
4	2.94	3.32	3.50	4.00	4.50	4.88	7.50
5	3.44	3.82	4.00	4.50	5.00	6.12	9.00
6	3.44	3.82	4.06	4.62	5.50	6.75	10.75
8	3.94	4.32	4.62	5.25	6.38	8.38	12.50
10	3.94	4.56	4.88	6.00	7.25	10.00	16.50
12	4.44	5.06	5.38	6.12	7.88	11.12	18.25
14	4.94	5.56	5.88	6.50	8.38	11.75	
16	4.94	5.69	6.00	7.00	8.50	12.25	
18	5.44	6.19	6.50	7.25	9.00	12.88	
20	5.62	6.32	6.62	7.50	9.75	14.00	
22	5.82	6.44	6.75	6.88	...	...	
24	5.94	6.56	6.88	8.00	11.50	16.00	

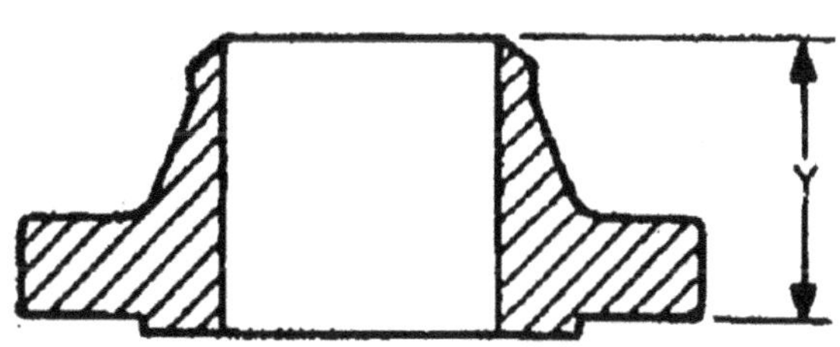

Gate Valves (See Note)
Solid Wedge, Double Disc, and Conduit
End-to-End Dimensions "A" Per ASME B16.10 (2017 Edition)
Flanged and Buttwelding Ends

Nominal Size		Pressure Class											
		150		300	600		900		1500		2500		
NPS	DN	Flanged End	Butt Weld		Long Pattern	Short Pattern	Long Pattern	Short Pattern	Long Pattern	Short Pattern	Long Pattern	Short Pattern
¼	8	4.00	4.00	...	6.50	...	...	...	...	...	...	...
3/8	10	4.00	4.00	...	7.50	...	...	...	...	...	...	...
½	15	4.25	4.25	5.00	8.50	5.25	...	...	...	...	10.38	...
¾	20	4.62	4.62	6.00	9.00	5.75	...	...	...	...	10.75	...
1	25	5.00	5.00	6.50	9.50	6.00	10.00	5.50	10.00	5.50	12.12	7.31
1 ¼	32	5.50	5.50	7.00	9.00	5.75	11.00	6.50	11.00	6.50	13.75	9.12
1 ½	40	6.50	6.50	7.50	9.50	6.00	12.00	7.00	12.00	7.00	15.12	9.12
2	50	7.00	8.50	8.50	11.50	7.00	14.50	8.50	14.50	8.50	17.75	11.00
2½	65	7.50	9.50	9.50	13.00	8.50	16.50	10.00	16.50	10.00	20.00	13.00
3	80	8.00	11.12	11.12	14.00	10.00	15.00	12.00	18.50	12.00	22.75	14.50
4	100	9.00	12.00	12.00	17.00	12.00	18.00	14.00	21.50	16.00	26.50	18.00
5	125	10.00	15.00	15.00	20.00	15.00	22.00	17.00	26.50	19.00	31.25	21.00
6	150	10.50	15.88	15.88	22.00	18.00	24.00	20.00	27.75	22.00	36.00	24.00
8	200	11.50	16.50	16.50	26.00	23.00	29.00	26.00	32.75	28.00	40.25	30.00
10	250	13.00	18.00	18.00	31.00	28.00	33.00	31.00	39.00	34.00	50.00	36.00
12	300	14.00	19.75	19.75	33.00	32.00	38.00	36.00	44.50	39.00	56.00	41.00
14	350	15.00	22.50	22.50	35.00	35.00	40.50	39.00	49.50	42.00		44.00
16	400	16.00	24.00	24.00	39.00	39.00	44.50	43.00	54.50	47.00		49.00
18	450	17.00	26.00	26.00	43.00	43.00	48.00		60.50	53.00		55.00
20	500	18.00	28.00	28.00	47.00	47.00	52.00		65.50	58.00		
22	550	20.00	30.00	43.00	51.00	51.00						
24	600	20.00	32.00	31.00	55.00	55.00	61.00		76.50			
26	650	22.00	34.00	49.00	57.00	57.00						
28	700	24.00	36.00	53.00	61.00	61.00						
30	750	24.00	36.00	55.00	65.00	65.00						
32	800	28.00	38.00	60.00	70.00	70.00						
34	850	30.00	40.00	64.00	76.00	76.00						
36	900	28.00	40.00	68.00	82.00	82.00						
40	1200	32.00										

Not all valve types are available in all sizes, refer to valve manufacturer information and/or ASME 816.10.

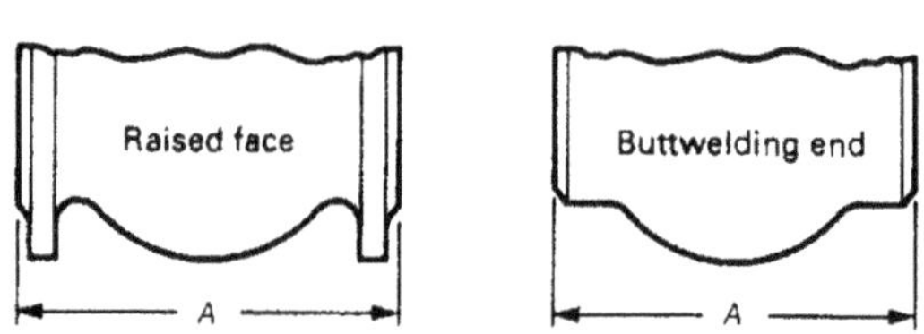

Globe, Lift Check and Swing Check Valves (See note.)
End-to-End Dimensions "A" Per ASME B16.10 (2017 Edition)
Flanged and Buttwelding Ends

Nominal Size		Pressure Class									
		150	300	600		900		1500		2500	
NPS	DN			Long Pattern	Short Pattern	Long Pattern	Short Pattern	Long Pattern	Short 3l Pattern	Long Pattern	Short Pattern
¼	8	4.00									
¾	10	4.00									
½	15	4.25	6.00	6.50				8.50		10.38	
¾	20	4.62	7.00	7.50		9.00		9.00		10.75	
1	25	5.00	8.00	8.50	5.25	10.00		10.00		12.12	
1¼	32	5.50	8.50	9.00	5.75	11.00		11.00		13.75	
1½	40	6.50	9.00	9.50	6.00	12.00		12.00		15.12	
2	50	8.00	10.50	11.50	7.00	14.50		14.50	8.50	17.75	11.00
2½	65	8.50	11.50	13.00	8.50	16.50	10.00	16.50	10.00	20.00	13.00
3	80	9.50	12.50	14.00	10.00	15.00	12.00	18.50	12.00	22.75	14.50
4	100	11.50	14.00	17.00	12.00	18.00	14.00	21.50	16.00	26.50	18.00
5	125	13.00	15.75	20.00	15.00	22.00	17.00	26.50	19.00	31.25	21.00
6	150	14.00	17.50	22.00	18.00	24.00	20.00	27.75	22.00	36.00	24.00
8	200	19.50	21.00	26.00	23.00	29.00	26.00	32.75	28.00	40.25	30.00
10	250	24.50	24.50	31.00	28.00	33.00	31.00	39.00	34.00	50.00	36.00
12	300	27.50	28.00	33.00	32.00	38.00	36.00	44.50	39.00	56.00	41.00
14	350	31.00	33.00	35.00		40.50	39.00	49.50	42.00		
16	400	36.00	34.00	39.00		44.50	43.00	54.50	47.00		
18	450	38.50	38.50	43.00		48.00		60.50			
20	500	38.50	40.00	47.00		52.00		65.50			
22	550	42.00	44.00	51.00							
24	600	51.00	53.00	55.00		61.00		76.50			
26	650	51.00	53.00	57.00							
28	700	57.00	59.00	63.00							
30	750	60.00	62.75	65.00							
36	900	77.00	87.00	82.00							

Not all valve types are available in all sizes, refer to valve manufacturer information and/or ASME B16.10.

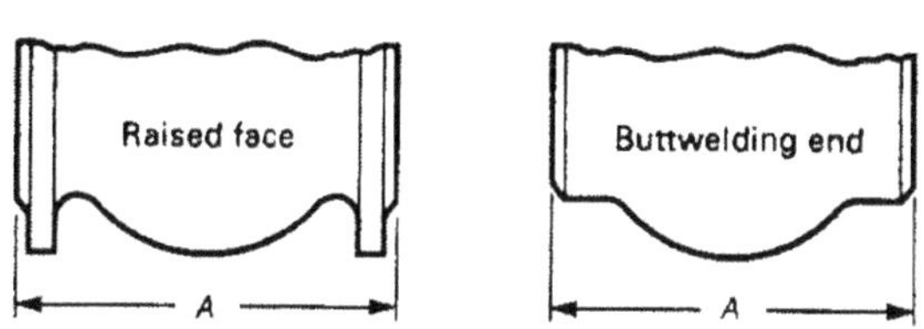

Angle and Lift Check (See Note)
End-to-End Dimensions "D" Per ASME B16.10 (2017 Edition)
Flanged and Butt welding Ends

Nominal Size		Pressure Class							
		150	300	600		900		1500	2500
NPS	DN			Long Pattern	Short Pattern	Long Pattern	Short Pattern	Long Pattern	Long Pattern
¼	8	2.00							
¾	10	2.00							
½	15	2.25	3.00	3.25				4.25	5.19
¾	20	2.50	3.50	3.75		4.50		4.50	5.38
1	25	2.75	4.00	4.25		5.00		5.00	6.06
1¼	32	3.00	4.25	4.50		5.50		5.50	6.88
1½	40	3.25	4.50	4.75		6.00		6.00	7.56
2	50	4.00	5.25	5.75	4.25	7.25		7.25	8.88
2½	65	4.25	5.75	6.50	5.00	8.25		8.25	10.00
3	80	4.75	6.25	7.00	6.00	7.50	6.00	9.25	11.38
4	100	5.75	7.00	8.50	7.00	9.00	7.00	10.75	13.25
5	125	7.00	7.88	10.00	8.50	11.00	8.50	13.25	15.62
6	150	8.00	8.75	11.00	10.00	12.00	10.00	13.88	18.00
8	200	9.75	10.50	13.00		14.50	13.00	16.38	20.12
10	250	12.25	12.25	15.50		16.50	15.50	19.50	25.00
12	300	13.75	14.00	16.50		19.00	18.00	22.25	28.00
14	350	15.50				20.25	19.50	24.75	
16	400	18.00				26.00			
18	450					29.00			
20	500					32.50			
22	550								
24	600					39.00			

Not all valve types are available in all sizes, refer to valve manufacturer information and/or ASME 816.10.

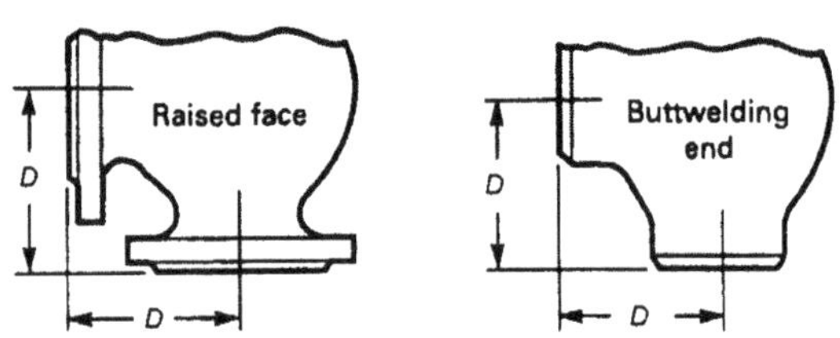

Table C-1 Thermal Expansion Data

A = Mean Coefficient of Thermal Expansion, 10^{-6} in./in./°F } in Going From 70°F to Indicated Temperature [Note (1)]
B = Linear Thermal Expansion, in./100 ft

Material	Coefficient	Temperature Range 70°F to																
		−325	−150	−50	70	200	300	400	500	600	700	800	900	1,000	1,100	1,200	1,300	1,400
Group 1 carbon and low alloy steels [Note (2)]	*A*	5.5	5.9	6.2	6.4	6.7	6.9	7.1	7.3	7.4	7.6	7.8	7.9	8.1	8.2	8.3	8.4	8.4
	B	−2.6	−1.6	−0.9	0	1.0	1.9	2.8	3.7	4.7	5.7	6.8	7.9	9.0	10.1	11.3	12.4	14.7
Group 2 low alloy steels [Note (3)]	*A*	6.0	6.5	6.7	7.0	7.3	7.4	7.6	7.7	7.8	7.9	8.0	8.1	8.2	8.3	8.4	8.4	8.5
	B	−2.9	−1.7	−1.0	0	1.1	2.0	3.0	4.0	5.0	6.0	7.0	8.1	9.2	10.3	11.4	12.5	13.5
5Cr–1Mo steels	*A*	5.6	6.0	6.2	6.4	6.7	6.9	7.0	7.1	7.2	7.2	7.3	7.4	7.5	7.6	7.6	7.7	7.8
	B	−2.7	−1.6	−0.9	0	1.0	1.9	2.8	3.7	4.6	5.5	6.4	7.4	8.4	9.3	10.3	11.4	12.4
9Cr–1Mo steels	*A*	5.0	5.4	5.6	5.8	6.0	6.2	6.3	6.4	6.5	6.6	6.7	6.8	6.9	7.0	7.1	7.2	7.2
	B	−2.4	−1.4	−0.8	0	0.9	1.7	2.5	3.3	4.1	5.0	5.9	6.8	7.7	8.7	9.7	10.6	11.6
Straight chromium stainless steels																		
12Cr to 13Cr steels	*A*	5.1	5.5	5.7	5.9	6.2	6.3	6.4	6.5	6.5	6.6	6.7	6.7	6.8	6.8	6.9	6.9	7.0
	B	−2.4	−1.5	−0.8	0	1.0	1.7	2.5	3.3	4.2	5.0	5.8	6.7	7.6	8.5	9.4	10.2	11.1
15Cr to 17Cr steels	*A*	4.5	4.9	5.1	5.3	5.5	5.7	5.8	5.9	6.0	6.1	6.2	6.2	6.3	6.4	6.4	6.5	6.5
	B	−2.1	−1.3	−0.7	0	0.9	1.6	2.3	3.0	3.8	4.6	5.4	6.2	7.0	7.9	8.7	9.5	10.4
27Cr steels	*A*	4.3	4.7	4.9	5.0	5.2	5.2	5.3	5.4	5.4	5.5	5.6	5.7	5.7	5.8	5.9	5.9	6.0
	B	−2.0	−1.2	−0.7	0	0.8	1.4	2.1	2.8	3.5	4.2	4.9	5.6	6.4	7.2	8.0	8.7	9.6
Austenitic stainless steels (304, 305, 316, 317, 321, 347, 348, 19-9DL, XM-15, etc.)	*A*	7.5	8.0	8.2	8.5	8.9	9.2	9.5	9.7	9.9	10.0	10.1	10.2	10.3	10.4	10.6	10.7	10.8
	B	−3.6	−2.1	−1.2	0	1.4	2.5	3.8	5.0	6.3	7.5	8.8	10.2	11.5	12.9	14.3	15.8	17.2
Other austenitic stainless steels (309, 310, 315, XM-19, etc.)	*A*	7.1	7.6	7.8	8.2	8.5	8.7	8.9	9.1	9.2	9.3	9.4	9.5	9.6	9.7	9.8	9.9	10.1
	B	−3.4	−2.0	−1.1	0	1.3	2.4	3.5	4.7	5.8	7.0	8.2	9.5	10.7	12.0	13.3	14.7	16.1
Gray iron	*A*	...	...	...	...	5.8	5.9	6.1	6.3	6.5	6.7	6.8	7.0	7.2	...	...	...	...
	B	...	...	...	0	0.9	1.6	2.4	3.2	4.1	5.0	6.0	7.0	8.0	...	...	...	...
Ductile cast iron	*A*	...	4.9	5.3	5.7	6.0	6.3	6.6	6.8	7.0	7.1	7.3	7.4	7.5	...	...	...	...
	B	...	−1.3	−0.8	0	0.9	1.7	2.6	3.5	4.5	5.4	6.4	7.3	8.4	...	...	...	...

Thermal Expansion Data (Cont'd)

A = Mean Coefficient of Thermal Expansion, 10^{-6} in./in./°F } in Going From 70°F to Indicated Temperature [Note (1)]
B = Linear Thermal Expansion, in./100 ft

Material	Coefficient	Temperature Range 70°F to −325	−150	−50	70	200	300	400	500	600	700	800	900	1,000	1,100	1,200	1,300	1,400
Monel (67Ni–30Cu) N04400	*A*	5.8	6.8	7.2	7.7	8.1	8.3	8.5	8.7	8.8	8.9	8.9	9.0	9.1	9.1	9.2	9.2	9.3
	B	−2.7	−1.8	−1.0	0	1.3	2.3	3.4	4.5	5.6	6.7	7.8	9.0	10.1	11.3	12.4	13.6	14.8
Nickel alloys N02200 and N02201	*A*	5.3	6.0	6.3	6.6	7.2	7.5	7.7	7.9	8.0	8.2	8.3	8.4	8.5	8.6	8.7	8.8	8.9
	B	−2.7	−1.7	−1.0	0	1.1	2.1	3.1	4.1	5.1	6.2	7.3	8.4	9.5	10.7	11.8	13.0	14.2
Nickel alloy N06022	*A*	...	...	...	6.9	6.9	6.9	6.9	7.0	7.0	7.2	7.3	7.5	7.7	7.9	8.1	8.3	8.5
	B	...	...	...	0	1.1	1.9	2.7	3.6	4.5	5.4	6.4	7.5	8.6	9.8	11.0	12.2	13.6
Nickel alloy N06600	*A*	5.5	6.1	6.4	6.8	7.1	7.3	7.5	7.6	7.8	7.9	8.0	8.2	8.3	8.4	8.6	8.7	8.9
	B	−2.6	−1.6	−0.9	0	1.1	2.0	3.0	3.9	5.0	6.0	7.0	8.1	9.3	10.4	11.6	12.9	14.2
Nickel alloy N06625	*A*	...	...	...	6.7	7.1	7.2	7.3	7.4	7.4	7.5	7.6	7.7	7.9	8.0	8.2	8.4	8.5
	B	...	...	...	0	1.1	2.0	2.9	3.8	4.7	5.6	6.6	7.7	8.8	9.9	11.1	12.3	13.6
Nickel alloys N08800 and N08810	*A*	5.9	6.9	7.4	7.9	8.4	8.6	8.8	8.9	9.0	9.1	9.2	9.3	9.4	9.5	9.6	9.7	9.8
	B	−2.8	−1.7	−1.1	0	1.3	2.4	3.5	4.6	5.7	6.9	8.1	9.3	10.5	11.8	13.0	14.4	15.7
Nickel alloy N08825	*A*	...	...	7.2	7.5	7.7	7.9	8.0	8.1	8.2	8.3	8.4	8.5	8.6	...	...	...	...
	B	...	...	−1.0	0	1.2	2.2	3.2	4.2	5.2	6.3	7.4	8.5	9.6	...	...	...	...
Nickel alloy N10276	*A*	...	...	...	6.0	6.3	6.5	6.7	6.9	7.1	7.2	7.4	7.5	7.6	7.7	7.8	7.9	8.0
	B	...	...	...	0	1.0	1.8	2.7	3.6	4.5	5.5	6.4	7.5	8.5	9.5	10.6	11.7	12.8
Copper alloys C1XXXX series	*A*	7.7	8.7	9.0	9.3	9.6	9.7	9.8	9.9	10.0	...	...	...	...	...	...	...	...
	B	−3.7	−2.3	−1.3	0	1.5	2.7	3.9	5.1	6.4	...	...	...	...	...	...	...	...
Bronze alloys	*A*	8.4	8.8	9.2	9.6	10.0	10.1	10.2	10.3	10.4	10.5	10.6	10.7	10.8	10.9	11.0	...	...
	B	−4.0	−2.3	−1.3	0	1.6	2.8	4.0	5.3	6.6	8.0	9.3	10.7	12.1	13.5	14.9	...	...
Brass alloys	*A*	8.2	8.5	9.0	9.3	9.8	10.0	10.2	10.5	10.7	10.9	11.2	11.4	11.6	11.9	12.1	...	...
	B	−3.9	−2.2	−1.3	0	1.5	2.8	4.1	5.4	6.8	8.2	9.8	11.4	13.0	14.7	16.4	...	...
Copper–nickel (70Cu–30Ni)	*A*	6.7	7.4	7.8	8.1	8.5	8.7	8.9	9.1	9.2	9.2	...	...	...	...	...	...	...
	B	−3.2	−2.0	−1.1	0	1.3	2.4	3.5	4.7	5.8	7.0	...	...	...	...	...	...	...

A = Mean Coefficient of Thermal Expansion, 10^{-6} in./in./°F
B = Linear Thermal Expansion, in./100 ft
} **in Going From 70°F to Indicated Temperature [Note (1)]**

Material	Coefficient	Temperature Range 70°F to −325	−150	−50	70	200	300	400	500	600	700	800	900	1,000	1,100	1,200	1,300	1,400
Aluminum alloys	A	9.9	10.9	11.6	12.1	13.0	13.3	13.6	13.9	14.2	...	...	...	...	...	...	...	...
	B	−4.7	−2.9	−1.7	0	2.0	3.7	5.4	7.2	9.0	...	...	...	...	...	...	...	...
Titanium alloys (Grades 1, 2, 3, 7, and 12)	A	...	...	4.5	4.6	4.7	4.8	4.8	4.9	4.9	5.0	5.1	...	...	...	...	...	...
	B	...	...	−0.6	0	0.7	1.3	1.9	2.5	3.1	3.8	4.5	...	...	...	...	...	...

NOTES:

(1) These data are for information and it is not to be implied that materials are suitable for all the temperature ranges shown.

(2) Group 1 alloys (by nominal composition):

Carbon steels (C, C–Si, C–Mn, and C–Mn–Si)	3Cr–1Mo
C–½Mo	½Ni–½Mo–V
½Cr–⅕Mo–V	½Ni–½Cr–¼Mo–V
½Cr–¼Mo–Si	¾Ni–½Mo–Cr–V
½Cr–½Mo	¾Ni–½Mo–⅓Cr–V
½Cr–½Ni–¼Mo	¾Ni–½Cu–Mo
¾Cr–½Ni–Cu	¾Ni–½Cr–½Mo–V
¾Cr–¾Ni–Cu–Al	¾Ni–1Mo–¾Cr
1Cr–⅕Mo	1Ni–½Cr–½Mo
1Cr–⅕Mo–Si	1¼Ni–1Cr–½Mo
1Cr–½Mo	1¾Ni–¾Cr–¼Mo
1Cr–½Mo–V	2Ni–¾Cr–¼Mo
1¼Cr–½Mo	2Ni–¾Cr–⅓Mo
1¼Cr–½Mo–Si	2½Ni
1¾Cr–½Mo–Cu	3½Ni
2Cr–½Mo	3½Ni–1¾Cr–½Mo–V
2¼Cr–1Mo	

(3) Group 2 alloys (by nominal composition):

Mn–V	Mn–½Mo–¼Ni
Mn–¼Mo	Mn–½Mo–½Ni
Mn–½Mo	Mn–½Mo–¾Ni

APPENDIX

IV

A PRACTICAL GUIDE TO EXPANSION JOINTS

This appendix is largely a copy of the publication, "A Practical Guide to Expansion Joints" by the Expansion Joint Manufacturers Association, Inc. 25 North Broadway, Tarrytown, NY 10591. It is reproduced, courtesy of EJMA.

IV-1 WHAT ARE EXPANSION JOINTS

IV-1.1 DEFINITION OF AN EXPANSION JOINT

For purposes of this publication, the definition of an expansion joint is "any device containing one or more metal bellows used to absorb dimensional changes such as those caused by thermal expansion or contraction of a pipe-line, duct, or vessel."

IV-1.2 EXPANSION DEVICES (FIG. IV-1.1)

Bellows expansion joints provide an alternative to pipe expansion loops and slip-type joints.

IV-1.3 MANUFACTURING METHODS

There are several methods of manufacturing bellows, the two most common bellows types are:

1. FORMED BELLOWS - (Fig. IV-1.2.1) Whether formed hydraulically or mechanically, this is by far the most common type available today. Formed from a thin-walled tube, a formed bellows contains only longitudinal welds, and exhibits significant flexibility. Formed bellows of varying shapes are made from single or multiple plies of suitable material (usually stainless steel) ranging in thickness from 0.004" to 0.125" and greater, and in diameters of 3/4" to over 12 ft. These bellows are usually categorized according to convolution shape (Fig. IV-1.2.1 and IV-1.4).
2. FABRICATED BELLOWS - Made by welding together a series of thin gage diaphragms or discs (Fig. IV-1.2.2 and IV-1.4). Fabricated bellows are usually made of heavier gage material than formed bellows and are therefore able to withstand higher pressures.

Due to the variations in manufacturing and design configurations, the number, height, and pitch of the convolutions should never be specified in the purchase of an expansion joint as a condition of design, but should be left to the decision of the manufacturer. These dimensions will vary from one manufacturer to another and in particular will depend on the specific design conditions and the manufacturer's own rating procedure which has been established on the basis of laboratory tests and field experience.

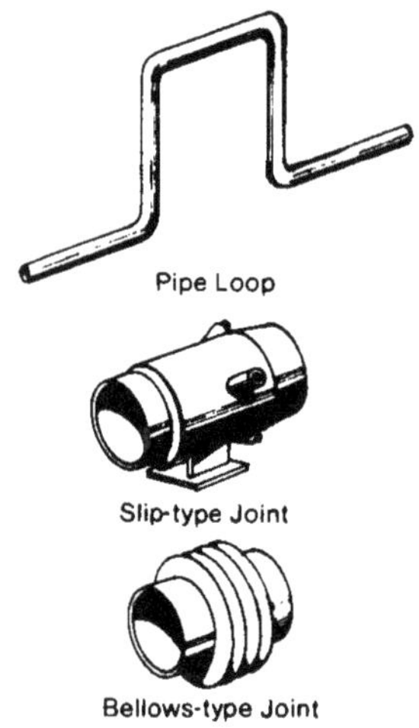

FIG. IV-1.1
EXPANSION DEVICES

IV-1.4 DESIGN VARIABLES

If an expansion joint is to fulfill its intended function safely and reliably, it must be kept in mind that it is a highly specialized product. Interchangeability is rare in expansion joints, and in a real sense, each unit is custom-made for an intended application. It becomes necessary then, to supply the expansion joint manufacturer with accurate information regarding the conditions of design that the expansion joint will be subjected to in service. Because many of these design conditions interact with each other, designing a bellows becomes much like assembling a jigsaw puzzle; one needs all of the pieces (design conditions) before a clear picture of an expansion joint design can emerge. The following is a listing of the basic design conditions that should be supplied to the manufacturer when specifying an expansion joint.

IV-1.4.1 Size. Size refers to the diameter of the pipeline (or dimensions of the duct in the case of rectangular joints) into which the expansion joint is to be installed. The size of an expansion joint affects its pressure-retaining capabilities, as well as its ability to absorb certain types of movements.

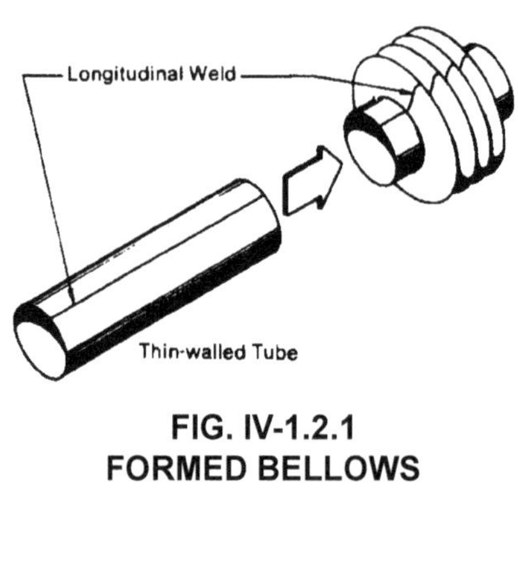

FIG. IV-1.2.1
FORMED BELLOWS

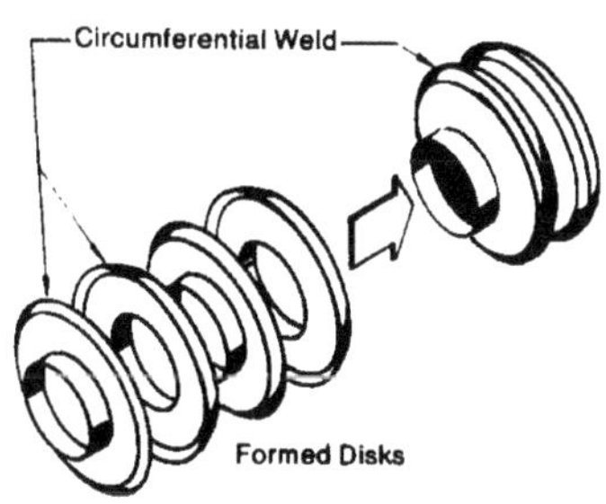

FIG. IV-1.2.2
FABRICATED BELLOWS

Note: When specifying pipe sizes, misunderstandings often result due to the confusing array of nominal pipe sizes (NPS) and different pipe schedules. To eliminate any misconceptions, pipe sizes should be supplied to the expansion joint manufacturer in terms of the actual outside diameter and the actual wall thickness of the pipe. The outside diameters, schedules, and wall thickness of pipe sizes up thru 36 inches can be found in the American Society of Mechanical Engineers (ASME B36.10).

IV-1.4.2 Flowing Medium.

The substances that will come in contact with the expansion joint should be specified. In some cases, due to excessive erosion, or corrosion potential, or in cases of high viscosity, special materials and accessories should be specified. When piping systems containing expansion joints are cleaned periodically, the cleaning solution must be compatible with the bellows materials.

IV-1.4.3 Pressure.

Pressure is possibly the most important factor determining expansion joint design. Minimum and maximum anticipated pressure should be accurately determined. If a pressure test is to be performed, this pressure should be specified as well. While the determination of pressure requirements is important, care should be exercised to insure that these specified pressures are not increased by unreasonable safety factors as this could result in a design which may not adequately satisfy other performance characteristics.

IV-1.4.4 Temperature.

The operating temperature of the expansion joint will affect its pressure capacity, allowable stresses, cycle life, and material requirements. All possible temperature sources should be investigated when determining minimum and maximum temperature requirements. In so doing, however, it is important that only those temperatures occurring at the expansion joint location itself be specified. Specifying temperatures remote from the expansion joint may unnecessarily result in the need for special materials at additional expense.

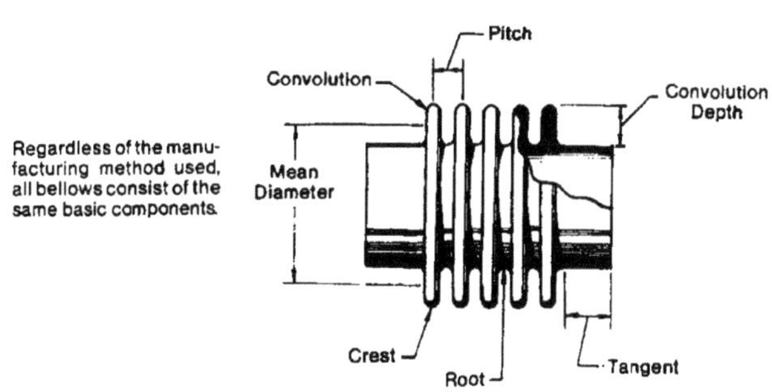

FIG. IV-1.3
BELLOWS COMPONENTS

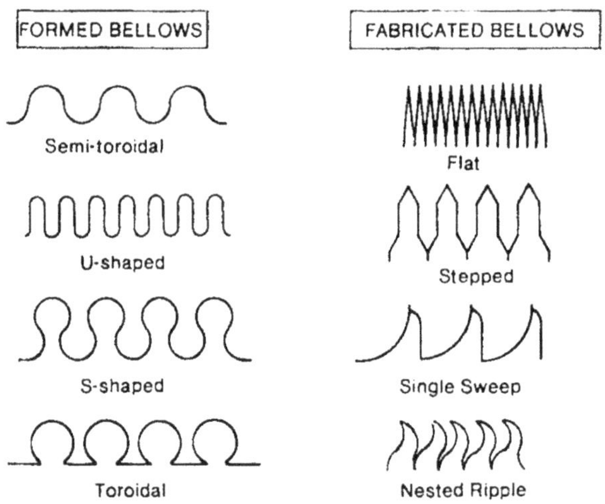

FIG. IV-1.4
CONVOLUTION SHAPES

IV-1.4.5 MOTION.

Movements due to temperature changes or mechanical motion to which the expansion joint will be subjected must be specified. (Methods for determining movements are presented in Section IV-2.) In addition, other extraneous movements, such as wind loading or installation misalignment must be considered. The various dimensional changes which an expansion joint is required to absorb, such as those resulting from thermal changes in a piping system, are as follows (Fig. IV-1.5):

Axial Motion is motion occurring parallel to the center line of the bellows and can be either extension or compression.

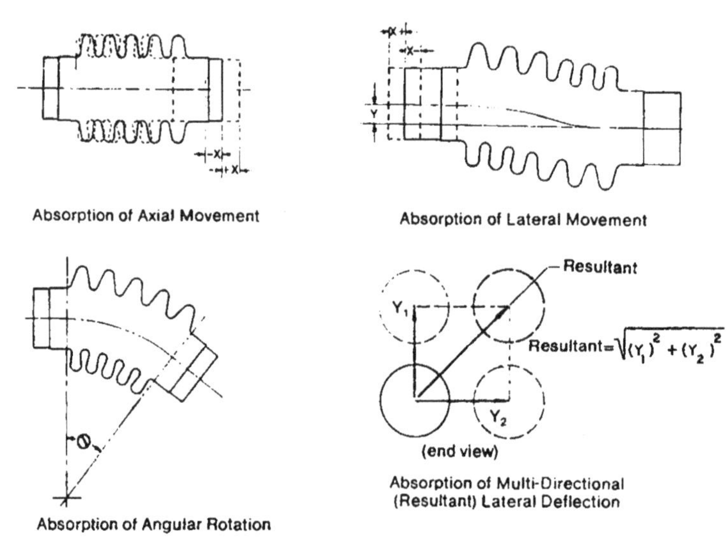

FIG. IV-1.5
TYPES OF EXPANSION JOINT MOVEMENT

In addition to Axial, Lateral, and Angular Movements, an expansion joint may be subjected to Torsional motion, or twisting. Torsion imposes severe stresses on the expansion joint and all such cases should be referred to the manufacturer.

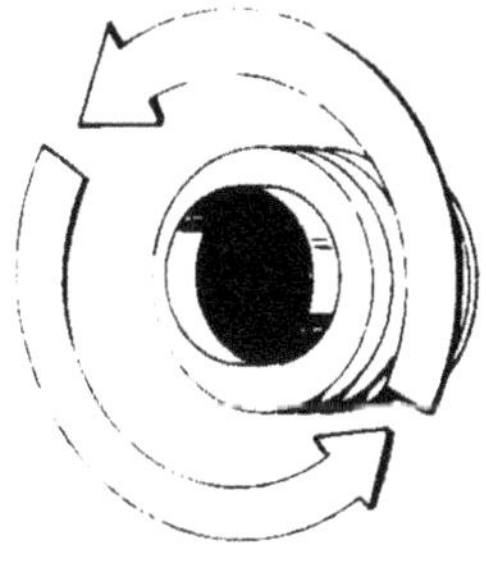

FIG. IV-1.6
TORSIONAL MOVEMENT

Lateral Deflection, or offset, is motion which occurs perpendicular, or at right angles to the centerline of the bellows. Lateral deflection can occur along one or more axis simultaneously.

Angular Rotation is the bending of an expansion joint along its centerline.

IV-1.5 EXPANSION JOINT ACCESSORIES

The basic unit of every expansion joint is the bellows. By adding additional components, expansion joints of increasing complexity and capability are created which are suitable for a wide range of applications.

Refer to Figure IV-1.7 and the following list for a description of the basic expansion joint accessories.

1. BELLOWS - The flexible element of an expansion joint, consisting of one or more convolutions, formed from thin material.
2. LINER - A device which minimizes the effect on the inner surface of the bellows by the fluid flowing through it. Liners are primarily used in high velocity applications to prevent erosion of the inner surface of the bellows and to minimize the likelihood of flow inducted vibration. Liners come in single, tapered, or telescoping configurations according to the application requirements. An expansion joint, if provided with liners, must be installed in the proper orientation with respect to flow direction. Liners are sometimes referred to as internal sleeves.
3. COVER - A device used to provide external protection to the bellows from foreign objects, mechanical damage, and/or external flow. The use of a cover is strongly recommended to all applications. A cover is sometimes referred to as a shroud.

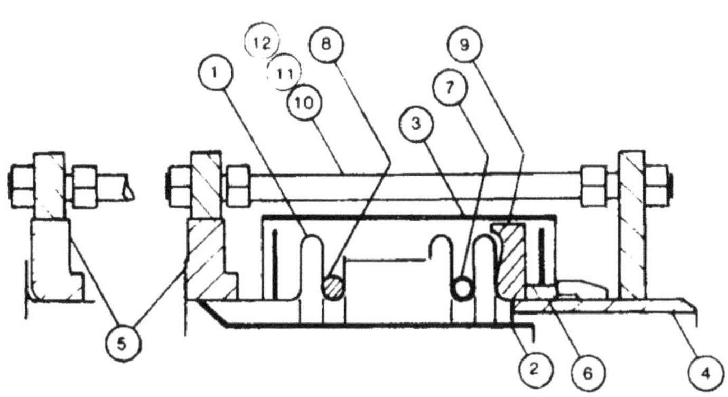

FIG. IV-1.7
EXPANSION JOINT ACCESSORIES

4. WELD END - The ends of an expansion joint equipped with pipe for weld attachment to adjacent equipment or piping. Weld ends are commonly supplied beveled for butt welding.
5. FLANGED END - The ends of an expansion joint equipped with flanges for the purpose of bolting the expansion joint to the mating flanges of adjacent equipment or piping.
6. COLLAR - A ring of suitable thickness which is used to reinforce the bellows tangent, or cuff, from bulging due to pressure.
7. HOLLOW REINFORCING RINGS - Devices used on some expansion joints, fitting snugly in the roots of the convolutions. The primary purpose of these devices is to reinforce the bellows against internal pressure. Hollow rings are usually formed from a suitable pipe or tubing section.
8. SOLID ROOT RINGS - Identical in function to hollow root rings, but formed from solid bar stock for greater strength.
9. EQUALIZING RINGS - "T" Shaped in cross-section, these rings are made of gray iron, steel, stainless steel or other suitable alloys. In addition to resisting internal pressure, equalizing rings limit the amount of compression movement per convolution.
10. CONTROL RODS - Devices, usually in the form of rods or bars, attached to the expansion joint assembly whose primary function is to distribute the applied movement between the two bellows of a universal expansion joint. Control rods are NOT designed to restrain bellows pressure thrust.
11. LIMIT RODS - Devices, usually in the form of rods or bars, attached to the expansion joint assembly whose primary function is to restrict the bellows movement range (axial, lateral and angular) during normal operation. In the event of a main anchor failure, limit rods are designed to prevent bellows over-extension or over-compression while restraining the full pressure loading and dynamic forces generated by the anchor failure.
12. TIE RODS - Devices, usually in the form of rods or bars, attached to the expansion joint assembly whose primary function is to continuously restrain the full bellows pressure thrust during normal operation while permitting only lateral deflection. Angular rotation can be accommodated only if two tie rods are used and located 90° from the direction of rotation.
13. PANTOGRAPHIC LINKAGES - A scissors-like device. A special form of control rod attached to the expansion joint assembly whose primary function is to positively distribute the movement equally between the two bellows of the universal joint throughout its full range of movement. Pantographic linkages, like control rods, are NOT designed to restrain pressure thrust.

IV-1.6 TYPES OF EXPANSION JOINTS

There are several different types of expansion joints. Each is designed to operate under a specific set of design conditions. The following is a listing of the most basic types of expansion joint designs, along with a brief description of their features and application requirements. A more in-depth discussion of specific expansion joint applications appears in Section IV-2.2.

All of the following expansion joint types are available in both circular and rectangular models

1. SINGLE EXPANSION JOINT. The simplest form of expansion joint; of single bellows construction, it absorbs all of the movement of the pipe section into which it is installed.
2. DOUBLE EXPANSION JOINT. A double expansion joint consists of two bellows jointed by a common connector which is anchored to some rigid part of the installation by means of an anchor base. The anchor base may be attached to the common connector either at installation or time of manufacturer. Each bellows of a double expansion joint functions independently as a single unit. Double bellows expansion joints should not be confused with universal expansion joints.
3. UNIVERSAL EXPANSION JOINT. A universal expansion joint is one containing two bellows joined by a common connector for the purpose of absorbing any combination of the three (3) basic movements. (Section IV-1.4.5) A universal expansion joint is used in cases to accommodate greater amounts of lateral movement than can be absorbed by a single expansion joint.

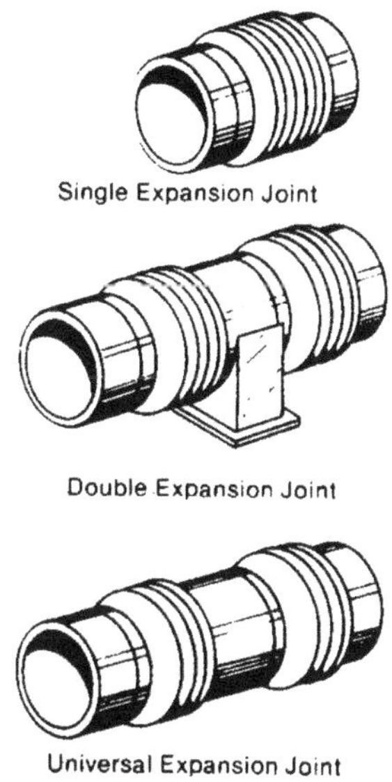

FIG. IV-1.8
EXPANSION JOINT TYPES

4. UNIVERSAL TIED EXPANSION JOINTS. Tied universal expansion joints are used when it is necessary for the assembly to eliminate pressure thrust forces from the piping system. In this case the expansion joint will absorb lateral movement and will not absorb any axial movement external to the tied length.

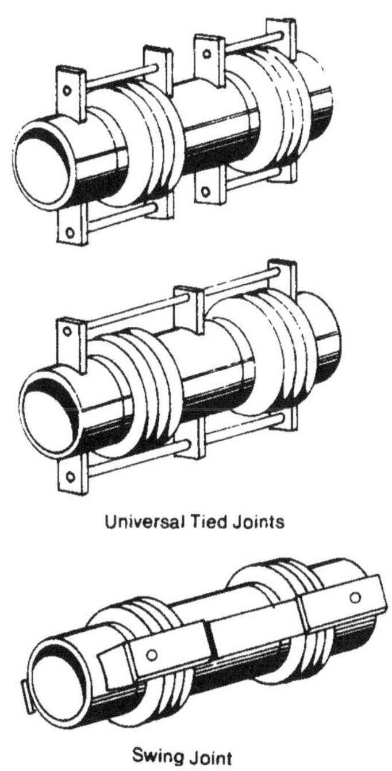

FIG. IV-1.8.1
EXPANSION JOINT TYPES (CONT.)

Hinged Expansion Joint

Gimbal Expansion Joint

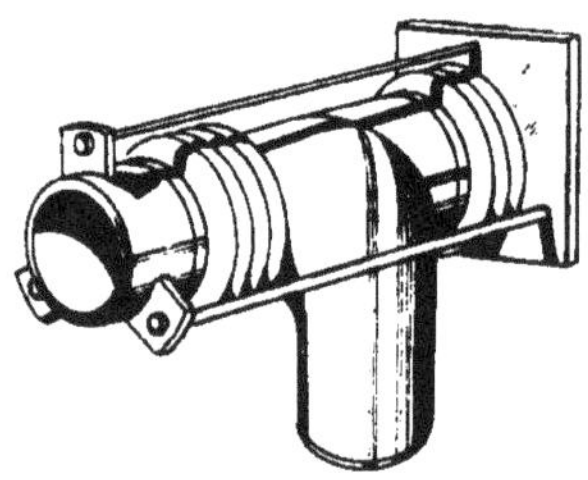

Pressure-Balanced Expansion Joint

FIG. IV-1.8.2
EXPANSION JOINT TYPES (CONT.)

5. SWING EXPANSION JOINT. A swing expansion joint is designed to absorb lateral deflection and/ or angular rotation in one plane only by the use of swing bars, each of which is pinned at or near the ends of the unit.
6. HINGED EXPANSION JOINT. A hinged expansion joint contains one bellows and is designed to permit angular rotation in one plane only by the use of a pair of pins running through plates attached to the expansion joint ends. Hinged expansion joints should be used in sets of 2 or 3 to function properly.
7. GIMBAL EXPANSION JOINT. A gimbal expansion joint is designed to permit angular rotation in any plane by the use of 2 pairs of hinges affixed to a common floating gimbal ring.

   NOTE: Expansion joints Types 4, 5, 6, and 7 are normally used for restraint of pressure thrust forces. (Section IV-2.1.3.2)
8. PRESSURE-BALANCED EXPANSION JOINT. A pressure-balanced expansion joint is designed to absorb axial movement and/or lateral deflection while restraining the bellows pressure thrust force (Section IV-2.1.3.2) by means of the devices interconnecting the flow bellows with an opposed bellows also subjected to line pressure. This type of joint is installed where a change of direction occurs in a run of pipe.

IV-2 USING EXPANSION JOINTS

IV-2.1 SYSTEM PREPARATION

When thermal growth of a piping system has been determined to be a problem, and the use of an expansion joint is indicated, the most effective expansion joint should be selected as a solution. When using expansion joints it should be kept in mind that movements are not eliminated, but are only directed to certain points where they can best be absorbed by the expansion joint. The methods used and the steps that must be taken in preparing a piping system for the introduction of an expansion joint are covered in this section. As will be seen, many factors must be considered to insure the proper functioning of an expansion joint.

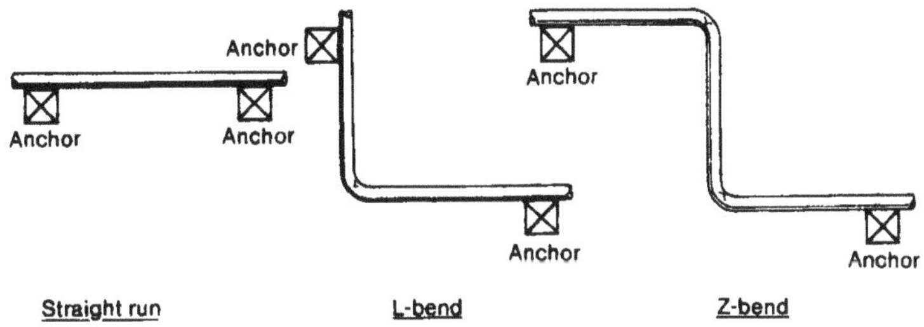

FIG. IV-2.1
STANDARD PIPE SECTIONS

IV-2.1.1 Simplify the System.

The first step in the selection of expansion joints is to choose the tentative location of the pipe anchors. The purpose of these anchors is to divide a piping system into simple, individual expanding sections. Since thermal growth cannot be restrained, it becomes the function of the pipe anchors to limit and control the amount of movement that the expansion joints, installed between them, must absorb. It is generally advisable to begin with the assumption that using single or double expansion joints in straight axial movement will provide the most simple and economical layout. *Never install more than one 'single' type joint between any two anchors in a straight run of pipe.*

In general, piping systems should be anchored so that a complex system is divided into several sections which conform to one of the configurations shown in Fig. IV-2.1.

Major pieces of equipment such as turbines, pumps, compressors, etc. may function as anchors provided they are designed for this use. Fig. IV-2.2 illustrates a hypothetical piping system in which expansion joints are to be installed. The following sections will describe expansion joint application methods as applied to this system.

IV-2.1.2 Calculating Thermal Growth.

Step 1 in determining thermal movement is to choose tentative anchor positions. The goal here is to divide a system into several straight-line sections exhibiting only axial movement. Referring to Fig. IV-2.3 demonstrates this method.

The pump and two tanks are logical anchor positions. The addition of anchors at pts. B and C result in the desired series of straight-line piping legs. The next step is to determine the actual change in length of each of the piping legs due to changes in temperature.

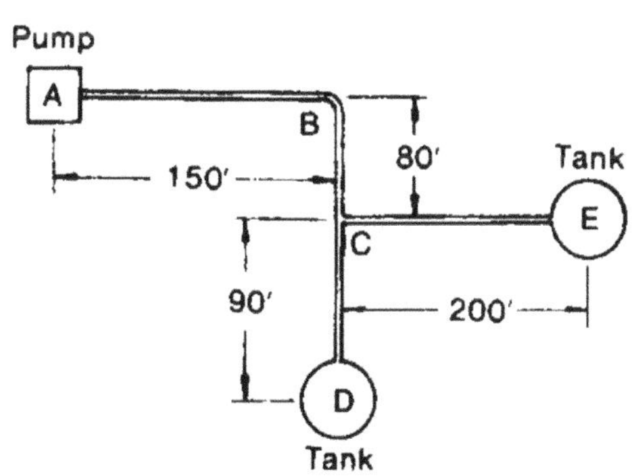

FIG. IV-2.2
TYPICAL PIPING LAYOUT

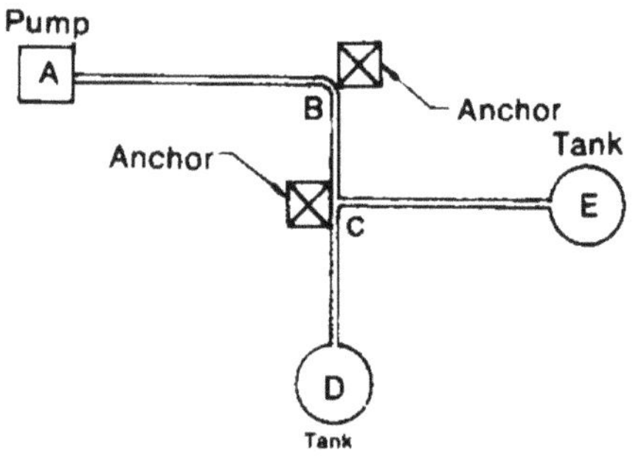

FIG. IV-2.3
ANCHOR POSITIONS

In determining thermal movement, all potential sources of temperature must be considered. Usually, the temperature of the flow medium is the major source of dimensional changes, but in certain extreme cases, ambient temperature outside the system can contribute to thermal movement. Movements caused by other sources such as mechanical movements or wind loading, must also be considered.

IV-2.1.3 Pipe Anchors and Forces.

Pipe anchors, their attachments, and the structures to which they are attached, must be designed to withstand all of the forces acting upon them. In addition to the normal pipe system forces, anchors in systems which contain expansion joints are subjected to additional forces. Two significant forces which are unique to expansion joint systems are spring force and pressure thrust force.

IV-2.1.3.1 Spring Force Spring force is the force required to deflect an expansion joint a specific amount. It is based on operating conditions and manufacturing methods and materials. Spring force imparts a resisting force throughout the system much as a spring would when compressed or otherwise deflected. (Fig. IV-2.4) In order for an expansion joint to operate properly, this spring force must be restrained by anchors.

The magnitude of spring force is determined by the expansion joint's spring rate, and by the amount of movement to which the expansion joint is subjected. The spring rate varies for each expansion joint and is based upon the physical dimensions and material of a specific expansion joint.

A simple example will illustrate the magnitude of spring force for an expansion joint installed in a 24 inch diameter pipe system. The axial spring rate is given as: 1568 lbs./in. If the expansion joint deflects ½ in. axially, the spring force is:

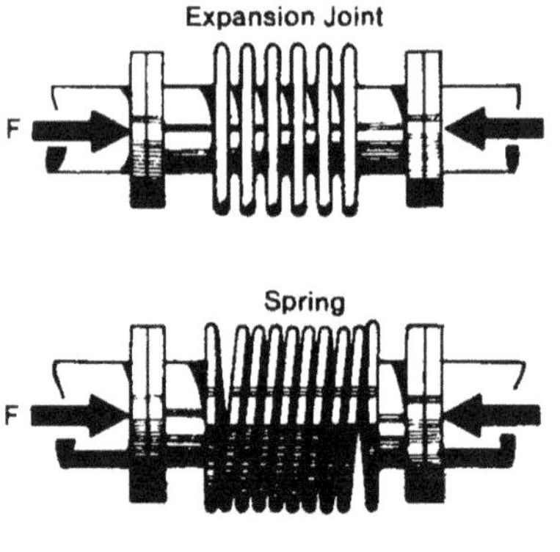

FIG. IV-2.4
SPRING FORCE

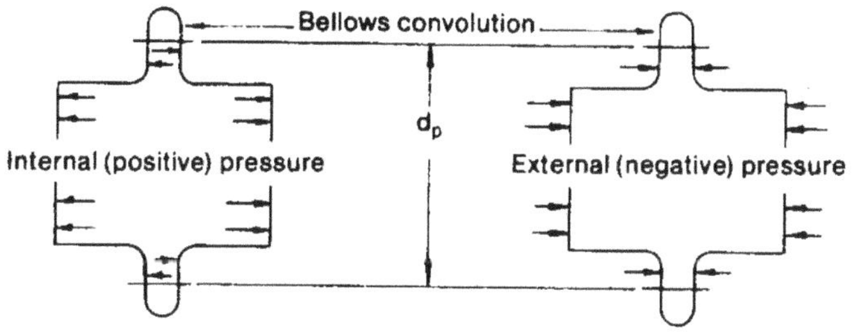

FIG. IV-2.5
PRESSURE THRUST

IV-2.1.3.2 Pressure Thrust Force Pressure thrust force is often misunderstood. It is a condition created by the installation of a flexible unit, such as an expansion joint, into a rigid piping system which is under pressure. Pressure thrust force is a function of the system pressure and mean diameter (Fig. IV-2.5) of the bellows. Mean diameter is determined by the height, or span, of the bellows and can vary from unit to unit. Mean diameter (d_p) is usually greater than the pipe diameter. Fig IV-2.5 illustrates the effect of pressure on a bellows convolution. In cases of internal, or positive pressure, the convolutions are pushed apart, causing the bellows to extend or increase in length while the opposite is observed in cases of external or negative pressure. The force required to maintain the bellows at it's proper length is equal to this pressure thrust and can be significantly higher than all other system forces combined. See Fig. (IV-2.6) for further details of pressure thrust effects.

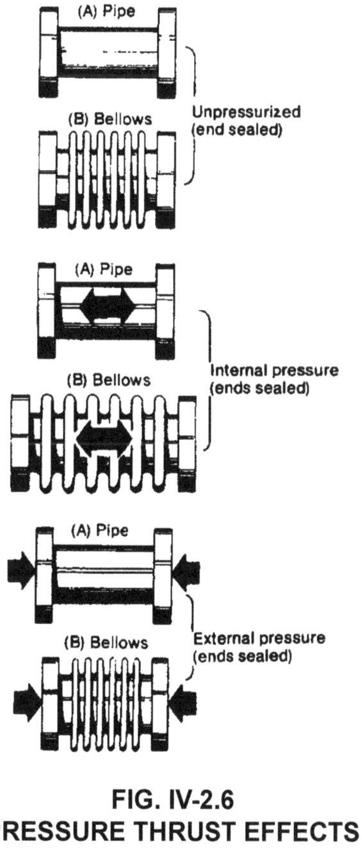

FIG. IV-2.6
PRESSURE THRUST EFFECTS

IV-2.1.3.3 Determining Pressure Thrust Force The magnitude of pressure thrust force (Fs) in lbs., is determined by the following equation:

$$Fs = Pa$$

where

(P) is the pressure (psig) and
(a) is the effective area of the expansion joint (in2)

The following example will illustrate the effect of pressure thrust. Suppose that an expansion joint is to be installed in a 24 in. diameter pipe system which operates a 150 psig pressure. From the manufacturer of the expansion joint the effective area is found to be 560 in^2.

$$a = \frac{\pi (d_p)^2}{4}$$

The pressure thrust force (Fs) is found to be:

$$Fs = Pa \text{ or } Fs = (150)(560) = 84{,}000 \text{ lbs.}$$

Note that when determining pressure thrust forces, the pressure value used in Fs = Pa should equal the maximum anticipated pressure that the system is likely to experience. For this reason, any pressure test must be considered. Be careful not to specify unrealistic safety factors which could lead to over-design of the anchors at additional cost.

The force is transmitted from the ends of the expansion joint along the pipe. Referring to Fig. IV-2.6, a comparison of the effect of pressure thrust is shown between a section of rigid pipe (A) and a flexible bellows (B).

In an unpressurized state, there is no pressure thrust force transmitted by A or B and their lengths remain the same.

Under internal pressure, a force is acting upon both A and B that is equal to the pressure times the effective areas of A & B. The force of this thrust is totally absorbed by the rigid walls of the pipe (A) and its length remains unchanged. The length of (B) will tend to increase under internal pressure since the flexible expansion joint is unable to resist the force. It therefore yields, or stretches.

In cases of external pressure or vacuum conditions, the effect of internal pressure is reversed and the expansion joint will tend to compress.

To stop the expansion joint from extending or compressing due to pressure thrust, main anchors must be located at some point on each side of the joint to withstand the forces of pressure thrust and keep the expansion joint at its proper length. Tie rods, hinges and/or gimbals may also be used to restrain pressure thrust force by attaching the ends of the expansion joint to each other. In this case, these restraints prevent the bellows from absorbing any axial movement external to the expansion joint. Pressure-balanced expansion joints restrain pressure thrust and absorb axial movement external to the joint.

As can be seen, the forces exerted on a piping system due to the presence of expansion joints can be quite significant. In the preceding example the spring force was found to be 784 lbs., while the pressure thrust force was equal to 84,000 lbs.—a combined force of 84,784 lbs! One can see that proper anchoring of a piping system containing expansion joints is vital.

IV-2.1.3.4 Main Anchors A main anchor must be designed to withstand the forces and moments imposed upon it by each of the pipe sections to which it is attached. In the case of a pipe section containing one or more unrestrained expansion joints, these will consist of the full bellows thrust due to pressure, media flow, the forces and/or moments required to deflect the expansion joint or joints their full rated movement, and the frictional forces caused by the pipe guides, directional anchors and supports. In certain applications it may be necessary to consider the weight of the pipe and its contents and any other forces and/or moments resulting from wind loading etc. In systems containing expansion joints, main anchors are installed at any of the following locations:

(A) At a change of direction of flow:

(B) Between two expansion joints of different sizes installed in the same straight run:

(C) At the entrance of a side branch containing an expansion joint into the main line:

(D) Where a shut-off or pressure reducing valve is installed in a pipe run between two expansion joints:

(E) At a blind end of pipe:

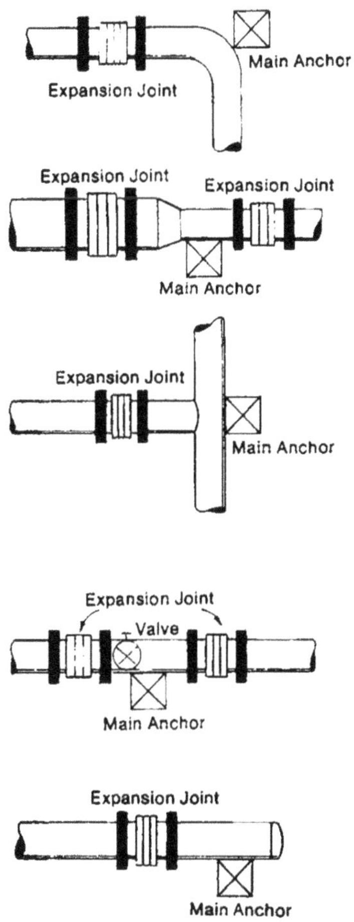

FIG. IV-2.7

IV-2.1.3.5 Intermediate Anchors *Intermediate anchors are not intended to withstand pressure thrust force.* This force is absorbed by main anchors, by devices on the expansion joint such as tie rods, swing bars, hinges, gimbals, etc., or, as in the case of a pressure-balanced or double expansion joint, is balanced by an equal pressure force acting in the opposite direction. An intermediate anchor must withstand all of the non-pressure forces acting upon it by each of the pipe sections to which it is attached. In the case of a pipe section containing an expansion joint, these forces will consist of the force required to move the expansion joint and the frictional forces caused by the pipe guides.

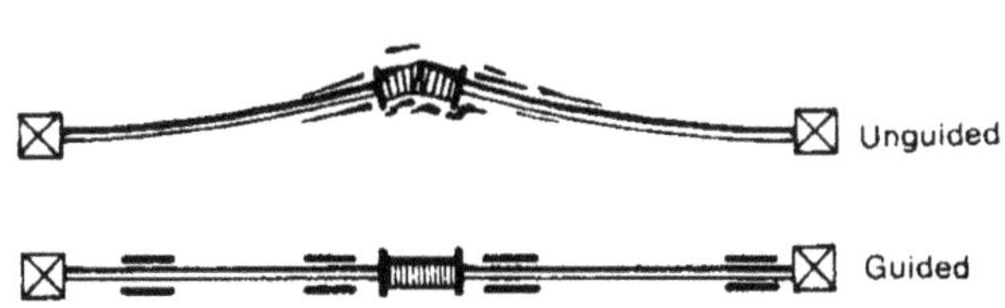

FIG. IV-2.8.1
PIPE GUIDES AND STABILITY

IV-2.1.4 Pipe Guides and Supports.

Correct alignment of the pipe adjoining an expansion joint is of vital importance to its proper function. Although expansion joints are designed and built for a long and satisfactory life, maximum service will be obtained only when the pipeline has the recommended number of guides and is anchored and supported in accordance with good piping practice. Proper supporting of the pipeline is required not only to support the pipe itself, but also to provide support at each end of the expansion joint. Pipe guides are necessary to insure proper alignment of movement to the expansion joint and to prevent buckling of the line. Buckling is caused by a combination of the expansion joint flexibility and the internal pressure loading on the pipe which causes it to act like a column loaded by the pressure thrust of the expansion joint. (Fig. IV-2.8.1).

IV-2.1.4.1 Pipe Guide Application When locating pipe guides for applications involving axial movement only, it is generally recommended that the expansion joint be located near an anchor, and that the first guide be located a maximum of 4 pipe diameters away from the expansion joint. This arrangement will provide proper movement guiding as well as proper support for each end of the expansion joint. The distance between the first and second guide should not exceed 14 pipe diameters, and the distance between the remaining anchors should be determined from paragraph IV-2.2.4.2. of the EJMA Standards.

Common types of pipe guides are shown in Fig. IV-2.9. For systems subjected to lateral motion or angular rotation, directional anchors or planar guides may be required. Examples of these guides are shown in Section IV-2.2.

The recommendations given for pipe anchors and guides represent the minimum requirements for controlling pipelines which contain expansion joints and are intended to protect the expansion joint and pipe system from abuse and failure. However, additional pipe supports are often required between guides in accordance with standard piping practice.

IV-2.1.4.2 Pipe Support Application A pipe support is any device which permits free movement of the piping while carrying the dead weight of the pipe and any valves or attachments. Pipe supports must also be capable of carrying the live weight. Pipe supports cannot be substituted for pipe alignment or planar guides. Pipe rings, U-bolts, roller supports, spring hanger etc., are examples of conventional pipe support devices.

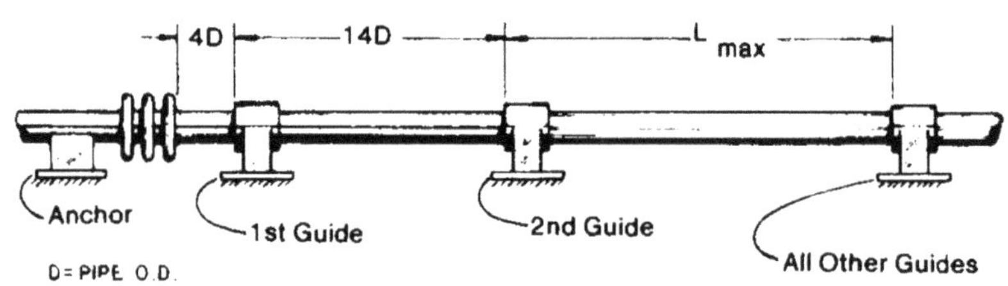

FIG. IV-2.8.2
PROPER PIPE GUIDE LOCATION

IV-2.2 EXPANSION JOINT APPLICATIONS

The following examples demonstrate common types of expansion joint applications, some dealing with simple axial movement and others with more complex and specialized applications.

IV-2.2.1 Axial Movement.

Figure IV-2.10 typifies good practice in the use of a single expansion joint to absorb axial pipeline expansion. Note the use of just one expansion joint between anchors, the nearness of the joint to an anchor and the spacing of the pipe guides.

Standard Pipe Alignment Guide

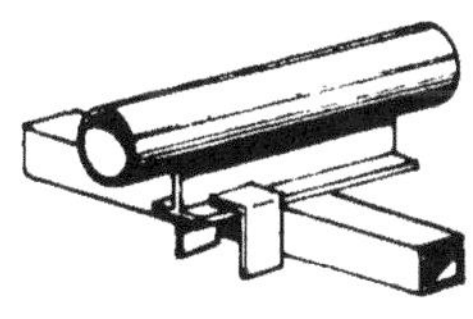

Tee Guide

Strap Guide

FIG. IV-2.9
COMMON PIPE GUIDES

Fig. IV-2.11 typifies good practice in the use of a double expansion joint to absorb axial pipeline expansion. Note the addition of the intermediate anchor which, in conjunction with the two main anchors effectively divides this section into two distinct expanding segments, so that there is only one expansion joint between anchors.

Fig. IV-2.12 typifies good practice in the use of a pressure-balanced expansion joint (Section IV-1.6, Paragraph 8) to absorb axial movement. Note that the expansion joint is located at a change of direction of the piping and that the elbow and the end of the pipe are secured by intermediate anchors. Since the pressure thrust is absorbed by the expansion joint itself, and only the force required to deflect the expansion joint is imposed on the piping, a minimum of guiding is required. Guiding near the expansion joint as shown will frequently suffice. In long, small-diameter pipelines, additional guiding may be necessary.

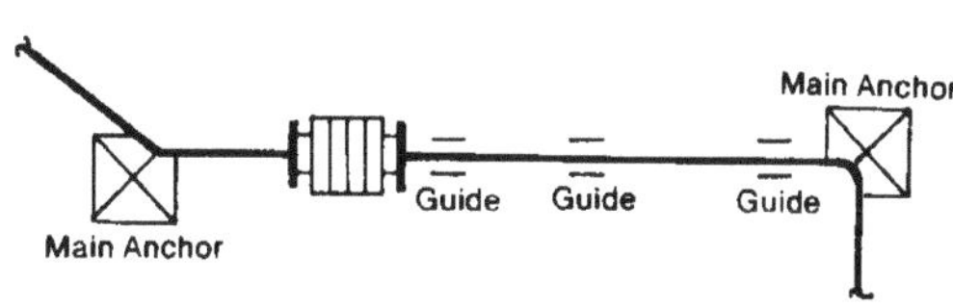

FIG. IV-2.10
SINGLE EXPANSION JOINT APPLICATION

FIG. IV-2.11
DOUBLE EXPANSION JOINT APPLICATION

IV-2.2.2 Combined Movements, Lateral Deflection, and Angular Rotation.

Because it offers the lowest expansion joint cost, the single expansion joint is usually considered first for any application. Fig. IV-2.13 shows a typical application of a single expansion joint used to absorb lateral deflection, as well as axial compression. The joint is located near the end of the long pipe run with main anchors at each end and guides properly spaced for both movement control and protection of the piping against buckling. In this case, as opposed to the similar configuration in Fig. IV-2.10, the anchor at left is a directional main anchor which, while absorbing the main anchor loading in the direction of the expansion joint axis, permits the thermal expansion of the short leg to act upon the expansion joint as lateral deflection. Because the main anchor loading (pressure thrust) exists only in the leg containing the expansion joint, the anchor at the end of the short piping leg is an intermediate anchor.

Figure IV-2.14 represents modifications of Figure IV-2.10 in which the main anchors at either end of the expansion joint are replaced by tie rods. Where the piping configuration permits, the use of the tie rods adjusted to prevent axial movement frequently simplifies and reduces the cost of the installation. Because of these tie rods, the expansion joint is not capable of absorbing any axial movement other than its own thermal expansion (See Section IV-2.1.3.3). The thermal expansion in the short leg is imposed as deflection on the long piping leg.

Fig. IV-2.15 shows a tied universal expansion joint used to absorb lateral deflection in a single plane 'Z' bend. The thermal movement of the horizontal lines is absorbed as lateral deflection by the expansion joint. Both anchors need only to be intermediate anchors since the pressure thrust is restrained by the tie rods. Where possible, the expansion joint should fill the entire offset leg so that its expansion is absorbed within the tie rods as axial movement, and bending in the horizontal lines is minimized.

Figure IV-2.16 shows a typical application of a tied universal expansion joint in a 3-plane 'Z' bend. Since the universal expansion joint can absorb lateral deflection in any direction, the two horizontal legs may lie at any angle in the horizontal plane.

Figure IV-2.17 shows a case of pure angular rotation using a pair of hinge joints in a single-plane 'Z' bend. Since the pressure thrust is restrained by the hinges, only intermediate anchors are required. The axial expansion of the pipe leg between the expansion joints is imposed on the two horizontal legs if there is sufficient flexibility.

If not, a 3 hinge system such as in Fig. IV-2.18 may be required. The thermal expansion of the offset leg is absorbed by joints (A), (B), & (C). Expansion joint (B) must be capable of absorbing the total of the rotation of (A) & (C). Hence, it is frequently necessary that the center expansion joint contain a greater number of convolutions than those at either end.

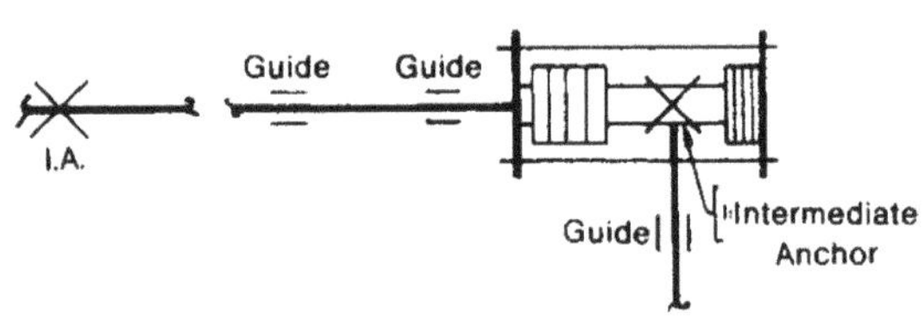

FIG. IV-2.12
PRESSURE-BALANCED APPLICATION

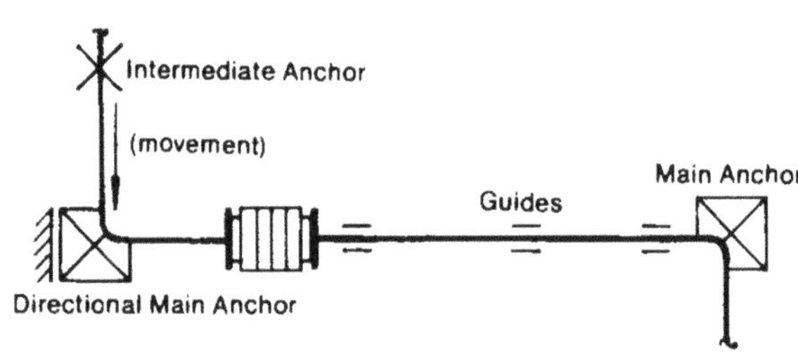

FIG. IV-2.13
COMBINED AXIAL AND LATERAL MOVEMENT

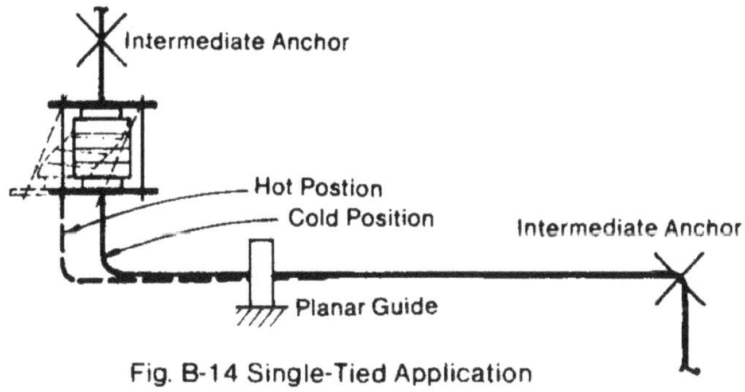

FIG. IV-2.14
SINGLE-TIED APPLICATION

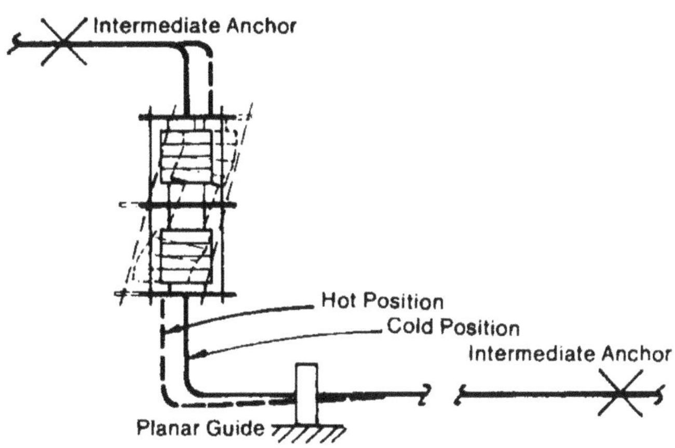

FIG. IV-2.15
UNIVERSAL-TIED APPLICATION

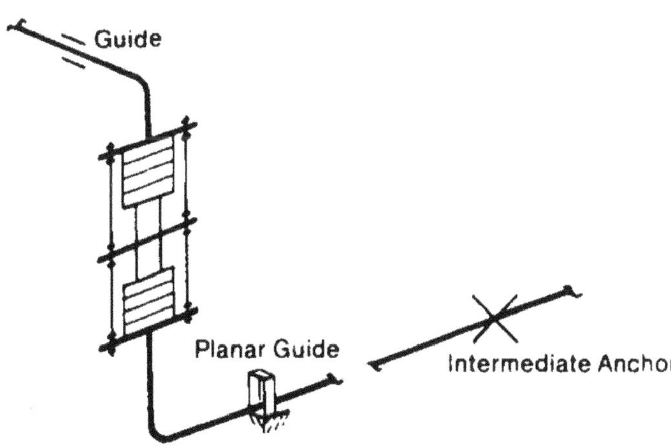

FIG. IV-2.16
UNIVERSAL-TIED APPLICATION

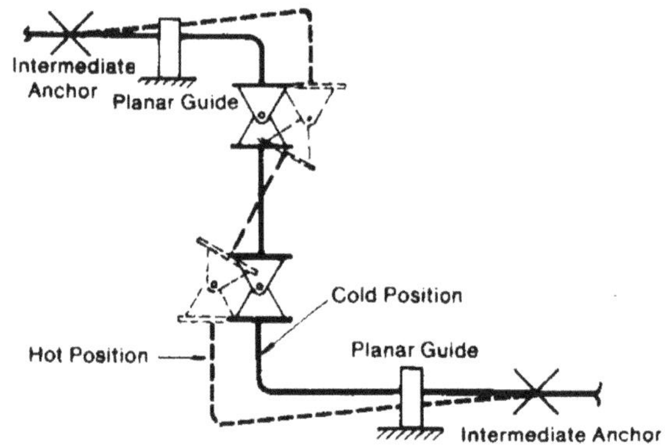

FIG. IV-2.17
TWO-HINGE APPLICATION

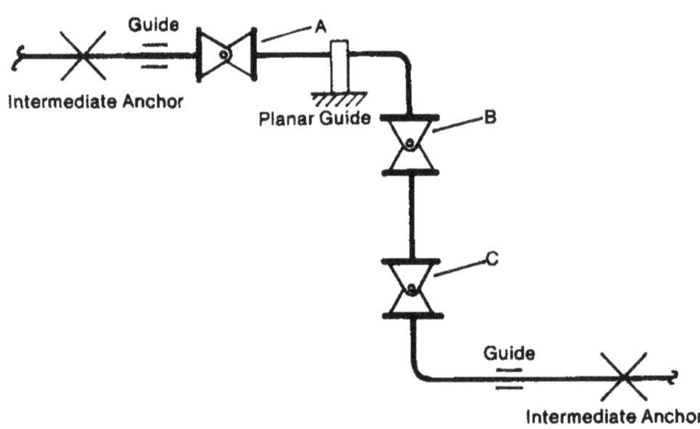

FIG. IV-2.18
THREE-HINGE APPLICATION

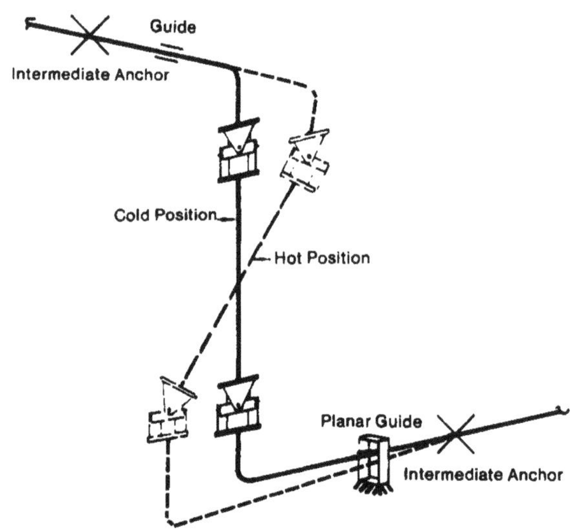

FIG. IV-2.19
TWO-GIMBAL APPLICATION

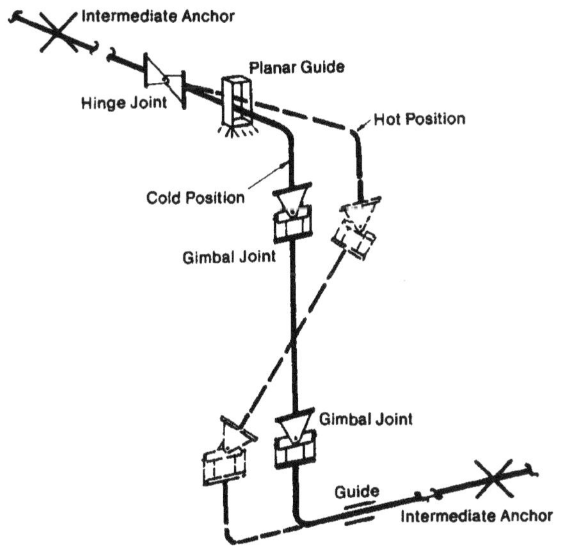

FIG. IV-2.20
ONE HINGE/TWO GIMBAL APPLICATION

Just as hinged expansion joints may offer great advantages in single-plane applications, gimbal expansion joints are designed to offer similar advantages in multi-plane systems. The ability of gimbal expansion joints to absorb angular rotation in any plane is frequently applied by utilizing two such units to absorb lateral deflection.

Figure IV-2.19 shows a case of pure angular rotation using a pair of gimbal expansion joints in a multi-plane 'Z' bend. Since the pressure thrust is restrained by the gimbal assemblies, only intermediate anchors are required.

In applications where the flexibility of the two horizontal piping legs is not sufficient to absorb the axial growth of the pipe section between the two gimbal joints in Figure IV-2.19, the addition of a hinge joint, as shown in Figure IV-2.20, is indicated. Since the expansion of the offset leg takes place in only one plane, the use of the simpler hinge expansion joint is justified.

IV-3 EXPANSION JOINT HANDLING, INSTALLATION, AND SAFETY RECOMMENDATIONS

IV-3.1 SHIPPING AND HANDLING

Responsible manufacturers of expansion joints take every design and manufacturing precaution to assure the user of a reliable product. The installer and the user have the responsibility to handle, store, install, and apply the expansion joints in a way which will not impair the quality built into them.

IV-3.1.1 Shipping Devices.

All manufacturers should provide some way of maintaining the face-to-face dimension of an expansion joint during shipment and installation. These usually consist of overall bars or angles welded to the flanges or weld ends at the extremities of the expansion joint. Washers or wooden blocks between equalizing rings are also used for this purpose. (Fig. IV-3.1) Do not remove these shipping devices until all expansion joints, anchors, and guides in the system have been installed. Shipping devices manufactured by members of the Expansion Joint

Manufacturers Association, Inc., are usually painted yellow or otherwise distinctively marked as an aid to the installer. The shipping devices must be removed prior to start-up or testing of the system. (Ref. Section IV-3.4)

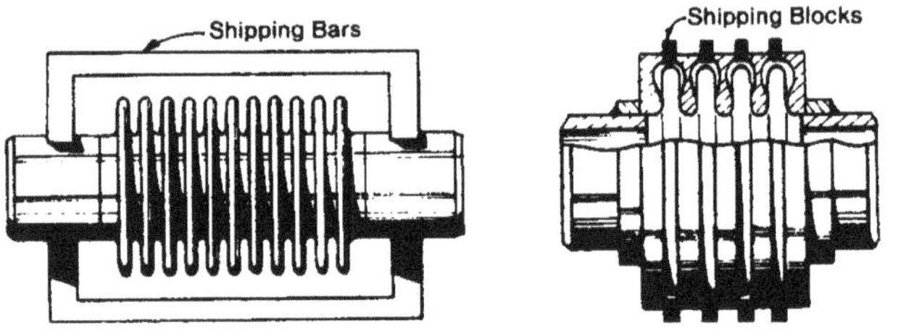

FIG. IV-3.1
SHIPPING DEVICES MAINTAIN OVERALL LENGTH DURING SHIPMENT AND INSTALLATION

IV-3.1.2 Storage.

Some conditions of outdoor storage may be detrimental, and where possible, should be avoided; preferably, the storage should be in a cool, dry area. Where this cannot be accomplished, the expansion joint manufacturer should be so advised either through the specifications or purchase contract. Care must be taken to avoid mechanical damage such as caused by stacking, bumping or dropping. For this reason, it is strongly suggested that covers be specified on all expansion joints to protect the bellows. Certain industrial and natural atmospheres can be detrimental to some bellows materials. If expansion joints are to be stored or installed in such atmospheric environments, materials should be specified which are compatible with these environments.

IV-3.1.3 Installation Instructions.

Expansion joints are shipped with documents which furnish the installer with instructions covering the installation of the particular expansion joint. These documents should be left with the expansion joint until installation is completed.

IV-3.2 INSTALLATION

Metal bellows-type expansion joints have been designed to absorb a specified amount of movement by flexing of the thin-gage bellows. If proper care is not exercised in the installation of the expansion joint, cycle life and pressure capacity could be reduced, leading to premature failure and damage to the piping system.

It is important that the expansion joint be installed at the length specified by the manufacturer. They should never be extended or compressed in order to make up for deficiencies in length, nor should they be offset to accommodate misaligned pipe.

Remember that a bellows is designed to absorb motion by flexing. The bellows, therefore, must be sufficiently thick to withstand the design pressure, while being thin enough to absorb the required flexing. Optimum design will always require a bellows to be of thinner material than virtually any other component in the piping system in which it is installed. The installer must recognize this relative fragility of the bellows and take every possible measure to protect it during installation. Avoid denting, weld spatter, arc strikes or the possibility of allowing foreign matter to interfere with the proper flexing of the bellows. It is highly recommended that a cover be specified for every expansion joint. The small cost of a cover is easily justified when compared to the cost of replacing a damaged bellows element. With reasonable care during storage, handling, and installation, the user will be assured of the reliability designed and built into the expansion joint.

IV-3.3 DO'S AND DON'T'S - INSTALLATION AND HANDLING

The following recommendations are included to avoid the most common errors that occur during installation. When in doubt about an installation procedure, contact the manufacturer for clarification before attempting to install the expansion joint.

DO'S...

Do . . . Inspect for damage during shipment such as: dents, broken hardware, water marks on carton, etc.
Do . . . Store in a clean, dry area where it will not be exposed to heavy traffic or damaging environment.
Do . . . Use only designated lifting lugs when provided.
Do . . . Make the piping system fit the expansion joint. By stretching, compressing, or offsetting the joint to fit the piping, the expansion joint may be overstressed when the system is in service.
Do . . . Leave one flange loose on the adjacent piping when possible, until the expansion joint has been fitted into position. Make necessary adjustments of this loose flange before welding.
Do . . . Install the joint with the arrow pointing in the direction of the flow.
Do . . . Install single vanstone liners pointing in the direction of flow. Be sure also to install a gasket between a vanstone liner and flange.
Do . . . In case of telescoping liner, install the smallest I.D. liner pointing in the direction of flow.
Do . . . Remove all shipping devices after the installation is complete and before any pressure test of the fully installed system.
Do . . . Remove any foreign material that may have become lodged between the convolutions.
Do . . . Refer to the proper guide spacing and anchoring recommendations in Section IV-2.1.3 & IV-2.1.4.

DON'T'S . . .

Don't . . . Drop or strike expansion joint.
Don't . . . Remove the shipping bars until the installation is complete.
Don't . . . Remove any moisture-absorbing desiccant bags or protective coatings until ready for installation.
Don't . . . Use hanger lugs or shipping bars as lifting lugs.
Don't . . . Use chains or any lifting device directly on the bellows or bellows cover. Don't. . . Allow weld spatter to hit unprotected bellows.
Don't . . . Use cleaning agents which contain chlorides. Don't. . . Use steel wool or wire brushes on bellows.
Don't . . . Force or rotate one end of an expansion joint for alignment of bolt holes. Bellows are not ordinarily capable of absorbing torsion.
Don't . . . Hydrostatic pressure test or evacuate the system before proper installation of all guides and anchors.
Don't . . . Use shipping bars to restrain the pressure thrust during testing.
Don't . . . Use pipe hangers as guides.
Don't . . . Exceed the manufacturers rated test pressure of the expansion joint.

Caution: The manufacturers warranty may be void if improper installation procedures have been used.

IV-3.4 SAFETY RECOMMENDATIONS

This section was prepared in order to better inform the user of those factors which many years of experience have shown to be essential for the successful installation and performance of piping systems containing bellows expansion joints.

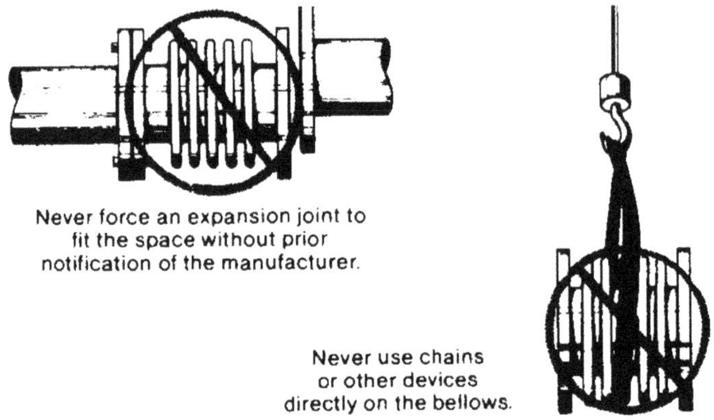

FIG. IV-3.2

IV-3.4.1 Inspection Prior to Start-up or Pressure Test.

Expansion joints are usually considered to be non-repairable items and generally do not fall into the category for which maintenance procedures are required. However, immediately after the installation is complete a careful visual inspection should be made of the entire piping system to insure that there is no evidence of damage, with particular emphasis on the following:

1. Are anchors, guides, and supports installed in accordance with the system drawings?
2. Is the proper expansion joint in the proper location?
3. Are the expansion joint's flow direction and pre-positioning correct?
4. Have all of the expansion joint shipping devices been removed?
5. If the system has been designed for a gas, and is to be tested with water, has provision been made for proper support of the additional dead weight load on the piping and expansion joint? Some water may remain in the bellows convolutions after the test. If this is detrimental to the bellows or system operation, means should be provided to remove this water.
6. Are all guides, pipe supports and the expansion joints free to permit pipe movement?
7. Has any expansion joint been damaged during handling and installation?
8. Is any expansion joint misaligned? This can be determined by measuring the joint overall length, inspection of the convolution geometry, and checking clearances at critical points on the expansion joint and at other points in the system.
9. Are the bellows and other movable portions of the expansion joint free of foreign material?

IV-3.4.2 Inspection During and Immediately after Pressure Test - Warning:

Extreme care must be exercised while inspecting any pressurized system or component.
A visual inspection of the system should include the following:

1. Evidence of leaking or loss of pressure.
2. Distortion or yielding of anchors, expansion joint hardware, the bellows and other piping components.
3. Any unanticipated movement of the piping due to pressure.
4. Evidence of instability in the bellows.
5. The guides, expansion joints, and other movable parts of the system should be inspected for evidence of binding.
6. Any evidence of abnormality or damage should be reviewed and evaluated by competent design authority.

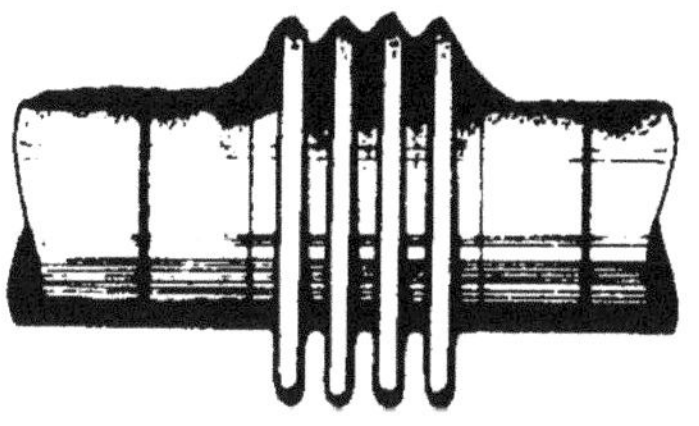

FIG. IV-3.3

IV-3.4.3 Periodic In service Inspection Warning:

Extreme care must be exercised while inspection any pressurized system or component.

1. Immediately after placing the system in operation, a visual inspection should be conducted to insure that the thermal expansion is being absorbed by the expansion joints in the manner for which they were designed.
2. The bellows should be inspected for evidence of unanticipated vibration.
3. A program of periodic inspection should be planned and conducted throughout the operating life of the system. The frequency of these inspections should be determined by the service and environmental conditions involved. Such inspections can spot the more obvious potential problems such as external corrosion, loosening of threaded fasteners, and deterioration of anchors, guides and other hardware.

IT MUST BE UNDERSTOOD THAT THIS INSPECTION PROGRAM, WITHOUT ANY OTHER BACKUP INFORMATION, CANNOT GIVE EVIDENCE OF DAMAGE DUE TO FATIGUE, STRESS CORROSION OR GENERAL INTERNAL CORROSION. THESE CAN BE THE CAUSE OF SUDDEN FAILURES AND GENERALLY OCCUR WITHOUT ANY VISIBLE OR AUDIBLE WARNING.

4. When any inspection reveals evidence of malfunction, damage, or deterioration, this should be reviewed by competent design authority for resolution. Additionally, any changes in the system operating conditions such as pressure, temperature, movement, flow, velocity, etc. that may adversely affect the expansion joint should be reported to and evaluated by a competent design authority.

IV-3.5 KEY TO SYMBOLS USED

IV-3.6 SPECIFICATION AND ORDERING GUIDE

This guide has emphasized that the most reliable and safe bellows expansion joint installations are the result of a high degree of understanding between the user and the manufacturer. The prospect for successful long-life performance of piping systems containing expansion joints begins with the submittal of accurate data by the system designer.

When preparing specifications for expansion joints, it is imperative that the system designer completely review the piping system layout and provide the necessary data relating to the selection of an expansion joint: flow medium, pressure, temperature, movements, etc.

To aid the user in the specification and ordering of expansion joints, the Expansion Joint Manufacturers Association has provided the following expansion joint specification sheets. Complete data, when submitted in accordance with these specifications, will help in obtaining an optimum design. Other data that must be considered but which is not covered, may be included in the "remarks" section of the specification sheets.

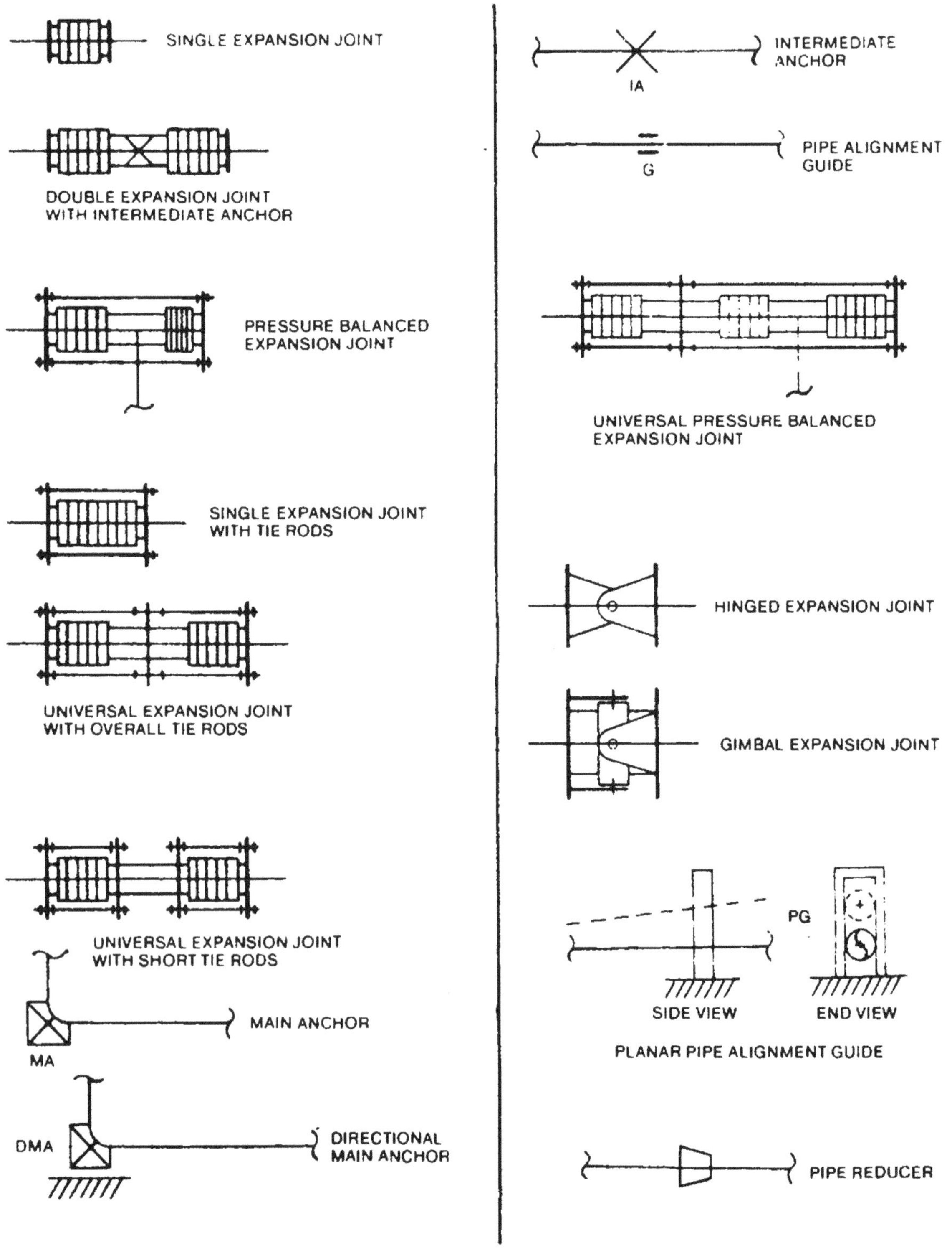
SINGLE EXPANSION JOINT
DOUBLE EXPANSION JOINT WITH INTERMEDIATE ANCHOR
PRESSURE BALANCED EXPANSION JOINT
SINGLE EXPANSION JOINT WITH TIE RODS
UNIVERSAL EXPANSION JOINT WITH OVERALL TIE RODS
UNIVERSAL EXPANSION JOINT WITH SHORT TIE RODS
MAIN ANCHOR
MA
DMA
DIRECTIONAL MAIN ANCHOR
IA
INTERMEDIATE ANCHOR
G
PIPE ALIGNMENT GUIDE
UNIVERSAL PRESSURE BALANCED EXPANSION JOINT
HINGED EXPANSION JOINT
GIMBAL EXPANSION JOINT
PG
SIDE VIEW
END VIEW
PLANAR PIPE ALIGNMENT GUIDE
PIPE REDUCER

APPENDIX A-1

STANDARD EXPANSION JOINT SPECIFICATION SHEET

COMPANY:	DATE / /
	SHEET OF
PROJECT:	INQUIRY NO.
	JOB NO.

	ITEM NO./EJ TAG NO.		EJMA PAGE REFERENCE			
1	QUANTITY					
2	NOMINAL SIZE/I.D./O.D. (IN.)					
3	EXPANSION JOINT TYPE		1			
4a	FLUID INFORMATION	MEDIUM GAS/LIQUID	5, 6, 147			
4b		VELOCITY (FT./SEC)	77			
4c		FLOW DIRECTION				
5	DESIGN PRESSURE, PSIG		6, 19,			
6	TEST PRESSURE, PSIG		83, 135			
7a	TEMPERATURE	DESIGN (°F)				
7b		MAXIMUM/MINIMUM (°F)	6, 13			
7c		INSTALLATION (°F)				
8a	MAXIMUM INSTALLATION MOVEMENT	AXIAL COMPRESSION (IN.)				
8b		AXIAL EXTENSION (IN.)	6, 7, 8,			
8c		LATERAL (IN.)	141			
8d		ANGULAR (DEG.)				
9a	MAXIMUM DESIGN MOVEMENTS	AXIAL COMPRESSION (IN.)				
9b		AXIAL EXTENSION (IN.)				
9c		LATERAL (IN.)	6, 7, 13,			
9d		ANGULAR (DEG.)	47			
9e		NO. OF CYCLES				
10a	OPERATING FLUCTUATIONS	AXIAL COMPRESSION (IN.)				
10b		AXIAL EXTENSION (IN.)				
10c		LATERAL (IN.)	84			
10d		ANGULAR (DEG.)				
10e		NO. OF CYCLES				
11a	MATERIALS OF CONSTRUCTION	BELLOWS	5, 6, 45			
11b		LINERS	77, 78			
11c		COVER	3, 7, 72			
11d		PIPE SPECIFICATION				
11e		FLANGE SPECIFICATION	3, 43			
12	RODS (TIE/LIMIT/CONTROL)		3, 4, 41			
13	PANTOGRAPHIC LINKAGE		4			
14	ANCHOR BASE (MAIN/INTERMEDIATE)		1, 2, 17			
15a	DIMENSIONAL LIMITATIONS	OVERALL LENGTH (IN.)				
15b		OUTSIDE DIAMETER (IN.)				
15c		INSIDE DIAMETER (IN.)				
16a	SPRING RATE LIMITATIONS	AXIAL (LBS./IN.)				
16b		LATERAL (LBS./IN.)	54			
16c		ANGULAR (IN.-LBS./DEG.)				
17	INSTALLATION POSITION HORIZ./VERT.		8, 141			
18a	QUALITY ASSURANCE REQUIREMENTS	BELLOWS WELD NDE — LONG. SEAM	133			
18b		BELLOWS WELD NDE — ATTACH.				
18c		PIPING NDE				
18d		DESIGN CODE REQRD.				
18e		PARTIAL DATA REQRD.				
18f						
18g						
19	VIBRATION AMPLITUDE/FREQUENCY					

SAMPLE

APPENDIX A-2

SUPPLEMENTAL SPECIFICATION SHEET

To be used with Standard Expansion Joint Specification Sheet

Company ______________________ Date ______________________

______________________ Proposal No. ______________________

Project ______________________ Inquiry/Job No. ______________________

______________________ Sheet ________ of ________

		ITEM NO.				
20.		PURGE, INSTRUMENTATION CONNECTION				
21a.	SPECIAL FLANGE DESIGN	FACING				
21b.		O.D. (IN.)				
21c.		I.D. (IN.)				
21d.		THICKNESS (IN.)				
21e.		B.C. DIAMETER (IN.)				
21f.		NO. HOLES				
21g.		SIZE HOLES				
21h.		HOLE ORIENTATION				

ISOMETRIC PIPING SKETCH:

APPENDIX

V

Conversion Factors

CONVERSION FACTORS
(Courtesy of Anvil International)

Multiply	by	To Obtain
Absolute viscosity (poise)	1	Gram/second centimeter
Absolute viscosity (centipoise)	0.01	Poise
Acceleration due to gravity (*g*)	32.174	feet/second^2
	980.6	Centimeters/second^2
Acres	0.4047	Hectares
	10	Square Chains
	43,560	Square Feet
	4047	Square Meters
	0.001562	Square Miles
	4840	Square Yards
	160	Square Rods
Acre-feet	43,560	Cubic Feet
	325,851	Gallons (US)
	1233.49	Cubic Meters
	1,233,490	Liters
Acre-feet/hour	726	Cubic feet/Minute
	5430.86	Gallons Minute
Angstroms	10^{-10}	Meters
Ares	0.01	Hectares
	1076.39	Square Feet
	0.02471	Acres
Atmospheres	76.0	Cms of Hg at 32°F
	29.921	Inches of Hg at 32° F
	33.94	Feet of Water at 62° F
	10,333	Kgs Square meter
	14.6963	Pounds Square inch
	1.058	Tons/Square foot
	1013.15	Millibars
	235.1408	Ounces/Square inch
Bags of cement	94	Pounds of cement
Barrels of oil	42	Gallons of oil (US)
Barrels of cement	376	Pounds of cement
Barrels (not legal)	31	Gallons (US)
or	31.5	Gallons (US)
Board feet	144 x 1 in.*	Cubic inches
Boiler horse power	33,479	BTU/hour
	9.803	Kilowatts
	34.5	pounds of water evaporated/ hour at 212° F
BTU	252.016	Calories (gm)
	0.252	Calories (Kg)
	778.26	Foot pounds
	0.0003927	Horse power hours
	1055.1	Joules
	107.5	Kilogram meters
	0.0002928	Kilowatt hours
BTU/Cu foot	8.89	Calories (Kg)/Cu meter at 32° F
BTU/Hr/ft^2/°F/ft	0.00413	Cal (gm)/Sec/cm^2/°C/cm
	1.49	Cal (Kg)/ Hr/M^2/°C/Meter
BTU/minute	12.96	Foot pounds/second
	0.02356	Horse Power
	0.01757	Kilowatts

* For thickness less than 1 in. use actual thickness in decimals of an inch.

Multiply	by	To Obtain
BTU/minute	17.57	Watts
BTU/pound	0.556	Calories (Kg)/Kilogram
Bushels	2150.4	Cubic inches
	35.24	Liters
	4	Pecks
	32	Quarts (dry)
Cables	120	Fathoms
Calories (gm)	0.003968	BTU
	0.001	Calories (Kg)
	3.088	Foot pounds
	1.558×10^{-6}	Horse power hours
	4.185	Joules
	0.4265	Kilogram meters
	1.1628×10^{-6}	Kilowatt hours
	0.0011628	Watt hours
Cal (gm)/sec/cm^2/°C/cm	242.13	BTU/Hr/ft^2/°F/ft
Calories (Kg)	3.968	BTU
	1000	Calories (gm)
	3088	Foot pounds
	0.001558	Horse power hours
	4185	Joules
	426.5	Kilograms meters
	0.0011628	Kilowatt hours
	1.1628	Watt hours
Calories (Kg) Cu meter	0.1124	BTU/Cu foot at 0° C
Cal (Kg)/Hr/M^2/°C/M	0.671	BTU Hr/ft^2/°F/foot
Calories (Kg)/Kg	1.8	BTU pound
Calories (Kg) minute	51.43	Foot pounds/second
	0.09351	Horse power
	0.06972	Kilowatts
Carats (diamond)	200	Milligram
Centares (Centiares)	1	Square meters
Centigram	0.01	Grams
Centiliters	0.01	Liters
Centimeters	0.3937	Inches
	0.032808	Feet
	0.01	Meters
	10	Millimeters
Centimeters of Hg at 32° F	0.01316	Atmospheres
	0.4461	Feet of water at 62° F
	136	Kgs/Square meter
	27.85	Pounds/Square foot
	0.1934	Pounds/Square inch
Centimeters/second	1.969	Feet/minute
	0.03281	Feet/second
	0.036	Kilometers/hour
	0.6	Meters/minute
	0.02237	Miles/hour
	0.0003728	Miles/minute
Centimeters/second^2	0.03281	Feet/second^2
Centipoise	0.000672	Pounds/sec foot
	2.42	Pounds/hour foot
	0.01	Poise
Chains (Gunter's)	4	Rods
	66	Feet
	100	Links

CONVERSION FACTORS (CONTINUED)
(Courtesy of Anvil International)

Multiply	by	To Obtain
Cheval-vapeur	1	Metric horse power
	75	Kilogram meters/second
	0.98632	Horse power
Circular inches	10^6	Circular mils
	0.7854	Square inches
	785,400	Square mils
Circular mils	0.7854	Square mils
	10^{-6}	Circular inches
	7.854×10^{-5}	Square inches
Cubic centimeters	3.531×10^{-5}	Cubic feet
	0.06102	Cubic inches
	10^{-6}	Cubic meters
	1.308×10^{-6}	Cubic yards
	0.0002642	Gallons (US)
	0.001	Liters
	0.002113	Pints (liq. US)
	0.001057	Quarts (Liq. US)
	0.0391	Ounces (fluid)
Cubic feet	28.320	Cubic centimeters
	1728	Cubic inches
	0.02832	Cubic meters
	0.03704	Cubic yards
	7.48052	Gallons (US)
	28.32	Liters
	59.84	Pints (liq. US)
	29.92	Quarts (liq. US)
	2.296×10^{-5}	Acre feet
	0.803564	Bushels
Cubic feet of water	62.4266	Pounds at 39.2° F
	62.3352	Pounds at 62° F
Cubic feet/minute	472	Cubic centimeters/sec
	0.1247	Gallons (US)/second
	0.472	Liters/second
	62.34	Pounds water/min at 62°F
	7.4805	Gallons (US)/minute
	10,772	Gallons/24 hours
	0.033058	Acre feet/24 hours
Cubic feet/second	646.317	Gallons (US)/24 hours
	448.831	Gallons/minute
	1.98347	Acre feet/24 hours
Cubic inches	16.387	Cubic centimeters
	0.0005787	Cubic feet
	1.639×10^{-5}	Cubic meters
	2.143×10^{-5}	Cubic yards
	0.004329	Gallons (US)
	0.01639	Liters
	0.03463	Pints (liq. US)
	0.01732	Quarts (liq. US)
Cubic meters	10^6	Cubic centimeters
	35.31	Cubic feet
	61.023	Cubic inches
	1.308	Cubic yards
	264.2	Gallons (US)
	1000	Liters
	2113	Pints (liq. US)
	1057	Quarts (liq. US)
Cubic yards	764,600	Cubic centimeters
	27	Cubic feet
	46,656	Cubic inches
	0.7646	Cubic meters
	202	Gallons (US)
	764.6	Liters
	1616	Pints (liq. US)

Multiply	by	To Obtain
	807.9	Quarts (liq. US)
Cubic yards/minute	0.45	Cubic feet/second
	3.367	Gallons (US)/second
	12.74	Liters/second
Cubit	18	Inches
Days (mean)	1440	Minutes
	24	Hours
	86,400	Seconds
Days (sidereal)	86,164.1	Solar seconds
Decigrams	0.1	Grams
Deciliters	0.1	Liters
Decimeters	0.1	Meters
Degrees (angle)	60	Minutes
	0.01745	Radians
	3600	Seconds
Degrees F [less 32]	0.5556	Degrees C
Degrees F	1 [plus 460]	Degrees F above absolute 0
Degrees C	1.8 [plus 32]	Degrees F
	1 [plus 273]	Degrees C above absolute 0
Degrees/second	0.01745	Radians/second
	0.1667	Revolutions/minute
	0.002778	Revolutions/second
Dekagrams	10	Grams
Dekaliters	10	Liters
Dekameters	10	Meters
Diameter (circle)	3.14159265359	Circumference
(approx)	3.1416	
(approx)	3.14	
(approx)	$\frac{22}{7}$	
Diameter (circle)	0.88623	Side of equal square
	0.7071	Side of inscribed square
Diameter³ (sphere)	0.5236	Volume (sphere)
Diam (major) x diam (minor)	0.7854	Area of ellipse
Diameter² (circle)	0.7854	Area (circle)
Diameter² (sphere)	3.1416	Surface (sphere)
Diam (inches) x RPM	0.262	Belt speed ft/minute
Digits	0.75	Inches
Drams (avoirdupois)	27.34375	Grains
	0.0625	Ounces (avoir.)
	1.771845	Grams
Fathoms	0.16667	Feet
Feet	30.48	Centimeters
	12	Inches
	0.3048	Meters
	1/3	Yards
	0.06061	Rods
Feet of water at 62	0.029465	Atmospheres
	0.88162	Inches of Hg at 32° F
	62.3554	Pounds/square foot
	0.43302	Pounds/square inch
	304.44	Kilogram/sq meter
Feet/minute	0.5080	Centimeters/second
	0.01667	Feet/second
	0.01829	Kilometers/hour

CONVERSION FACTORS (CONTINUED)
(Courtesy of Anvil International)

Multiply	by	To Obtain
Feet/minute	0.3048	Meters/minute
	0.01136	Miles/hour
Feet/Second	30.48	Centimeters/second
	1.097	Kilometers/hour
	0.5921	Knots
	18.29	Meters/minute
	0.6818	Miles/hour
	0.01136	Miles/minute
Feet/second^2	30.48	Centimeters/second^2
	0.3048	Meters/second^2
Flat of a hexagon	1.155	Distance across corners
Flat of a square	1.414	Distance across corners
Foot pounds	0.00128492	BTU
	0.32383	Calories (gm)
	0.0003238	Calories (Kg)
	5.05×10^{-7}	Horse power hours
	1.3558	Joules
	0.13826	Kilogram meters
	3.766×10^{-7}	Kilowatt hours
	0.0003766	Watt hours
Foot pounds/minute	0.001286	BTU/minute
	0.01667	Foot pounds/second
	3.03×10^{-5}	Horse power
	0.0003241	Calories (Kg)/minute
	2.26×10^{-5}	Kilowatts
Foot pounds/second	0.07717	BTU/minute
	0.001818	Horse power
	0.01945	Calories (Kg)/minute
	0.001356	Kilowatts
Furlong	40	Rods
	220	Yards
	660	Feet
	0.125	Miles
	0.2042	Kilometers
Gallons (Imperial)	277.42	Cubic inches
	4.543	Liters
	1.20095	Gallons (US)
Gallons (US)	3785	Cubic centimeters
	0.13368	Cubic feet
	231	Cubic inches
	0.003785	Cubic meters
	0.004951	Cubic yards
	3.785	Liters
	8	Pints (liq. US)
	4	Quarts (liq. US)
	0.83267	Gallons (Imperial)
	3.069×10^{-6}	Acre feet
Gallons (US) of water at 62° F	8.333	Pounds of water
Gallons (US) of water/ minute	6.0086	Tons of water/24 hours
Gallons(US)/minute	0.002228	Cubic feet/second
	0.13368	Cubic feet/minute
	8.0208	Cubic feet/hour
	0.06309	Liters/second
	3.78533	Liters/minute
	0.0044192	Acre feet/24 hours
Grains	1	Grains (avoirdupois)
	1	Grains (apothecary)
	1	Grains (troy)
	0.0648	Grams
	0.0020833	Ounces (troy)

Multiply	by	To Obtain
	0.0022857	Ounces (avoir.)
Grains/gallon (US)	17.118	Parts/million
	142.86	Pounds/million gallons (US)
Grams	980.7	Dynes
	15.43	Grains
	0.001	Kilograms
	1000	Milligrams
	0.03527	Ounces (avoir.)
	0.03215	Ounces (troy)
	0.002205	Pounds
Grams/centimeters	0.00521	Pounds/inch
Grams/cubic centimeter	62.45	Pounds/cubic foot
	0.03613	Pounds/cubic inch
	4.37	Grains/100 cubic ft
Grams/liter	58.405	Grains/gallon (US)
	8.345	Pounds/100 gallons (US)
	0.062427	Pounds/cubic foot
	1000	Parts/million
Gravity (*g*)	32.174	Feet/second^2
	980.6	Centimeters/second^2
Hand	4	Inches
	10.16	Centimeters
Hectares	2.471	Acres
	107,639	Square feet
	100	Ares
Hectograms	100	Grams
Hectoliters	100	Liters
Hectometers	100	Meters
Hectowatts	100	Watts
Hogshead	63	Gallons (US)
	238.4759	Liters
Horse power	42.44	BTU/minute
	33,000	Foot pounds/minute
	550	Foot pounds/second
	1.014	Metric horse power (Cheval vapeur)
	10.7	Calories (Kg)/min
	0.7457	Kilowatts
	745.7	Watts
Horse power (boiler)	33,479	BTU/hour
	9.803	Kilowatts
	34.5	Pounds of water evaporated/hour at 212° F
Horse power hours	2546.5	BTU
	641,700	Calories (gm)
	641.7	Calories (Kg)
	1980198	Foot pounds
	2688172	Joules
	273,740	Kilogram meters
	0.7455	Kilowatt hours
	745.5	Watt hours
Inches	2.54	Centimeters
	0.08333	Feet
	1000	Mils
	12	Lines
	72	Points
Inches of Hg at 32° F	0.03342	Atmospheres
	345.3	Kilograms/square meter
	70.73	Pounds/square foot
	0.49117	Pounds/square inch
	1.1343	Feet of water at 62° F

CONVERSION FACTORS (CONTINUED)
(Courtesy of Anvil International)

Multiply	by	To Obtain
Inches of Hg at 32° F	13.6114	Inches of water at 62° F
	7.85872	Ounces/square inch
Inches of water at 62° F	0.002455	Atmospheres
	25.37	Kilograms/square meter
	0.5771	Ounces/square inch
	5.1963	Pounds/square foot
	0.03609	Pounds/square inch
	0.07347	Inches of Hg at 32° F
Joules	0.00094869	BTU
	0.239	Calories (gm)
	0.000239	Calories (Kg)
	0.73756	Foot pounds
	3.72×10^{-7}	Horse power hours
	0.10197	Kilogram meters
	2.778×10^{-7}	Kilowatt hours
	0.0002778	Watt hours
	1	Watt second
Kilograms	980,665	Dynes
	2.205	Pounds
	0.001102	Tons (short)
	1000	Grams
	35.274	Ounces (avoir.)
	32.1507	Ounces (troy)
Kilogram meters	0.009302	BTU
	2.344	Calories (gm)
	0.002344	Calories (Kg)
	7.233	Foot pounds
	3.653×10^{-6}	Horse power hours
	9.806	Joules
	2.724×10^{-6}	Kilowatt hours
	0.002724	Watt hours
Kilograms/ cubic meter	0.06243	Pounds/ cubic foot
Kilograms/meter	0.6720	Pounds/foot
Kilograms/sq centimeter	14.223	Pounds/sq inch
	1	Metric atmosphere
Kilogram/sq meter	9.678×10^{-5}	Atmospheres
	0.003285	Feet of water at 62° F
	0.002896	Inches of Hg at 32° F
	0.2048	Pounds/square foot
	0.001422	Pounds/square inch
	0.007356	Centimeters of Hg at 32° F
Kiloliters	1000	Liters
Kilometers	100,000	Centimeters
	1000	Meters
	3281	Feet
	0.6214	Miles
	1094	Yards
Kilometers/hour	27.78	Centimeters/second
	54.68	Feet/minute
	0.9113	Feet/second
	16.67	Meters/minute
	0.6214	Miles/hour
	0.5396	Knots
Kilometers/hr/sec	27.78	Centimeters/sec/sec
	0.9113	Feet/sec/sec
	0.2778	Meters/sec/sec
Kilowatts	56.92	BTU/minute
	44,250	Foot pounds/minute
	737.6	Foot pounds/second
	1.341	Horse power
	14.34	Calories (Kg)/min
	1000 Watts	
Kilowatt hours	3413	BTU
Kilowatt hours	859,999	Calories (gm)
	858.99	Calories (Kg)
	2,655,200	Foot pounds
	1.341	Horse power hours
	3,600,000	Joules
	367,100	Kilogram meters
	1000	Watt hours
Knots	1	Nautical miles/hour
	1.1516	Miles/hour
	1.8532	Kilometers/hours
Leagues	3	Miles
Lines	0.08333	Inches
Links	7.92	Inches
Liters	1000	Cubic Centimeters
	0.03531	Cubic feet
	61.02	Cubic inches
	0.001	Cubic meters
	0.001308	Cubic yards
	0.2642	Gallons (US)
	0.22	Gallons (Imp)
	2.113	Pints (liq. US)
	1.057	Quarts (liq. US)
	8.107×10^{-7}	Acre Feet
	2.2018	Pounds of water at 62° F
Liter/minute	0.0005886	Cubic feet/second
	0.004403	Gallons (US)/second
	0.26418	Gallons (US)/minute
Meters	100	Centimeters
	3.281	Feet
	39.37	Inches
	1.094	Yards
	0.001	Kilometers
	1000	Millimeters
Meters/minute	1.667	Centimeters/second
	3.281	Feet/minute
	0.05468	Feet/second
	0.06	Kilometers/hour
	0.03728	Miles/hour
Meters/second	196.8	Feet/minute
	3.281	Feet/second
	3.6	Kilometers/hour
	0.06	Kilometers/minute
	2.237	Miles/hour
	0.03728	Miles/minute
Microns	10^{-6}	Meters
	0.001	Millimeters
	0.03937	Mils
Mils	0.001	Inches
	0.0254	Millimeters
	25.4	Microns
Miles	160,934	Centimeters
	5280	Feet
	63,360	Inches
	1.609	Kilometers
	1760	Yards
	80	Chains
	320	Rods
	0.8684	Nautical miles
Miles/hour	44.70	Centimeters/second
	88	Feet/minute
	1.467	Feet/second
	1.609	Kilometers/hours
	0.8684	Knots
	26.82	Meters/minute

CONVERSION FACTORS (CONTINUED)
(Courtesy of Anvil International)

Multiply	by	To Obtain
Miles/minute	2682	Centimeters/second
	88	Feet/second
	1,609	Kilometers/minute
	60	Miles/hour
Millibars	0.000987	Atmosphere
Milliers	1000	Kilograms
Milligrams	0.01	Grams
	0.01543	Grains
Milligrams/liter	1	Parts/million
Milliliters	0.001	Liters
Million gals/24 hours	1.54723	Cubic feet/second
Millimeters	0.1	Centimeters
	0.03937	Inches
	39.37	Mils
	1000	Microns
Miner's inches	1.5	Cubic feet/minute
Minutes (angle)	0.0002909	Radians
Nautical miles	6080.2	Feet
	1.516	Miles
Ounces (avoirdupois)	16	Drams (avoir.)
	437.5	Grains
	0.0625	Pounds (avoir.)
	28.349527	Grams
	0.9115	Ounces (troy)
Ounces (fluid)	1.805	Cubic inches
	0.02957	Liters
	29.57	Cubic centimeters
	0.25	Gills
Ounces (troy)	480	Grains
	20	Pennyweights (troy)
	0.08333	Pounds (troy)
	31.103481	Grams
	1.09714	Ounces (avoir.)
Ounces/square inch	0.0625	Pounds/square inch
	1.732	Inches of water at 62° F
	4.39	Centimeters of water at 62° F
	0.12725	Inches of Hg at 32° F
	0.004253	Atmospheres
Palms	3	Inches
Parts/million	0.0584	Grains/gallon (US)
	0.07016	Grains/gallon (Imp)
	8.345	Pounds/million gal (US)
Pennyweights (troy)	24	Grains
	1.55517	Grams
	0.05	Ounces (troy)
	0.0041667	Pounds (troy)
Pints (liq. US)	4	Gills
	16	Ounces (fluid)
	0.5	Quarts (liq. US)
	28.875	Cubic inches
	473.1	Cubic centimeters
Pipe	126	Gallons (US)
Points	0.01389	Inches
Poise	0.0672	Pounds/sec foot
	242	Pounds/hour foot
	100	Centipoise
Poncelots	100	Kilogram meters/second

Multiply	by	To Obtain
	1.315	Horse power
Pounds (avoirdupois)	16	Ounces (avoir.)
	256	Drams (avoir.)
	7000	Grains
	0.0005	Tons (short)
	453.5924	Grams
	1.21528	Pounds (troy)
	14.5833	Ounces (troy)
Pounds (troy)	5760	Grains
	240	Pennyweights (troy)
	12	Ounces (troy)
	373.24177	Grams
	0.822857	Pounds (avoir.)
	13.1657	Ounces (avoir.)
	0.00036735	Tons (long)
	0.00041143	Tons (short)
	0.00037324	Tons (metric)
Pounds of water at 62° F		0.01604 Cubic feet
	27.72	Cubic inches
	0.120	Gallons (US)
Pounds of water/min. at 62° F	0.0002673	Cubic feet/second
Pounds/cubic foot	0.01602	Grams/cubic centimeter
	16.02	Kilograms/cubic meter
	0.0005787	Pounds/cubic inch
Pounds/cubic inch	27.68	Grams/cubic centimeter
	27.680	Kilograms/cubic meter
	1728	Pounds/cubic foot
Pounds/foot	1.488	Kilograms/meter
Pounds/inch	178.6	Grams/centimeter
Pounds/hour foot	0.4132	Centipoise
	0.004132	Poise grams/sec cm
Pounds/sec foot	14.881	Poise grams/sec cm
	1488.1	Centipoise
Pounds/square foot	0.016037	Feet of water at 62° F
	4.882	Kilograms/square meter
	0.006944	Pounds/square inch
	0.014139	Inches of Hg at 32° F
	0.0004725	Atmospheres
Pounds/square inch	0.068044	Atmospheres
	2.30934	Feet of water at 62° F
	2.0360	Inches of Hg at 32° F
	703.067	Kilograms/square meter
	27.912	Inches of water at 62° F
Quadrants (angular)	90	Degrees
	5400	Minutes
	324,000	Seconds
	1.751	Radians
Quarts (dry)	67.20	Cubic inches
Quarts (liq.US)	2	Pints (liq. US)
	0.9463	Liters
	32	Ounces (fluid)
	57.75	Cubic inches
	946.3	Cubic centimeters
Quintal, Argentine	101.28	Pounds
Brazil	129.54	Pounds
Castile, Peru	101.43	Pounds
Chile	101.41	Pounds
Metric	220.46	Pounds
Mexico	101.47	Pounds

CONVERSION FACTORS (CONTINUED)
(Courtesy of Anvil International)

Multiply	by	To Obtain
Quires	25	Sheets
Radians	57.30	Degrees
	3438	Minutes
	206.186	Seconds
	0.637	Quadrants
Radians/second	57.30	Degrees/second
	0.1592	Revolutions/second
	9.549	Revolutions/minute
Radians/second^2	573.0	Revolutions/minute^2
	0.1592	Revolutions/second^2
Reams	500	Sheets
Revolutions	360	Degrees
	4	Quadrants
	6.283	Radians
Revolutions/minute	6	Degrees/second
	0.1047	Radians/second
	0.01667	Revolutions/second
Revolutions/minute^2	0.001745	Radians/second^2
	0.0002778	Revolutions/second^2
Revolutions/second	360	Degrees/second
	6.283	Radians/second
	60	Revolutions/minute
Revolutions/second^2	6.283	Radians/second^2
	3600	Revolutions/minute^2
Rods	16.5	Feet
	5.5	Yards
Seconds (angle)	4.848×10^{-6}	Radians
Sections	1	Square miles
Side of a square	1.4142	Diameter of inscribed circle
	1.1284	Diameter of circle with equal area
Span	9	Inches
Square centimeters	0.001076	Square feet
	0.1550	Square inches
	0.0001	Square meters
	100	Square millimeters
Square feet	2.296×10^{-5}	Acres
	929.0	Square centimeters
	144	Square inches
	0.0929	Square meters
	3.587×10^{-8}	Square miles
	0.1111	Square yards
Sqaure inches	6.452	Square centimeters
	0.006944	Square feet
	645.2	Square millimeters
	1.27324	Circular inches
	1,273,239	Circular mils
	1,000,000	Square mils
Square kilometers	247.1	Acres
	10,760,000	Square feet
	1,000,000	Square meters
	0.3861	Square miles
	1,196,000	Square yards
Square meters	0.0002471	Acres
	10.764	Square feet
	1.196	Square yards
	1	Centares
Square miles	640	Acres
	27,878,400	Square feet

Multiply	by	To Obtain
Square Miles	2.590	Square kilometers
	259	Hectares
	3,097,600	Square yards
	102,400	Square rods
	1	Sections
Square millimeters	0.01	Square centimeters
	0.00155	Square inches
	1550	Square mils
	1973	Circular mils
Square mils	1.27324	Circular mils
	0.0006452	Square millimeters
	10^{-6}	Square inches
Square yards	0.0002066	Acres
	9	Square feet
	0.8361	Square meters
	3.228×10^{-7}	Square miles
Stere	1	Cubic meters
Stone	14	Pounds
	6.35029	Kilograms
Tons (long)	1016	Kilograms
	2240	Pounds
	1.12	Tons (short)
Tons (metric)	1000	Kilograms
	2205	Pounds
	1.1023	Tons (short)
Tons (short)	2000	Pounds
	32,000	Ounces
	907.185	Kilograms
	0.90718	Tons (metric)
	0.89286	Tons (long)
Tons of refrigeration	12,000	BTU/hour
	288,000	BTU/24 hours
Tons of water/24 hours at 62° F	83.33	Pounds of water/hour
	0.16510	Gallons (US)/minute
	1.3263	Cubic feet/hour
Watts	0.05692	BTU/minute
	44.26	Foot pounds/minute
	0.7376	Foot pounds/second
	0.001341	Horse power
	0.01434	Calories (Kg)/minute
	0.001	Kilowatts
	1	Joule/second
Watt hours	3.413	BTU
	860.5	Calories (gm)
	0.8605	Calories (Kg)
	2655	Foot pounds
	0.001341	Horse power hours
	3600	Joules
	367.1	Kilogram meters
	0.001	Kilowatt hours
Watts/square inch	8.2	BTU/square foot/minute
	6373	Foot pounds/sq ft/minute
	0.1931	Horse power/ square foot
Yards	91.44	Centimeters
	3	Feet
	36	Inches
	0.9144	Meters
	0.1818	Rods
Year (365 days)	8760	Hours

References

API 6D, *Pipeline Valves*, American Petroleum Institute.

API 526, *Flanged Steel Pressure Relief Valves,* American Petroleum Institute.

API 570, *Piping Inspection Code: Inspection Repair, and Alteration of In-Service Piping Systems,* American Petroleum Institute.

API 594, *Check Valves: Flanged, Lug, Wafer, and Butt-welding,* American Petroleum Institute.

API 599, *Metal Plug Valves—Flanged, Threaded and Welding Ends,* American Petroleum Institute.

API 600, *Steel Gate Valves - Flanged and Butt-welding Ends, Bolted Bonnets,* American Petroleum Institute.

API 602, *Gate, Globe, and Check Valves for Sizes DN 100 (NPS 4) and Smaller for the Petroleum and Natural Gas Industries,* American Petroleum Institute.

API 603, *Corrosion-resistant, Bolted Bonnet Gate Valves - Flanged and Butt-welding Ends,* American Petroleum Institute.

API 608, *Metal Ball Valves—Flanged, Threaded, and Welding Ends,* American Petroleum Institute.

API 609, *Butterfly Valves: Double Flanged, Lug- and Water-Type,* American Petroleum Institute.

API 610, *Centrifugal Pumps for Petroleum, Petrochemical and Natural Gas Industries,* American Petroleum Institute.

API 611, *General Purpose Steam Turbines for Petroleum, Chemical, and Gas Industry Services,* American Petroleum Institute.

API 612, *Petroleum, Petrochemical and Natural Gas Industries-Steam Turbines-Special-purpose Applications,* American Petroleum Institute.

API 617, *Axial and Centrifugal Compressors and Expander-compressors,* American Petroleum Institute.

API 619, *Rotary-Type Positive Displacement Compressors for Petroleum, Chemical, and Gas Industry Services,* American Petroleum Institute.

API 661, Petroleum, Petrochemical, and Natural Gas Industries - Air-cooled Heat Exchangers, American Petroleum Institute.

API 686, *Recommended Practice for Machinery Installation and Installation Design,* American Petroleum Institute.

ASME B 1.20.1, *Pipe Threads, General Purpose (Inch),* American Society of Mechanical Engineers.

ASME B16.1, *Gray Iron Pipe Flanges and Flanged Fittings, Classes 25, 125, and 250*, American Society of Mechanical Engineers

ASME B16.3, *Malleable Iron Threaded Fittings,* American Society of Mechanical Engineers.

ASME B16.4, *Gray Iron Threaded Fittings,* American Society of Mechanical Engineers.

ASME B16.5, *Pipe Flanges and Flanged Fittings,* American Society of Mechanical Engineers.

ASME B16.9, *Factory-Made Wrought Buttwelding Fittings,* American Society of Mechanical Engineers.

ASME B16.10, *Face-to-Face and End-to-End Dimensions of Valves,* American Society of Mechanical Engineers.

ASME B16.ll, *Forged Steel Fittings, Socket-Welding and Threaded,* American Society of Mechanical Engineers.

ASME B16.14, *Ferrous Pipe Plugs, Bushings, and Locknuts with Pipe Threads,* American Society of Mechanical Engineers.

ASME B16.15, *Cast Bronze Threaded Fittings, Classes 125 and 250,* American Society of Mechanical Engineers.

ASME B16.18, *Cast Copper Alloy Solder Joint Pressure Fittings,* American Society of Mechanical Engineers.

ASME B 16.22, *Wrought Copper and Copper Alloy Solder Joint Pressure Fittings,* American Society of Mechanical Engineers.

ASME B 16.24, *Cast Copper Alloy Pipe Flanges and Flanged Fittings and Valves: Classes 150, 300, 600, 900, 1500, and 2500,* American Society of Mechanical Engineers.

ASME B 16.26, *Cast Copper Alloy Fittings for Flared Copper Tubes,* American Society of Mechanical Engineers.

ASME B 16.34, *Valves—Flanged, Threaded, and Welding End,* American Society of Mechanical Engineers.

ASME B16.36, *Orifice Flanges,* American Society of Mechanical Engineers.

ASME B16.39, *Malleable Iron Threaded Pipe Unions, Classes 150, 250, and 300,* American Society of Mechanical Engineers.

ASME B 16.42, *Ductile Iron Pipe Flanges and Flanged Fittings, Classes 150 and 300,* American Society of Mechanical Engineers.

ASME B 16.47, *Large Diameter Steel Flanges, NPS 26 Through NPS 60,* American Society of Mechanical Engineers.

ASME B 16.48, *Steel Line Blanks,* American Society of Mechanical Engineers.

ASME B 18.2.1, *Square, Hex, Heavy Hex, and Askew Head Bolts and Hex, Heavy Hex, Hex Flange, Lobed Head, and Lag Screws (Inch Series),* American Society of Mechanical Engineers.

ASME B 18.2.2, *Nuts for General Applications: Machine Screw Nuts, Hex, Square, Hex Flange, and Coupling Nuts (Inch Series),* American Society of Mechanical Engineers.

ASME B18.31.2, Continuous Thread Stud, Double-End Stud, and Flange Bolting Stud (Stud Bolt) (Inch Series). American Society of Mechanical Engineers.

ASME B31.1, *Power Piping,* American Society of Mechanical Engineers.

ASME B31.3, *Process Piping,* American Society of Mechanical Engineers.

ASME B31.4, *Pipeline Transportation Systems for Liquid and Slurries,* American Society of Mechanical Engineers.

ASME B31.5, *Refrigeration Piping and Heat Transfer Components,* American Society of Mechanical Engineers.

ASME B31.8, *Gas Transmission and Distribution Piping Systems,* American Society of Mechanical Engineers.

ASME B31.9, *Building Services Piping,* American Society of Mechanical Engineers. ASME B31.ll,*Slurry Piping,* American Society of Mechanical Engineers.

ASME B31.12, *Hydrogen Piping and Pipelines,* American Society of Mechanical Engineers.

ASME B31E, *Standard for the Seismic Design and Retrofit of Above-Ground Piping Systems,* American Society of Mechanical Engineers.

ASME B31G, Manual for Determining the Remaining Strength of Corroded Pipelines, American Society of Mechanical Engineers.

ASME B31J, Stress Intensification Factors (i-Factors), Flexibility Factors (k-Factors), and Their Determination for Metallic Piping Components, American Society of Mechanical Engineers.

ASME B31P, Standard Heat Treatments for Fabrication Processes, American Society of Mechanical Engineers.

ASME B31T, Standard Toughness Requirements for Piping, American Society of Mechanical Engineers.

ASME Boiler and Pressure Vessel Code, Section I, *Power Boilers,* American Society of Mechanical Engineers.

ASME Boiler and Pressure Vessel Code, Section II, *Materials,* Part A, *Ferrous Material Specifications,* American Society of Mechanical Engineers.

ASME Boiler and Pressure Vessel Code, Section II, *Materials,* Part B, *Nonferrous Material Specifications,* American Society of Mechanical Engineers.

ASME Boiler and Pressure Vessel Code, Section II, *Materials,* Part D, *Properties,* American Society of Mechanical Engineers.

ASME Boiler and Pressure Vessel Code, Section III, *Rules for Construction of Nuclear Power Plant Components,* American Society of Mechanical Engineers.

ASME Boiler and Pressure Vessel Code, Section V, *Nondestructive Examination,* American Society of Mechanical Engineers.

ASME Boiler and Pressure Vessel Code, Section VIII, Division 1, *Pressure Vessels,* American Society of Mechanical Engineers.

ASME Boiler and Pressure Vessel Code, Section VIII, Division 2, *Pressure Vessels, Alternative Rules,* American Society of Mechanical Engineers.

ASME Boiler and Pressure Vessel Code, Section VIII, Division 3, *Pressure Vessels, Alternative Rules for Construction of High Pressure Vessels,* American Society of Mechanical Engineers.

ASME Boiler and Pressure Vessel Code, Section VIII, Divisions 1 and 2, Code Case 2286, *Alternative Rules for Determining Allowable Compressive Stresses for Cylinders, Cones, Spheres, and Formed Heads,* American Society of Mechanical Engineers.

ASME Boiler and Pressure Vessel Code, Section IX, *Welding and Brazing Qualifications,* American Society of Mechanical Engineers.

ASME PCC-1, *Guidelines for Pressure Boundary Bolted Flange Joint Assembly,* American Society of Mechanical Engineers.

ASME PCC-2, *Repair of Pressure Equipment and Piping,* American Society of Mechanical Engineers.

ASME Standard OM-S/G, *Standards and Guides for Operation and Maintenance of Nuclear Power Plants,* American Society of Mechanical Engineers.

ASTM A53, *Standard Specification for Pipe, Steel, Black and Hot-Dipped, Zinc-Coated, Welded and Seamless,* ASTM International.

ASTM A106, *Standard Specification for Seamless Carbon Steel Pipe for High-Temperature Service,* ASTM International.

ASTM B828, *Standard Practice for Making Capillary Joints by Soldering of Copper and Copper Alloy Tube and Fittings*, ASTM International.

ASTM C582, *Standard Specification for Contact-Molded Reinforced Thermosetting Plastic (RTP) Laminates for Corrosion Resistant Equipment,* ASTM International.

ASTM D1599, *Standard Test* Method *for Resistance to Short-Time Hydraulic Pressure of Plastic Pipe, Tubing, and Fittings,* ASTM International.

ASTM B61, *Standard Specification for Steam or Valve Bronze Castings,* ASTM International

ASTM B62, *Standard Specification for Composition Bronze or Ounce Metal Castings,* ASTM International.

ASTM D2837, *Test Method for Obtaining Hydrostatic Design Basis for Thermoplastic Pipe Materials or Pressure Design Basis for Thermoplastic Pipe Products,* ASTM International.

ASTM D2992, *Standard Practice for Obtaining Hydrostatic or Pressure Design Basis for "Fiberglass" (Glass-Fiber-Reinforced Thermosetting-Resin) Pipe and Fittings,* ASTM International.

ASTM F1476, *Performance of Gasketed Mechanical Couplings for Use in Piping Applications,* ASTM International.

ASTM F1548, *Performance of Fittings for Use with Gasketed Mechanical Couplings Used in Piping Applications*, ASTM International

AWS D10.10, *Recommended Practices for Local Heating of Welds in Piping and Tubing,* American Welding Society.

AWWA C110, *Ductile-Iron and Gray-Iron Fittings,* American Water Works Association.

AWWA C1 15, *Flanged Ductile-Iron with Ductile-Iron or Gray-Iron Threaded Flanges,* American Water Works Association.

AWWA C207, *Steel Pipe Flanges for Water Works Service, Sizes 4 Inch Through 144 Inch (100 mm Through 3,600 mm),* American Water Works Association.

AWWA C208, *Dimensions for Fabricated Steel Water Pipe Fittings,* American Water Works Association.

AWWA C500, *Metal-Seated Gate Valves for Water Supply Service,* American Water Works Association.

AWWA C504, *Rubber-Seated Butterfly Valves,* American Water Works Association.

Becht IV, C, 1980, "Root Bulge of Bellows," DOE Research and Development Report, ESG-DOE-13320, Rockwell International.

Becht IV, C, and Skopp, G., 1981a, "Stress Analysis of Bellows," *Metallic Bellows and Expansion Joints,* PVP-Vol. 51, American Society of Mechanical Engineers.

Becht IV, C, and Skopp, G., 1981b, "Root Bulge of Bellows," *Metallic Bellows and Expansion Joints,* PVP-Vol. 51, American Society of Mechanical Engineers.

Becht IV, C, Horton, P., and Skopp, G., 1981c, "Verification of Theoretical Plastic Ratchet Boundaries for Bellows," *Metallic Bellows and Expansion Joints,* PVP-Vol. 51, American Society of Mechanical Engineers.

Becht IV, C, 1983, "A Simplified Approach for Evaluating Secondary Stresses in Elevated Temperature Design," 83-PVP-51.

Becht IV, C, 1985, "Predicting Bellows Response by Numerical and Theoretical Methods," *A Decade of Progress in Pressure Vessels and Piping Technology*—7955, American Society of Mechanical Engineers.

Becht IV, C, 1988, "Elastic Follow-up Evaluation of a Piping System with a Hot Wall Slide Valve," *Design and Analysis of Piping, Pressure Vessels, and Components—1988,* PVP-Vol. 139, American Society of Mechanical Engineers.

Becht IV, C, 1989a, "Fatigue and Elevated Temperature Design of Bellows," *Metallic Bellows and Expansion Joints*—7959, PVP-Vol. 168, American Society of Mechanical Engineers.

Becht IV, C, 1989b, "Considerations in Bellows Thrust Forces and Proof Testing," *Metallic Bellows and Expansion Joints*—7959, PVP-Vol. 168, American Society of Mechanical Engineers.

Becht IV, C, Chen, Y., and Benteftifa, C, 1992, "Effect of Pipe Insertion on Slip-On Flange Performance," *Design and Analysis of Pressure Vessels, Piping, and Components*—7992, PVP-Vol. 235, American Society of Mechanical Engineers.

Becht IV, C, and Chen, Y, 2000, Span Limits for Elevated Temperature Piping," *Journal of Pressure Vessel Technology,* 122(2), pp. 121-124.

Becht IV, C, 1999, "Evaluation of Plastic Strain Concentration in Bellows," *Fracture, Design, Analysis of Pressure Vessels, Heat Exchangers, Piping Components, and Fitness for Service*—7999, PVP-Vol. 388, American Society of Mechanical Engineers.

Becht IV, C, 2000a, "The Effect of Bellows Convolution Profile on Stress Distribution and Plastic Strain Concentration," *Fitness for Service, Stress Classification and Expansion Joint Design,* PVP-Vol. 401, American Society of Mechanical Engineers.

Becht IV, C, 2000b, "An Evaluation of EJMA Stress Calculations for Unreinforced Bellows," *Fitness for Service, Stress Classification and Expansion Joint Design,* PVP-Vol. 401, American Society of Mechanical Engineers.

Becht IV, C, 2000c, "Fatigue of Bellows, A New Design Approach," *International Journal of Pressure Vessels and Piping,* 77(13), pp. 843-850.

Becht IV, C, and Hollinger, G., 2008, *Development of Basis Time-Dependent Allowable Stresses for Creep Regime in Section VIII Division 1,* STP-PT-024, ASME Standards Technology, LLC.

Becht IV, C, 2001, *Behavior of Bellows,* WRC Bulletin 466, Welding Research Council, Nov 2001.

Becht V, C, et al, 2014, "Sustained Stress Indices (SSI) in the B31.3 2010 Edition, ASME PVP2014-28267, 2014 PVP Conference Proceedings, American Society of Mechanical Engineers

Bednar, H., 1986, *Pressure Vessel Design Handbook,* Van Nostrand Reinhold, New York.

Bergman, E. O., 1960, "The New-Type Code Chart for the Design of Vessels Under External Pressure," *Pressure Vessel and Piping Design, Collected Papers 1927-1959,* American Society of Mechanical Engineers, pp. 647-654.

Biersteker, M., Dietemann, C, Sareshwala, S., and Haupt, R. W, 1991, "Qualification of Nonstandard Piping Product Form for ASME Code for Pressure Piping, B31 Applications," *Codes and Standards and Applications for Design and Analysis of Pressure Vessels and Piping Components,* PVP-Vol. 210-1, American Society of Mechanical Engineers.

Bijlaard, P. P., "Stresses from Local Loadings in Cylindrical Pressure Vessels," *Trans. A.S.M.E.,* ***11,*** 802-816(1955).

Bijlaard, P. P., "Stresses from Radial Loads in Cylindrical Pressure Vessels," *Welding Jnl,* **33**(12), Research supplement, 615-s to 623-s (1954).

Bijlaard, P. P., "Stresses from Radial Loads and External Moments in Cylindrical Pressure Vessel," *Ibid.,* **34**(12). Research supplement, 608-s to 617-s (1955).

Bijlaard, P. P., "Computation of the Stresses from Local Loads in Spherical Pressure Vessels or Pressure Vessel Heads," *Welding Research Council Bulletin* No. 34, (March 1957).

Bijlaard, P. P., "Local Stresses in Spherical Shells from Radial or Moment Loadings," *Welding Jnl,* **36**(5), Research Supplement, 240-s to 243-s (1957).

Bijlaard, P. P., "Stresses in a Spherical Vessel from Radial Loads Acting on a Pipe," *Welding Research Council Bulletin* No. 49, 1-30 (April 1959).

Bijlaard, P. P., "Stresses in a Spherical Vessel from External Moments Acting on a Pipe," *Ibid.,* No. 49, 31-62 (April 1959).

Bijlaard, P. P., "Influence of a Reinforcing Pad on the Stresses in a Spherical Vessel Under Local Loading," *Ibid.,* No. 49, 63-73 (April 1959).

Bijlaard, P. P., "Stresses in Spherical Vessels from Local Loads Transferred by a Pipe," *Ibid.,* No. 50, 1-9 (May 1959).

Bijlaard, P. P., "Additional Data on Stresses in Cylindrical Shells Under Local Loading," *Ibid.,* No. 50, 10-50 (May 1959).

Boardman, H. C, 1943, "Formulas for the Design of Cylindrical and Spherical Shells to Withstand Uniform Internal Pressure," *Water Tower,* 30, pp. 14-15.

Broyles, R., 1989, "Bellows Instability," *Metallic Bellows and Expansion Joints—1989,* American Society of Mechanical Engineers.

Broyles, R., 1994, "EJMA Design Equations," *Developments in a Progressing Technology*—7994, PVP-Vol. 279, American Society of Mechanical Engineers.

Copper Development Association, 1995, *The Copper Tube Handbook,* Copper Development Association, New York.

EJMA, *Standards of the Expansion Joint Manufacturer's Association,* EJMA, Tarrytown, New York.

Guidelines for Avoidance of Vibration Induced Fatigue Failure in Process Pipework, published by the Energy Institute.

Hinnant, C, and Paulin, T., 2008, "Experimental Evaluation of the Markl Fatigue Methods and ASME Piping Stress Intensification Factors," *Proceedings of PVP 2008,* The American Society of Mechanical Engineers.

Holt, M., 1960, "A Procedure for Determining the Allowable Out-of-Roundness for Vessels Under External Pressure," *Pressure Vessel and Piping Design, Collected Papers 1927-1959,* American Society of Mechanical Engineers, pp. 655-660.

ISO 15649, 2001, Petroleum and natural gas industries-piping.

Kobatake, K., et al, 1986, "Simplified Fatigue Life Evaluation Method of Bellows Expansion Joints at Elevated Temperature," *Fatigue and Fracture Assessment by Analysis and Testing,* PVP-Vol. 103, American Society of Mechanical Engineers.

Koves et. al in a paper *"Establishing Allowable Nozzle Loads"* PVP 2011-57100, American Society of Mechanical Engineers.

Langer, B. F., 1961, "Design of Pressure Vessels for Low Cycle Fatigue," *ASME Paper No. 61-WA-18,* American Society of Mechanical Engineers.

Markl, A., 1960a, "Fatigue Tests of Piping Components," *Pressure Vessel and Piping Design, Collected Papers, 1927-1959,* American Society of Mechanical Engineers, pp. 402-418.

Markl, A., 1960b, "Fatigue Tests of Welding Elbows and Comparable Double-Mitre Bends," *Pressure Vessel and Piping Design, Collected Papers, 1927-1959,* American Society of Mechanical Engineers, pp. 371-393.

Markl, A., 1960c, "Fatigue Tests on Flanged Assemblies," *Pressure Vessel and Piping Design, Collected Papers, 1927-1959,* American Society of Mechanical Engineers, pp. 91-101.

Markl, A., 1960d, Piping-Flexibility Analysis, *Pressure Vessel and Piping Design, Collected Papers, 1927-1959,* American Society of Mechanical Engineers, pp. 419-441.

Mershon, J., Mokhtarian, K., Ranjan G., and Rodabaugh, E., 1984, *Local Stresses in Cylindrical Shells Due To External Loadings on Nozzles—Supplement to WRC Bulletin No. 107,* Bulletin 297, Welding Research Council, New York.

MSS SP-42, *Corrosion Resistant Gate, Globe, Angle and Check Valves with Flanged and Butt Weld Ends (Classes 150, 300 & 600).*

MSS SP-43, *Wrought and Fabricated Butt-Welding Fittings for Low Pressure, Corrosion Resistant Applications,* Manufacturers Standardization Society.

MSS SP-44, *Steel Pipeline Flanges,* Manufacturers Standardization.

MSS SP-51, *Class 150LW Corrosion Resistant Cast Flanges and Flanged Fittings,* Manufacturers Standardization Society.

MSS SP-58, *Pipe Hangers and Supports—Materials, Design, Manufacture, Selection, Application, and Installation,* Manufacturers Standardization Society.

MSS SP-65, *High Pressure Chemical Industry Flanges and Threaded Stubs for Use with Lens Gaskets,* Manufacturers Standardization Society.

MSS SP-70, *Gray Iron Gate Valves, Flanged and Threaded Ends,* Manufacturers Standardization Society.

MSS SP-71, *Gray Iron Swing Check Valves, Flanged and Threaded Ends,* Manufacturers Standardization Society.

MSS SP-72, *Ball Valves with Flanged or Buttwelding Ends for General Service,* Manufacturers Standardization Society.

MSS SP-73, *Brazing Joints for Copper and Copper Alloy Pressure Fittings,* Manufacturers Standardization Society.

MSS SP-75, *High-Strength, Wrought, Butt-Welding Fittings,* Manufacturers Standardization Society.

MSS SP-78, *Gray Iron Plug Valves, Flanged and Threaded Ends*, Manufacturers Standardization Society.

MSS SP-79, *Socket-Welding Reducer Inserts,* Manufacturers Standardization Society.

MSS SP-80, *Bronze Gate, Globe, Angle and Check Valves,* Manufacturers Standardization Society.

MSS SP-81, *Stainless-Steel or Stainless-Steel-Lined, Bonnetless, Flanged, Knife Gate Valves,* Manufacturers Standardization Society.

MSS SP-83, *Class 3000 And 6000 Pipe Unions, Socket Welding and Threaded (Carbon Steel, Alloy Steel, Stainless Steels, And Nickel Alloys),* Manufacturers Standardization Society.

MSS SP-85, *Cast Iron Globe and Angle Valves, Flanged and Threaded Ends,* Manufacturers Standardization Society.

MSS SP-88, *Diaphragm Valves,* Manufacturers Standardization Society.

MSS SP-95, *Swage(d) Nipples and Bull Plugs,* Manufacturers Standardization Society.

MSS SP-97, *Integrally Reinforced Forged Branch Outlet Fittings—Socket Welding, Threaded, and Buttwelding Ends,* Manufacturers Standardization Society.

MSS SP-105, *Instrument Valves for Code Applications,* Manufacturers Standardization Society.

MSS SP-106, *Cast Copper Alloy Flanges and Flanged Fittings Class 125, 150, and 300,* Manufacturers Standardization Society.

MSS SP-119, *Factory-Made Wrought Belled End Pipe Fittings for Socket-Welding.* Manufacturers Standardization Society.

NEMA SM 23, *Steam Turbines for Mechanical Drive Service,* National Electrical Manufacturers Association.

Osweiller, F., 1989, "Design of an Expansion Joint by a Finite Element Program—Comparison with the EJMA Standards," *Metallic Bellows and Expansion Joints—1989,* PVP-Vol. 168, American Society of Mechanical Engineers.

Robinson, E., 1960, "Steam-Piping Design to Minimize Creep Concentrations," *Pressure Vessel and Piping Design, Collected Papers 1927-1959,* American Society of Mechanical Engineers, pp. 451-466.

Rodabaugh, E., and George, H., 1960, "Effect of Internal Pressure on Flexibility and Stress-Intensification Factors of Curved Pipe or Welding Elbows," *Pressure Vessel and Piping Design, Collected Papers 1927-1959,* American Society of Mechanical Engineers, pp. 467-477.

Rossheim, D. B., and Markl, A.R.C., (1960), "The Significance of, and Suggested Limits for, the Stress in Pipelines Due to the Combined Effects of Pressure and Expansion," *Pressure Vessels and Piping, Collected Papers, 1927-1959,* American Society of Mechanical Engineers, pp. 362-370.

SAE J513, *Refrigeration Tube Fittings—General Specifications,* Society of Automotive Engineers.

SAE J514, *Hydraulic Tube Fittings,* Society of Automotive Engineers.

SAE J518, *Hydraulic Flange Tube, Pipe, and Hose Connections, Four-Bolt Split Flanged Type,* Society of Automotive Engineers.

Saunders, H. E., and Windenburg, D., 1960, "Strength of Thin Cylindrical Shells Under External Pressure," *Pressure Vessel and Piping Design, Collected Papers 1927-1959,* American Society of Mechanical Engineers, pp. 600-611.

Short II, W. E., 1989, "Overview of Chapter VII, Nonmetallic Piping and Piping Lined with Non-metals in the ASME B31.3 Chemical Plant & Petroleum Refinery Piping Code," *Codes and Standards and Applications for Design and Analysis of Pressure Vessel and Piping Components—1989,* PVP-Vol. 161, American Society of Mechanical Engineers.

Short II, W. E., 1992, "Coverage of Non-Metals in the ASME B31.3 Chemical Plant and Petroleum Refinery Piping Code," *Journal of Process Mechanical Engineers,* **206,** pp. 67-72.

Short II, W. E., Leon, G. F., Widera, G. E. O., and Zui, C. G., 1996, *Literature Survey and Interpretive Study on Thermoplastic and Reinforced-Thermosetting-Resin Piping and Component Standards,* WRC 415, Welding Research Council.

Sims, J., 1986, "Development of Design Criteria for a High Pressure Piping Code," *High Pressure Technology—Design, Analysis, and Safety of High Pressure Equipment,* PVP-Vol. 110, D. P. Kendall, ed., American Society of Mechanical Engineers.

SNT-TC-1A, *Recommended Practice for Nondestructive Testing Personnel Qualification and Certification,* Society for Nondestructive Testing.

Takezono, S., 1971, "Fatigue Strength of Expansion Joints of Pressure Vessels," *Bulletin of the Japanese Society of Mechanical Engineers,* 14(76).

Tsukimori, K., et al, 1989, "Fatigue and Creep-Fatigue Life Prediction of Bellows," *Metallic Bellows and Expansion Joints—1989,* PVP-Vol. 168, American Society of Mechanical Engineers.

Wichman, K., Hopper A., and Mershon J., 1979, *Local Stresses in Spherical and Cylindrical Shells due to External Loadings,* Bulletin 107, Welding Research Council.

Windenburg, D., 1960, "Vessels Under External Pressure: Theoretical and Empirical Equations Represented in Rules for the Construction of Unfired Pressure Vessels Subjected to External Pressure," *Pressure Vessel and Piping Design, Collected Papers 1927-1959,* American Society of Mechanical Engineers, pp. 625-632.

Windenburg, D., and Trilling, C, 1960, "Collapse by Instability of Thin Cylindrical Shells Under External Pressure," *Pressure Vessel and Piping Design, Collected Papers 1927-1959,* American Society of Mechanical Engineers, pp. 612-624.

Yamamoto, S., et al, 1986, "Fatigue and Creep-Fatigue Testing of Bellows at Elevated Temperature," *Fatigue and Fracture Assessment by Analysis and Testing,* PVP-Vol. 103, American Society of Mechanical Engineers.

Yamashita, T, et al, 1989, "Ratchetting Mechanism of Bellows and Evaluation Method," *Metallic Bellows and Expansion Joints—1989,* PVP-Vol. 168, American Society of Mechanical Engineers.

INDEX

C

D

E

F

G

H

N

O

P

www.ingramcontent.com/pod-product-compliance
Ingram Content Group UK Ltd.
Pitfield, Milton Keynes, MK11 3LW, UK
UKHW050951290926
15470UKWH00005B/1756